HAVING ALL THE RIGHT CONNECTIONS

HAVING ALL THE RIGHT CONNECTIONS

Telecommunications and Rural Viability

Edited by
Peter F. Korsching, Patricia C. Hipple,
and Eric A. Abbott

Westport, Connecticut
London

Library of Congress Cataloging-in-Publication Data

Having all the right connections : telecommunications and rural viability / edited by
Peter F. Korsching, Patricia C. Hipple, and Eric A. Abbott.
      p.   cm.
    Includes bibliographical references and index.
    ISBN 0–275–96582–1 (alk. paper)
    1. Rural telecommunication—United States.   2. Rural development—United States.   I.
Korsching, Peter F.   II. Hipple, Patricia C., 1953–   III. Abbott, Eric A., 1945–
HE7775.H39   2000
384′.0973′091734—dc21   99–086113

British Library Cataloguing in Publication Data is available.

Library of Congress Catalog Card Number: 99–086113
ISBN: 0–275–96582–1

First published in 2000

Praeger Publishers, 88 Post Road West, Westport, CT 06881
An imprint of Greenwood Publishing Group, Inc.
www.praeger.com

Printed in the United States of America

The paper used in this book complies with the
Permanent Paper Standard issued by the National
Information Standards Organization (Z39.48–1984).

10 9 8 7 6 5 4 3 2 1

CONTENTS

ILLUSTRATIONS

FIGURES

TABLES

Acronyms and Abbreviations

AI	artificial intelligence
AOL	America Online
AP	Associated Press
ARPA	Advanced Research Projects Agency
ARPANET	Advanced Research Projects Agency Network
ASCII	American Standard Code for Information Interchange
ATA	American Telemedicine Association
AT&T	American Telephone and Telegraph
ATM	asynchronous transfer mode or automatic teller machine
BIDCO	Belmond Industrial Development Corporation
BOCs	Bell Operating Companies
bps	bits per second
Bps	bytes per second
CAD	computer-aided design
CAM	computer-aided manufacture
CASE	computer-aided software engineering
CATA	Cable Television Association
CD	compact disk
CD-ROM	compact disk, read-only memory
CIM	computer-integrated manufacture
CMC	computer-mediated communications
CME	continuing medical education
CNN	Cable News Network
Codec	coder/decoder
DBS	direct broadcasting by satellite
Dpi	dots per inch

DTN   data transmission network
DTMF   dual tone multi-frequency signaling

E911   enhanced 911
E-commerce electronic commerce
EDI   electronic data interchange
EFTPoS   electronic fund transfer at point of sale
E-mail   electronic mail
EVH   electronic village hall
EXNET   Extension Network
EXNET-IP   Extension Network-Internet protocol

FAX   facsimile
FCC   Federal Communications Commission
FMS   flexible manufacturing systems
FRED   Fund for Rural Education and Development
FTTH   fiber to the home

GAO   Government Accounting Office
GATS   General Agreement on Trade and Services
GATT   General Agreement on Tariffs and Trade
Gbps   gigabits per second
Gbyte   gigabyte
GDP   gross domestic product
GNP   gross national product
GTE   General Telephone and Electronics
GUI   graphic user interface

HBO   Home Box Office
HCFA   Health Care Financing Administration
HDTV   high-definition television

ICN   Iowa Communications Network
ICT   information and communication technology
IITF   Information Infrastructure Task Force
IMMC   Iowa Methodist Medical Center
INS   Iowa Network Services
IPR   intellectual property rights
IRS   Internal Revenue Service
IS   information systems
ISDN   Integrated Services Digital Network
ISP   Internet service provider
IT   information technology(ies)
ITA   Iowa Telecommunications Association
ITFS   instructional television fixed service
ITT   information and telecommunications technologies
IUB   Iowa Utilities Board
IV   interactive video

JIT	just-in-time system
JPEG	joint photographic experts group
Kbps	kilobits per second
Kbyte	kilobyte
LAN	local-area network
LATA	local access and transport area
LEC	local exchange carrier
MAN	metropolitan area network
Mbps	megabits per second
Mbyte	megabyte
MCI	Microwave Communications, Inc.
MCU	multipoint control unit
MFJ	Modified Final Judgment
Modem	MOdulator/DEModulator
MRTC	Midwest Rural Telemedicine Consortium
NECA	National Exchange Carriers Association
NII	national information infrastructure
NRECA	National Rural Electric Cooperative Association
NRTA	National Rural Telecommunications Association
NRTC	National Rural Telecommunications Cooperative
NTCA	National Telephone Cooperative Association
NTIA	National Telecommunications and Information Administration
OCR	optical character recognition
ONA	open network architecture
OPASTCO	Organization for the Protection and Advancement of Small Telephone Companies
OSI	open systems interconnection
OTA	Office of Technology Assessment
PACS	picture archiving and communications system
PANS	pretty amazing new stuff
PC	personal computer
PCN	personal communications network
P.O.	purchase order
POP	point of presence
PoS	point of sale
POTS	plain old telephone service
PSTN	public switched telephone network
RAM	random access memory
RAN	rural area network
RBOC	regional Bell operating company
REA	rural electrification act and Rural Electric Administration
RFD	rural free delivery

RHC   regional holding company
RIC   rural information center
RIITA   Rural Iowa Independent Telephone Association
ROM   read-only memory
RTB   Rural Telephone Bank
RTFC   Rural Telephone Finance Cooperative
RUS   Rural Utility Service

SAF   store and forward
SCSI   small computer systems interface
SMSA   standard metropolitan statistical area
SS7   Signaling System 7

Tbps   terabits per second
TCP/IP   transmission control protocol / Internet protocol
Telco   telecommunications company or telephone company
TQM   total quality management
TV   television
TVA   Tennessee Valley Authority

UPI   United Press International
USDA   United States Department of Agriculture
USF   Universal Service Fund
USTA   United States Telephone Association

VAN   value-added network
V-chip   violence-chip
VCR   videocassette recorder
VHF   very high frequency
VR   virtual reality

WAN   wide-area network
WTO   World Trade Organization
WWW   World Wide Web (the "Web")

Part I

Theory, Policy, and Practice: Setting the Stage for Rural Telecommunications

1

RURAL AMERICA AND THE INFORMATION AND COMMUNICATIONS REVOLUTION

Peter F. Korsching, Patricia C. Hipple, and Eric A. Abbott

INTRODUCTION

The development and introduction of new communications technologies, including enhanced telephone services, video-conferencing, teletext services, computers with modems, and local access Internet services, have raised important questions about their probable impacts on rural areas. Are they an answer to the problems of remoteness and isolation of rural communities? Or do they represent yet another way by which a well-connected urban network is poised to gain further advantage? Will telecommunications innovations become the boon or bane of rural communities? Which rural communities will benefit from development of telecommunications and which will experience the unexpected and detrimental consequences of these technologies? Are telecommunications technologies essential? Does their lack portend the further demise of rural communities?

A quarter of a century ago Daniel Bell (1973) predicted the coming of a post-industrial society in which a service economy, driven by information and communications technologies, would displace the manufacturing industrial sector as the primary source of employment. This shift, made possible by the convergence of communications technologies and computer technologies through digitization, is now well advanced. The seemingly limitless potentials of these technologies are generating futuristic visions focused upon concepts such as information society, information superhighway, virtual networks and communities, and cybersociety. From a practical standpoint, telecommunications technologies provide easy, inexpensive and efficient access to information, and therefore are critical to the American economy. Much as the development of railroad infrastructure in the 19th century and the air transportation and the interstate highway systems in the mid 20th century facilitated economic and social development, the telecommunications infrastructure is expected to facilitate economic and social development as we move into the 21st century.

Not all of America, however, benefits equally from new technologies. Technological infrastructure is of greatest benefit to those in close proximity to the system, and it least benefits, and indeed may adversely impact, those at a distance from the system. For example, many rural communities receive no direct benefits from the interstate highway system and face loss of trade and services because the highways open access to larger trade and service centers. So policy makers are realizing that there are no comprehensive and universally equitable technological fixes for rural social and economic problems. There is hope, however, in the strong potential of telecommunications technologies for breaking down distance barriers and returning isolated rural areas to mainstream America. But the promise of telecommunications depends on a number of factors, not the least of which is the availability of a telecommunications infrastructure. The story of one rural business owner (Case Study) illustrates the dilemmas, as well as the social and economic consequences of inadequate infrastructure.

CASE STUDY: LACKING ALL THE RIGHT CONNECTIONS

Colleen Dysart ran a successful home-based business, or "cottage industry," just outside Oklahoma City. An entrepreneur who designed jewelry and home decorative items, Colleen made most of her initial sales at art and craft shows, fairs, and festivals, but she maintained an active customer list, responding to requests for customized items. She did a sizeable portion of repeat sales via phone, facsimile (FAX), courier service, and the postal service. Although modest, her business was successful enough to provide Colleen's livelihood as well as support several part-time, seasonal workers in her community. Recent computer innovations seemed to promise future success as well. Colleen began with a personal computer for correspondence and record keeping, and she was poised to expand onto the Internet with her own Web page, potentially increasing her customer base and sales.

When she relocated with her husband to a sparsely populated corner of the state, Colleen did not foresee the impact it would have on her business. She had always worked well in relative isolation and her art was highly mobile, so she thought the business would be portable as well. Upon arrival at her new home, however, she learned that the technologies needed for business expansion were not so portable, for in this part of the state, residential telephone service was limited exclusively to party lines provided by a small rural telephone company.

Colleen's home-based business would have to share a four-resident party line. This meant she could not install a computer modem and was therefore precluded from doing business over the Internet. She could not install an 800 number for the convenience of her customers, nor could she use a FAX machine to expedite supplies or receive customer orders. She had no voice-mail capability, nor would her party line accommodate an answering machine. Even the most faithful customers had to contend with a seemingly perpetual busy signal while the party line was in use.

Not surprisingly, Colleen's sales dropped precipitously at the same time her costs escalated dramatically. The only way the business could survive was to "take it on the road," but the travel and accommodation expenses ate away at her profits and she was unable to maintain follow-up relationships with her customers. Although the local telephone company planned future service upgrades, Colleen's business could not survive the wait. As a result of inadequate infrastructure, she lost her investment and income, her employees lost their jobs, and her community lost needed revenue.

IMPORTANCE OF A RURAL TELECOMMUNICATIONS INFRASTRUCTURE

The adverse consequences to Colleen's business demonstrate the need for adequate telecommunications infrastructure to support commercial enterprises and thereby sustain rural communities' vitality and growth. The failure of Colleen's business meant the loss of several jobs as well as additional economic and social impacts on the community. Telecommunications infrastructure investments, notably in technologies that involve the use of computers to access databases or other online services, information exchange networks, and interactive technologies such as educational programs, classes, or conferences, are critically important to the economic vitality of rural communities (Fox, 1988). These technologies include:

- *Computers with modems* used to send and receive e-mail, access online databases and user groups, check inventories, place orders, or search for information from databases

- *Enhanced telephone service* such as wide-area telephone networks or other enhanced communication services that permit rapid access to databases, e-mail or telemarketing

- *Video-conferencing* systems that permit two-way interactive video via fiber-optic lines, or one-way video with audio feedback via satellites

- *Teletext services* used to receive information downloaded from a satellite or FM sideband and displayed on a computer terminal; these systems only receive information

- *Local connections* to fiber-optic lines and digital switches to support the implementation and use of innovative telecommunications technologies

The local development of such state-of-the-art telecommunications infrastructure is fundamental to a community's sustainable economic development, whether that development is based upon a strategy of grow your own, retention and expansion, or recruiting new business and industry from the outside (Allen, Johnson, Leistritz, Olsen, Sell, & Spilker, 1995; Schmandt, Williams, Wilson, & Strover, 1991). The growing importance of telecommunications infrastructure is confirmed by the decline in manufacturing jobs and an increase in service jobs. By the early 1990s, 71 percent of Americans held jobs in the service sector (Glasmeier & Howland, 1995). Among the most desired service sector jobs from the local community perspective are the producer services, or services that businesses perform as intermediate goods to other businesses such as consulting, marketing, and sales. These services employ workers at all skill levels with commensurate incomes, and their growth and ability to disperse geographically are made possible by computers and information technologies (Glasmeier & Howland, 1995; Dillman, 1991b).

Telecommunications technologies also provide opportunities for rural communities to cater to segments of the labor force identified as home-office workers, "lone eagles," and home-based businesses. Home-office workers, also known as telecommuters, have increased from less than 8 million in 1974, to 39 million in 1992, and will increase to an estimated 55 million in 1997 (Allen & Johnson, 1995). Home-office workers have jobs that, although tied to a particular location, provide some flexibility to work at home. Lone eagles like Colleen, however, are not constrained by any location requirements of their work other than the availability of the technological infrastructure to sell their products or services (Salant, Carley, & Dillman, 1996) and support their businesses. The literature predicts dramatic increases in the number of telecommunications supported home-based businesses or "electronic cottages" (Davidson, 1991; Christensen, 1989).

LESSONS FROM IOWA RESEARCH

Although modern telecommunications and the "information superhighway" are touted widely by the popular press as promising for revitalizing rural communities, the research literature is divided in its assessment of the prospects of telecommunications for rural areas. Iowa, a state often considered an innovator in the implementation of advanced telecommunications technologies, provides an example of opportunities and problems faced by rural communities in relation to telecommunications.

In Iowa, local telephone service is provided by 150 small, locally owned, independent companies and three large, absentee-owned corporations. The State of Iowa owns and operates a statewide fiber-optic system known as the Iowa Communications Network (ICN). With at least one point-of-presence in each of the 99 counties in Iowa, the ICN is intended primarily for educational purposes with legal users limited to educational institutions (K–12 schools, public and private colleges and universities), libraries, health care facilities, and state and federal government agencies. In the private sector, the three large telephone companies provide fiber-optic cable to the metropolitan areas, larger cities, and many rural towns in their service areas. The Iowa Network Services (INS), a consortium of 130 small, independent companies, operates a fiber-optic network that serves many rural communities. The Cooperative Extension Service has a wide-area network that links every county, area, and state Extension worker in Iowa, providing access to e-mail, the Internet, and the World Wide Web. Despite this seemingly wide availability of telecommunications technologies, however, there is little evidence that rural leaders and officials in Iowa are prepared to exploit the technologies and benefit from them.

Over the past four years a multidisciplinary team of researchers at Iowa State University has examined the potential of telecommunications technologies for use in community economic development. We began with case studies of four rural communities to identify the relevant issues of telecommunications-based economic development (Gregg, Abbott, & Korsching, 1996). Based on the literature in this area and what we learned from these case studies, we conducted statewide surveys of different sectors within rural communities deemed particularly relevant to the local adoption and use of telecommunications technologies. These community sectors included telephone companies, economic development organizations, municipal government, businesses, county offices of the Iowa Cooperative Extension Service, hospitals, libraries, community newspapers, and farm operators.[1] Each of these sectors has a unique position within the community as well as resources to promote telecommunications use. Each also experiences pressures that discourage or impede the adoption and sharing of telecommunications technologies. In conducting these surveys all members of the research team maintained a primary focus on the role of telecommunications in rural community economic development and several questions were standardized and included on all survey instruments. The larger share of the survey instruments, however, was related to sector-specific issues, and how those issues were defined and addressed was left largely to the judgment of team members with relevant disciplinary expertise. Not having a narrowly defined standard survey instrument for all sectors strengthened the research in that the results raised additional issues and provided insights which otherwise might have been missed.

To set the stage for the sector studies, we began by examining the policy and regulatory environment in which rural communities must function, with emphases on federal policy initiatives and their implications for rural telecommunications development. Three primary issues emerged: the tensions between public and private control, their implications for universal access, and their impact on continued competition and provision of affordable service for rural areas. Government involvement can both promote and inhibit community telecommunications initiatives. Forms of government involvement that promote telecommunications initiatives include financial assistance, enabling legislation, policy incentives, and the removal of regulatory barriers. For example, the federal Rural Utilities Service provides rural telephone companies with grants and loans for infrastructure development and service provision, and state right-of-way laws make it possible for telecommunications providers to have access to property necessary to expand their services. In contrast, local zoning laws, such as those that restrict the construction of antennas, can impede telecommunications development. Several outcomes are possible when government actually gets into the business of telecommunications. In Iowa, the state constructed a $500 million fiber-optic system to provide telecommunications services to public agencies across the state. But what the state characterizes as contributing to rural development may actually hinder it because private telecommunications providers lose their biggest rural customers and may have less incentive to invest in future system upgrades.

We might expect the providers of local telephone service to take the lead in telecommunications promotion and development in rural areas because they have telecommunications expertise, insights on the needs of the local community, access to resources, and the potential for increasing revenues by providing telecommunications development assistance. In particular, we might expect the largest providers—those with the greatest expertise, resources, and revenue potential—to be most involved in rural development efforts. Involvement in local development activities, however, was significantly greater for small, locally owned telephone companies than large, absentee-owned companies, regardless of whether that involvement was interaction with development organizations, assuming leadership in economic development, or providing financial support for economic development activities.

Of course, local telephone companies no longer have a monopoly in their service areas. The Telecommunications Act of 1996 opened the local exchange to competition from other entities, including the local community. Rural municipalities are becoming telecommunications entrepreneurs in an effort to make these technologies available and accessible to their residents, but this is occurring amid concern and controversy regarding the proper role of public investment in insuring a competitive environment. Local government administrators know that the Internet, high-speed data transfer, and excellent telephone service are critical for the long-term economic well being of their communities and the ability to conduct the business of government efficiently and effectively. Many, however, are unable to take advantage of these technologies because their rural communities have systems insufficient to accommodate the new services. A number of communities have countered this by initiating municipally owned telephone companies and using public revenues for telecommunications development. Referenda in more than thirty Iowa communities have approved the creation of such locally owned telecommunications companies.

We might expect local economic development organizations and their directors to provide leadership in applying telecommunications to development. In a survey of community economic development professionals in rural Iowa, however, most reported little or no contact with their local telephone companies. In addition, they were poorly informed about public efforts to develop a statewide telecommunications system, they lacked models of successful use of telecommunications technologies for promoting economic development, and they had little or no personal experience in using these technologies (Bultena & Korsching, 1996).

The case studies of four rural Iowa communities indicated that the impetus for business and organizational use of such technologies comes largely from sources outside the community, and there is little sharing of the technologies or information about them within the community (Gregg et al., 1996). Local organizations and institutions (government, main street businesses, the library, newspapers, hospitals, and schools) are adopting the technologies, but there are few attempts to organize these innovations on a communitywide basis or systematically plan for their use to benefit the community and improve services. In light of rural traditions of local autonomy and self-sufficiency built upon collaborative efforts among rural residents, this lack of cooperation and sharing raises important questions and concerns. Is there a changing ethos in rural America that precludes collaborative efforts in telecommunications development, or alternatively, do the characteristics of telecommunications technologies themselves impede communitywide adoption and use in rural areas? Do communities and organizations that have been successful in the dissemination, adoption, and use of telecommunications have unique characteristics that foster such development? And to what degree is it necessary for the impetus and capacity to adopt telecommunications technologies come from outside the community?

Rural business subsidiaries, as an illustration of this issue, may be directed by parent companies to adopt standardized computer capabilities and, indeed, may provide outside expertise to bring the local business online with headquarters and other subsidiaries. Linkages within and outside the organization also play a role in the adoption of telecommunications technologies by rural hospitals and health care providers. Although internal efforts for efficiency and cost savings may be an important incentive, it is more likely that forces external to rural hospitals impel telecommunications adoption. In this case it takes the form of internal versus external control and use. We found that the most common use of computers in rural hospitals, for example, is for electronic billing, ordering, and general record management, but a substantial proportion of computer adoption was mandated by insurers, licensers, governmental agencies, and other outside groups on which the rural hospital is dependent.

In addition to external impetus for telecommunications development by rural organizations, budget flexibility and the support of top management also were deciding factors in adoption of the technologies. This we discovered in studies of rural libraries and businesses, where achieving a critical mass of financial resources and clientele or users was key to telecommunications use. For libraries, critical mass took the form of annual budget and size of readership (number of library cards issued), whereas for businesses critical mass meant gross sales and the extent of vertical integration with parent organizations. Adoption of telecommunications technologies was greater in competitive industries and those with clearly seen advantages in adopting information technologies. For rural businesses, facsimile (FAX) is a major means of telecommunicating, followed by online data access using a modem. Word processing and accounting are the most popular applications. Lack of a local access number, however, is an inhibiting factor for Internet use. By way of contrast, nonadopters see little value in such technologies, or they are concerned that these technologies may be incompatible with their current business practices and value systems.

Surveys of local government leaders in Iowa's rural communities demonstrated that local governments see advantages in telecommunications development, and a growing number have e-mail, Internet access, access to the World Wide Web, and even municipal Web sites (homepages). Critical mass, in the form of community size and resource base, is a determining factor however, with larger government jurisdictions more likely to be active in adopting telecommunications technologies to conduct city business and more likely to promote telecommunications development within their communities to expand economic opportunity. But regardless of size, most jurisdictions are excited about prospects of telecommunications and many fear being left behind due to their inability to develop internal and local expertise. In addition to this paucity of technical expertise, other constraining factors are initial and continuing expense of technologies for small organizations, and the current inadequacy of local telecommunications infrastructure. These concerns become more pronounced as community size diminishes. Despite these fears, or perhaps because of them, some local jurisdictions are assuming public responsibility for economic development and playing an entrepreneurial role in telecommunications development.

For newspapers, use of innovative telecommunications, in particular the Internet and World Wide Web, were related to circulation size, in other words, readership and advertising base, and thus financial capital. Having the in-house technological base to use the Internet and Web, using computerized business and production technologies in the newspaper operation, using various online technologies, and operating a Web page all were related to a continuum from small weeklies to large weeklies to small dailies to large dailies. Also, respondents from dailies were more likely to perceive a central community role for the Internet and Web, and believe that telecommunications will strengthen the rural community rather than widen the gap between rural and urban communities.

Like other businesses, farms demonstrated increasing use of computer and telecommunications technologies, but both type and amount of use were a function of the scale of the farming operation. In large operations a greater percentage of use was for farm management purposes than in small farms. The uniqueness of the family farm as a business provides opportunity for nonfarm use of telecommunications technologies and use by other members of the family. Small farms had a great volume of use devoted to nonfarm purposes. Human capital seems to play an important role in the difference between types of technology applications. Education and managerial abilities of the farmer are strongly related to farm use. All farms experienced expanding use by other members of the farm family.

In a study of county Extension offices, we looked at Extension's online communication patterns before and after installation of a wide-area network. Despite initial reluctance on the part of Extension workers to approve the investment in telecommunications technologies, their subsequent adoption and diffusion proved quite successful and most are now enthusiastic supporters. Keys to winning their support, however, were high-quality, user friendly software, universal access to computers, a reliable system, and good training. While community development seeks to achieve telecommunications adoption and use through democratic means, the top-down planning and implementation approach used by Extension demonstrates that centralization of efforts can have benefits in reducing fragmentation, ensuring standardization, supporting training, and assuring compatibility of the system. It also demonstrates that the imposition of telecommunications technologies on reluctant users does not preclude creative and democratic use of the technologies once they are installed.

ECONOMIC DEVELOPMENT FOR COMMUNITIES

Economic development is defined as any effort to increase local capacity to generate income and employment to maintain, if not improve, the community's economic position (Summers, 1986). Economic development includes job creation, income enhancement, and the development of infrastructure to achieve those ends (Summers, 1986; Shaffer, 1989). Five basic strategies of economic development have been identified: (a) attracting new basic employers, also known as business recruitment or "smokestack chasing," (b) improving the efficiency of existing firms, also known as business retention and expansion; (c) improving the ability to capture dollars, such as marketing local products or services to the outside or adding value before exporting; (d) encouraging new business formation or "entrepreneurship", and (e) increasing aids and/or transfers received by the community, such as bringing in grants and state and federal financial assistance (Shaffer, 1989).

Development implies improvement, and economic development refers to both the process of achieving that improvement and the improvement itself. However, economic development may not be community development, particularly with regard to telecommunications. One common economic development strategy—attracting telemarketing to a community (which actually is a variation of smokestack chasing)—provides a case in point.

Telemarketing has readily established itself in rural communities. Given the proper telephone and computer infrastructure, telemarketing is not constrained by the remoteness of any particular location, and remote rural communities have aggressively sought to become telemarketing centers (Leistritz, 1993). Yet having a telemarketing firm may provide few benefits to a community if the employees work part-time at minimum wage. The company and its absentee owners or stockholders may benefit, but the local community will receive few, if any, benefits from this "economic development." Such a development strategy does little to enhance the human capital (knowledge, skills, talent, and commitment) of local community members and, indeed, it may siphon away social capital (the structure of supportive networks and relationships) by placing priority on relationships outside the community. In other words, rather than making an investment in community capital, such a development strategy depreciates community capital by de-skilling workers and interrupting local networks and relationships.

To ensure benefits for the community, community development and economic development should be integrated (Ryan, 1988). Development efforts should be focused on physical, financial, human, and social capital investments, rather than depreciation. Economic profit external to the community should not come at the expense of community capacity and potential; nor should it depreciate human and social capital within the community. Expanded human and social capital should be an integral part of efforts to improve infrastructure and introduce new revenue sources. "Community development can serve as a guide to economic development activities, making certain that the process is in a direction favorable to public rather than private concerns" (Ryan, 1988, p. 361). We might add, making certain the process is favorable to local networks and relationships.

A COMMUNITY DEVELOPMENT FRAMEWORK

Our research in Iowa confirms the importance of "having all the right connections" in terms of infrastructure. It also suggests that other, nontechnological factors may be equally important. Identifying these factors, determining their roles in telecommunications-based economic development, and incorporating them in a development program requires a conceptual framework. Although our primary focus is on economic development through telecommunications adoption, we use a community development framework in this research. A community development framework accentuates capacity-building within a community as the preferred goal, a process that is likely to balance economic benefits and costs against equally important social, aesthetic, and sentimental interests of community members. Community development includes deliberate efforts of a group of people in a community to improve their social, economic, or cultural situation (Christianson & Robinson, 1980). This definition suggests community development (a) is a process devoted to local improvement, (b) which is publicly oriented in terms of participation, implementation and resulting benefits, (c) whose participants are primarily local residents.

Numerous factors can facilitate or hinder the benefits of telecommunications technologies for community development (Schmandt et al.; 1991; Markus, 1987). Some of these factors are local, while others are external to the community. A community's wealth, assets, or resources—in a word, "capital"—are critical investments if significant community development is to result from telecommunications. And community capital exists in several forms. Most commonly recognized are the more tangible forms—physical capital and financial capital—but equally important are the less tangible forms—human capital and social capital.

Physical capital includes natural, physical, and material resources existing within a community or available to it. In our immediate context, physical capital includes telecommunications infrastructure—the telecommunications technologies owned and/or operated by organizations and individuals within the community. These include telephone systems, copper wiring and fiber-optic cabling, switching devices, satellite connections, computers with modems, networks, and software that enable these systems to function.

Financial capital consists of hard currency, savings, revenues, or instruments of credit available to the community for investment or speculation (Flora, 1995). In terms of telecommunications technologies, financial capital assumes numerous forms including retained earnings of providers, credit financing, government grants, low interest loans, and private venture capital, as well as personal financial resources for equipment purchases, maintenance, and training.

In contrast with material forms of capital which are measured using economic indicators, the intangible forms of capital are more difficult to measure, but nonetheless require critical levels of investment for successful telecommunications adoption and use. *Human capital* consists of the knowledge and skills possessed by community members or available to them that affect the adoption, adaptation, implementation, and reinvention for innovative applications of telecommunications technologies. Because the cumulative knowledge and skills are critical elements for community development, an important human capital is local leadership that is able to harness the knowledge, skills, abilities, and talents of community members.

Social capital resides within the networks of relationships between and among community members. It is a group rather than an individual possession, because social capital "inheres in the structure of relations between actors and among actors" (Coleman, 1988, p. S98). Social capital includes "features of social organization, such as networks, norms and trust that facilitate coordination and cooperation for mutual benefits" (Putnam, 1993, p. 35). Three components of social capital combine to form the basis of a collective conscience for community action: norms such as obligations, trust, and information (Coleman, 1988).

How can rural communities capture the benefits of physical capital, especially telecommunications infrastructure and technologies, for their social and economic revitalization? While adequate financial capital plays an obvious and major role in the process, human and social capital are also crucial to optimize the benefits of telecommunications technologies for rural communities. Visionary leadership able to mobilize local networks and capitalize on external linkages to technologies is particularly critical, as is the ability to glean information for action. Information acquisition can be a costly process both in time and money, but some information can be obtained through social relationships maintained for other purposes (Coleman, 1988). Human and social capitals are, in essence, resources invested for the development of a community's abilities to act collectively in meeting common needs (Flora & Flora, 1995; Ryan, Terry, & Besser, 1995).

Now, most decisions for implementing community development are made locally, but decision making and implementation are affected by linkages or relationships that individuals, businesses, agencies, organizations, and institutions within the community have with others outside the community as well as by linkages or relationships maintained locally (Warren, 1978). The former we refer to as vertical patterns or linkages, while the latter we refer to as horizontal patterns or linkages. Vertical relationships outside the community typically involve different hierarchical levels of power and authority, while horizontal relationships, in contrast, are at approximately the same hierarchical level, that is, within the community.

A community's horizontal pattern consists of formal and informal associations that maintain the structure and identity of the community and further its ability to perform necessary local functions. Much of the interaction that occurs within the horizontal pattern is specifically community oriented; that is, the principal actors are local residents, goals or interests of the actors are clearly identified with the locality, and the actions serve public rather than private interests (Wilkinson, 1991). When such community actions strengthen and improve relationships that facilitate local action on behalf of the community, these actions are termed *community development* (Wilkinson, 1972, 1986).

Community development thus entails strengthening working relationships or networks within local communities to build their capacity to address and resolve community problems and issues. The stronger these local networks, the greater the potential for successful community action and development (Wilkinson, 1991; Richards, 1984; Warren, 1978). This is not to imply that strengthening community networks, whether interpersonal or interorganizational, is the sole objective of community development. On the contrary, community development encompasses both structural change (i.e., strengthened relationships and networks) and task accomplishment (i.e., the concrete results of the action) (Kaufman, 1985). In other words, community development is a context, content, and outcome of local capacity building.

For community development to occur—for community structures to evolve into a singular entity that is capable of initiating and implementing action for the benefit of the larger community—individuals and organizations must set aside selfish interests and focus on their common interests. They must mutually define goals for their communities and work cooperatively to attain these goals (Coleman, 1988). This is rarely a spontaneous process. Rather it hinges on several factors, including the fund of local human capital or, more specifically, the capability, knowledge, skills, and commitment of local citizens, and the vision of local leaders (Garkovich, 1989). Community leadership mobilizes local capacity toward meeting community needs and solving collective problems. Through applications of human capital, community leaders mobilize social capital.

Garkovich (1989) identifies several functions community leaders perform in mobilizing local capacity (i.e., cumulative knowledge, talents, and skills) and network relationships to accomplish specific tasks that improve community welfare. First, leaders are visionary. They not only see immediate problems facing their communities, but they also have an image of what their communities can become through productively channeled activities. Second, leaders assist in community decision making by providing the knowledge base necessary for informed decisions. They know the problems and assets of their communities, understand the local effects of larger societal forces, and have information about sources of financial and technical assistance. Third, through their vision and knowledge about problems and resources, leaders identify potential and feasible action programs to address community problems and attain collective goals. Fourth, leaders manage group dynamics in the actions required for achieving community goals.

A fifth function that leaders perform for community development involves horizontal and vertical linkages. Leaders work to identify and procure resources necessary to initiate and sustain community actions. These resources may come from within or without the community. Garkovich (1989) states that the ability to mobilize action and identify resources depends upon "the strength and complexity of the networks that link local organizations together and those that tie the local community to external organizations and leadership" (p. 205). We have defined these ties as horizontal and vertical linkages. Horizontal linkages provide the community with the internal cohesion needed to solve its problems, while vertical linkages provide the community with access to information and resources not locally available.

Vertical relationships can potentially have negative impacts on communities in that they reduce the amount of control or autonomy that communities have over their own affairs and destiny, but they can also be positive, providing a mechanism through which new ideas and technologies are channeled into communities. Granovetter (1982, 1973) refers to this critical role of external linkages as the "strength of weak ties." Research has shown a relationship between a local community's share of economic growth and community leaders' contacts with elites in metropolitan centers (McGranahan, 1984). Leaders of entrepreneurial communities realize they cannot depend solely upon their own resources for development, but must gain access to additional needed resources, such as funding and management, through external networks (Flora & Flora, 1993).

CRITICAL MASS AND COMMUNITY ECONOMIC DEVELOPMENT

We have used an economic metaphor to conceptualize the four forms of community resources—physical, financial, human, and social capital—and stated that the community must invest in those resources to attain desirable development. But what level of investment is necessary to foster telecommunications-based community economic development? In other words, is there a threshold below which telecommunications development cannot succeed for want of a critical mass of physical infrastructure, financial backing, human talents, and social networks? Critical mass refers to the totality of individuals' actions to achieve a public good; a theory of critical mass "seeks to predict the probability, extent and effectiveness of group action in pursuit of that public good" (Markus, 1990, p. 201). Identifying such a threshold, or critical mass, could help communities identify whether, or at which point, they need to aggregate demand to achieve telecommunications development goals. Issues to be considered in relation to critical mass for each of the types of community capital include the following:

Physical Capital and Critical Mass. Modern complex technologies such as telecommunications technologies need a critical mass of infrastructure to support their adoption and use (Brown, 1981). One of the factors leading to the long-term success of Cabela's, a telephone order catalog house for hunting, fishing, and other outdoor equipment, is the installation of a point-of-presence (POP) at its Kearney facility. The POP offered Cabela's enhanced capability for its long-distance telemarketing and also helped reduce long-distance telephone costs. The Internet was made possible through the critical mass of telecommunications infrastructure the government developed in its effort to establish the USENET or ARPANET, the Internet's precursor. In Iowa, the ICN fiber-optic infrastructure has linked rural and urban hospitals, thus creating the potential of the ICN's use for telemedicine. Without first having the critical mass of telecommunications infrastructure that links people, organizations and communities, little adoption and use of telecommunications technologies can be expected.

Financial Capital and Critical Mass. Investment in telecommunications technologies represents unique challenges that affect access to financial capital. Rapidly changing technologies and the potential for rapid obsolescence create risk for the investor. The costs of technological upgrading may represent high volume outlays beyond the financial capital levels accessible to smaller telecommunications providers, thereby biasing participation. Moreover, investment in the technologies often represents software and intellectual or human capital more than hardware, which may discourage conventional lenders. Perhaps the most direct tie of financial capital to critical mass theory is that of assembling the potential demand to economically justify the up-front financial investment in the newer technologies. Smaller rural providers, especially, face this dilemma. Private-public consortiums may be needed to assemble an investment strategy that will justify the financial capital outlay.

Human Capital and Critical Mass. Telecommunications technologies, such as a microcomputer with modem, are often purchased by individuals and many different models and software can be found in rural communities. Despite this individuality of machine and software, research has found that adoption and use rates tend to be higher where a critical mass of knowledgeable people and users are present (Yarbrough, 1990). Case studies of communities also have found that a crucial element for setting the telecommunications process in motion is a local champion or visionary (Abbott, 1997). Even one such person in a community may begin the process in which continuously increasing numbers of locals find that, because the actions of others have made the technologies more feasible, they too want to participate (Oliver, Marwell, & Teireira, 1985).

Social Capital and Critical Mass. In rural development, community organizations and groups often play key roles in shaping community infrastructure investments, educational and public use decisions, and the general business climate for new telecommunications technologies. Building on Mancur Olson's *The Logic of Collective Action* (1965), considerable attention has been devoted to applying the principles of collective action to the problem of critical mass for new technologies (Markus, 1987; Oliver et al., 1985; Oliver & Marwell, 1988; Marwell, Oliver, & Prahl, 1988; Rogers, 1986). To what degree would community residents, especially members of the business community who might be in direct competition with each other, be willing to share telecommunications information and technologies with others with expectations and trust that those acts of sharing would be reciprocated? Such sharing should be more prevalent in a community with high levels of social capital, and, in turn, high levels of telecommunications innovation.

The case of Aurora, Nebraska, (Abbott, 1997) demonstrates how one community invested physical, financial, human, and social capital to achieve an effective telecommunications network for community economic development.

CASE STUDY: THE CASE OF AURORA, HAMILTON COUNTY, NEBRASKA

Following a 35-year history of community collaboration that includes creation of an industrial development corporation, community foundations, and local health facilities, Aurora and Hamilton County were well situated to take advantage of promising telecommunications innovations that facilitate future economic robustness. Citizens of Aurora have mobilized the resources needed to keep the community on "the leading edge of the information age." Community leaders have upgraded the telecommunications infrastructure, availed themselves of the latest information in technological innovations, and implemented an awareness and training campaign that opens access to telecommunications tools to any community resident who might want them.

Since 1988 the local telephone company, Hamilton Telecommunications, has provided digital phone service in every town in Hamilton County. In addition, for more than a decade fiber-optic cable has connected every town in the service area and provided links out of the community as well. Installation of these upgrades was costly however, with Hamilton Telecommunications hedging its investment against prospects for expanded service areas and the provision of new services to community entities as an outgrowth of community economic development efforts.

An initial impetus for this level of infrastructure investment came from a local business owner who founded a telemarketing firm to buffer against job losses that resulted from the farm crisis of the 1980s. His efforts not only created jobs within Aurora, but also resulted in increased demand and use of telephone lines. His investments multiplied within the community, allowing Hamilton Communications to upgrade services with the installation of fiber-optic cable.

The community took advantage of grants available for telecommunications development as well. Among these was a grant for special phone services for the deaf; this funding provided a springboard for the company to provide new digital switching equipment that could be used by the entire community. In response to increased use of the Internet (and concomitant escalating costs) by area schools, Hamilton Communications subsequently became an Internet service provider, serving Aurora and other communities beyond their original service area. Indeed, nine other telephone companies now use this Internet service, which permits Internet access without the expense of long-distance phone connections. In the course of this development, Hamilton Communications has grown from a company of 35 employees to some 250 today, their remuneration representing nearly $5 million in additional revenue to the community.

The story of Hamilton Communications demonstrates the key role to be played by local telephone companies in telecommunication-supported community and economic development. This small rural telco was able to mobilize extant resources within the community as well as attract resources from outside the community. While Hamilton Communications managers were the visionary force and champion of efforts, they, of course, did not do it alone.

Although much of the infrastructure was in place by the late 1980s, a 1993 meeting of community members was pivotal to the community's accelerated use of telecommunications. Community organizers recruited 175 residents to attend a town meeting to discuss the future of telecommunications for Aurora and Hamilton County. Representatives of the chamber of commerce, city council, school board, newspaper, churches, health service, and business and industry were among the citizens in attendance, and Don Dillman, national expert on telecommunications and community development, spoke to the assembly. University experts provided important assistance to the community as well. The meeting focused on community initiatives to take advantage of telecommunications, and was an exercise in community consensus building. This communitywide meeting and planning session was followed by general and targeted community surveys, and organization of the information technology task force. The task force developed an Information Technology Strategic Plan and its implementation further solidified community efforts.

Myriad activities followed development of the strategic plan: The school board wired all classrooms for Internet access and hired an information technology specialist. By 1997, 68 percent of Aurora students had computers at home; 25 percent of these had online access. E-mail also was made available to senior citizens through the school project. Two-way videoconferencing is planned. The local computer store teaches Internet classes as well as classes on a variety of software applications, and a local business has installed fiber-optic cable throughout its plant and disseminates information to the community. The city administrator developed a Web page to promote Aurora. The local library bought computers and provides free Internet access to library patrons. The telephone companies put the Web pages of Hamilton County businesses on their server free of charge. A local satellite downlink system is installed, and a local cable channel provides five scheduled programs each month. The local newspaper is online and continues to play a critical role in information and support of the project. University Extension provides training to farmers on the potentials of telecommunications. And a hands-on science center provides demonstrations on information technologies as an important resource to the community (Warren & Whitlow, 1997).

This case study of Aurora and Hamilton County, Nebraska, demonstrates that visionary leadership and collaboration are keys to telecommunications-based community economic development. It illustrates how critical mass, vertical-horizontal and internal-external linkages, a development "champion," and the synergies of community capacity-building factor into telecommunications adoption and use. The success of Aurora and Hamilton County also makes clear the importance of community organization, especially the complementarity of informal and formal organizations. And it offers hope of opportunities for rural youth as telecommunications resources, trainers, implementers, troubleshooters, and entrepreneurs who may creatively use telecommunications technologies in the near future as economic development tools. In Aurora and Hamilton County we see the creative and committed investment of physical, financial, human, and social capital to achieve a level of telecommunications development with substantial benefit to rural areas.

PURPOSE AND OBJECTIVES OF THE IOWA RESEARCH

Community organizations and institutions act as gatekeepers, disseminating technologies that might not otherwise reach individual households. Because new technologies often reach rural audiences indirectly through local organizations such as libraries, local government, county Extension offices, or health care providers, our research sought to examine these organizations and their influence on the adoption of new telecommunications technologies in rural communities. We sought to identify factors that contribute to successful adoption and use of telecommunications for rural community economic development, to assist local leaders, community service providers and businesses, community officials, and state policy makers capture the potential benefits of innovative telecommunications technologies for local economic development. We also sought to identify potential problems and pitfalls to be avoided by communities eager to implement telecommunications development strategies.

We took a manifold approach in this volume to achieve these goals. We gathered case studies and other research from throughout the Midwest to illustrate issues we raise. To broaden the geographic scope and relevance of the project we conducted the "Making Wise Choices" conference, which brought together community development practitioners, state policy makers, telecommunications providers, and rural development scholars from across the United States to share and highlight specific examples of rural community telecommunications development activities. The majority of our examples, indeed the majority of our expertise, however, have evolved from the experiences of Iowa communities. We conducted case studies of select rural communities in Iowa to identify current telecommunications use as well as the community networks and linkages that contributed to successful adoption and use. We then conducted statewide surveys of particular community sectors or institutions to determine current telecommunications technologies uses, visions for the technologies' application for community economic development, and needed education, technical and other support to utilize these technologies within local communities. We identified innovative projects that have enhanced telecommunications in rural communities and contributed significantly to community development efforts. Finally, in the discussion of findings, we highlight their program and policy relevance to provide guidance to local and state leaders and officials in using telecommunications technology in programs of rural community economic development.

This research is unique in providing detailed information on adoption and use of telecommunications within specific sectors of rural communities. While most previous research has focused on the community as the level of analysis, our focus is on particular sectors within the community, including telephone companies, local government, businesses, economic development professionals, local newspapers, public libraries, Extension to communities, hospitals and health care facilities, and farm operators. Specifically, we explore the roles sector representatives play and the attitudes they hold toward adoption and use of telecommunications technologies. In doing so, we highlight the innovative efforts within these sectors and discuss their contributions to the overall community development process. Although every state has unique and distinctive characteristics that define and shape its telecommunications environment, we believe the experiences of Iowa have implications and lessons for rural communities throughout the United States. Because Iowa is considered by many within *and* outside the state as a leader in telecommunications and, indeed, the state has implemented telecommunications policies and programs not found elsewhere, other states can benefit from Iowa's experiences not only for ideas to adopt but also for problems to avoid.

In the subsequent chapters we review the findings of our research and present community efforts and initiatives that show promise for application to other communities interested in implementing telecommunications for community economic development.

ORGANIZATION OF THIS VOLUME

This volume has three parts. Part I, which sets the stage for rural telecommunications with theory and policy, begins with an introduction to key telecommunications issues that confront rural America. We contextualized these issues within a community development framework, elaborating those elements of community capital central to organizing telecommunications development, namely human and social capital as well as physical and financial capital. By briefly providing key findings from our research and case studies from Oklahoma and Nebraska, we illustrated critical issues and demonstrated their consequence for rural communities. Chapter 2, by Dom Caristi, discusses the policy and regulatory environment for rural telecommunications, particularly federal policy, with warnings against oversimplified notions of government intervention as either good or bad for development. In Chapter 3, Peter F. Korsching and Sami El-Ghamrini provide the perspective of the organizations that historically have provided telecommunications services at the local level—the local telephone companies—and examine factors related to the innovativeness of these companies. Montgomery Van Wart, Dianne Rahm, and Scott Sanders observe in Chapter 4 that rural municipalities are increasingly assuming responsibility for community economic development, including the development of advanced telecommunications, when the private sector is unable or unwilling to make the commitment.

Part II is composed of six chapters, each chapter discussing the perspectives and actions of a community sector in its adoption and use of telecommunications. Issues raised by these discussions concern the significance of vertical and horizontal linkages, critical mass, strategies for aggregating service demand to lower costs, human and social capital requirements for successful telecommunications implementation, and the hoopla effect. The six sectors and the chapter authors are: business by G. Premkumar, newspapers by Eric A. Abbott and Walter E. Niebauer, libraries by Eric A. Abbott and Bridget Moser Pellerin, municipal governments by Erin K. Schreck and Patricia C. Hipple, hospitals and telemedicine by Patricia C. Hipple and Melody Ramsey, and farming by Eric A. Abbott, J. Paul Yarbrough, and Allan G. Schmidt.

In Part III we examine more closely what occurs within an organization as it implements a new telecommunications system, and the role of community economic development professionals in local telecommunications development. Eric A. Abbott and Jennifer L. Gregg discuss changes in communications patterns before and after implementation of a wide-area network in Iowa State Extension, and specific organization actions that led to success of the network. Brent Hales, Joy Gieseke, and Delfino Vargas-Chanes analyze the role of economic development professionals' perceptions, attitudes, and leadership in local telecommunications development in an effort to determine specific factors related to such development. In the final chapter, Peter F. Korsching, Eric A. Abbott, and Patricia C. Hipple discuss some of the myths and realities of telecommunications and their impact on rural community welfare. They also provide some lessons from our research for rural communities seeking to build a telecommunications infrastructure and promote telecommunications-based economic development.

NOTE

1. Although not including schools in this research may seem a glaring omission, we felt that there already is a large body of research on telecommunications in schools, especially distance learning, and that our resources would be more productively invested in exploring telecommunications use in sectors less researched.

2

POLICY INITIATIVES AND RURAL TELECOMMUNICATIONS

Dom Caristi

INTRODUCTION

Telecommunications development and deployment depend to a great extent on the policies adopted at the state and federal levels. Many Americans believe that things are regulated as they are because that is the only way it could possibly be done, when in fact nothing could be further from the truth. The United States has always operated under the assumption that telecommunications companies should be privately owned, which has not been the dominant worldwide model. Most other countries have owned all or part of their telecommunications systems and have only recently investigated the possibility of privatizing them.

While the United States has favored private over public telecommunications systems, it has had a bias toward monopoly-owned telecommunications systems. For most of our telecommunications history, American Telephone & Telegraph (AT&T) has operated as a natural monopoly for local and long distance telephone service. It has been less than 20 years since the United States recognized that monopolization of telephone service was not necessary for an efficient system. We are only now beginning to explore the extent to which competition can be introduced while still providing service to all.

Although competitive long distance providers flourished in the 1980s, the area where the least amount of competition has occurred has been in the local exchange loop. Despite the fact that most Americans can choose from many long distance providers, those same Americans are forced to use the one and only local exchange provider in their community. This is especially the case in rural communities, where the belief is that density would never be adequate to provide financial incentives for competition. The Modification of Final Judgment, the legal document that governed the breakup of AT&T, was based on Judge Greene's belief that "the exchange monopoly is a natural monopoly." Some contend that the rural telephone exchange will always be a monopoly, or that perhaps it *should* remain a monopoly (Selwyn, 1996, p. 87). With a low density of customers, economic efficiency might only be possible if one provider has all of the customers in a rural market. In larger markets, competitors that duplicate the hardware and software necessary to provide modern phone service are less efficient than providers who move telecommunications traffic all over one system. Assuming that one system is capable of handling all the traffic, a competitor who constructs a new system creates redundancy for which the customer inevitably pays. In a competitive market, the same number of ratepayers must subsidize two physical plants instead of one. The decades-old argument that local telephone exchanges must be monopolies is based on this efficiency model.

On the other hand, more than a decade before the Telecommunications Act, prominent commentators like economist Alfred Kahn and future Supreme Court Justice Stephen G. Breyer suggested that conditions might change to make competition possible in the future (Sidak & Spulber, 1997, p. 64). As Vogelsang and Mitchell (1997) demonstrate, belief that a natural monopoly has to exist in local telephone service is flawed for four reasons:

- it ignores the reality of multiple-line users such as businesses and apartment buildings;
- interconnection carriers can offer different services without having to duplicate the single-line connection to the consumer (for example, by leasing lines from an incumbent local telephone company);
- other services which already have a distribution network (such as cable television) do not have the huge capital investments associated with "start up" telephone operations; and
- wireless services do not operate on the same economies of scale as wireline service (p. 54).

The balance between public and private involvement in telecommunications is not ruled by some magic formula. Moreover, the formula likely changes given different technologies and market realities. The "ideal" mix of public and private involvement 100 years ago (if such existed) is unlikely to be appropriate for telecommunications development in the new millennium.

Government involvement can promote or inhibit the development of an industry. While a few might contend that any government involvement is inappropriate (Gasman, 1994), it is simplistic to believe that all government intervention is either good or bad for development. The telecommunications industry is no different: some government involvement can stimulate telecommunications development while other action can retard telecommunications growth.

Involvement can take many forms. If a government entity wants to become involved in telecommunications development, it can do so in one of three ways. First, the state might decide to provide *financial assistance* for telecommunications development, either in the form of direct subsidies or by providing low-cost loans for development. Second, the state can enact *favorable legislation* that creates an environment conducive to new telecommunications investment. Some states and locales create legislation that can impede telecommunications development, while others have laws that make telecommunications development less tedious. One example might be pole attachment rules. Some communities have rules that favor incumbent utilities, some have rules that favor new entrants, and others have no rules at all. Finally, the state can decide that telecommunications development is so important that the *government itself becomes involved in constructing the necessary infrastructure.* All of these forms have been utilized at one time or another. To further complicate the situation, government action might actually involve two forms simultaneously. For example, when a state assists a small rural telephone company by providing low interest loans and tax abatements, it is providing both direct financial assistance and creating favorable legislation to help the company get started. Both forms of involvement may be necessary: only one of the two might not provide an adequate boost for the fledgling company. An examination of the three forms of government involvement follows.

FINANCIAL ASSISTANCE

The Rural Electrification Act (REA) provided rural phone companies with financial assistance. This assistance was provided to nonprofit cooperatives, for-profit commercial companies, and even public bodies (Adams & Stephens, 1991, p. 16). There have been proponents of revised or expanded funding roles similar to the REA or railroad and highway construction programs where the federal government provided outright grants (Williams, 1997, p. 45). Currently, Pennsylvania provides grants to counties that want to bring distance education to their local schools. The State of Wisconsin has budgeted $100 million per year for four years to assist with bringing fiber-optic lines to schools throughout the state.

The administrative process for providing financial assistance has several models. The federal and state governments can provide funding directly or can make the money available in the form of low- or no-interest loans. The Rural Telephone Bank and Rural Utilities Services have provided loans to hundreds of independent telephone companies across the nation. The State of Wisconsin actually began providing loans to school districts and libraries in anticipation of the payments schools and libraries would receive from the Universal Service Fund (USF). To allow the institutions to begin upgrading their telecommunications facilities without having to wait for federal funding, the state lends up to 50 percent of the cost of a system's upgrade (Gamble-Risley, 1999, p. 34).

An interesting development in recent years has been government agencies forming consortia to deal with telecommunications providers. By aggregating telecommunications services, smaller agencies, which would otherwise not be entitled to volume discounts, are able to take advantage of rates previously available only to large customers (Harris, 1998b). This is particularly useful to rural communities that heretofore had no opportunity to avail themselves of lower rates. By aggregating telecommunications uses by schools, county government, law enforcement, and libraries, rural counties have enough volume to take advantage of discounts from providers.

Whether the formation of these consortia favors or disfavors telecommunications development is a matter of conjecture. Pacific Bell opposed a consortium in San Diego County because its rates were regulated, while those of its opponents were not. Eventually the Public Utilities Commission allowed Pacific Bell to bid below its tariff rate, which predictably caused quite a stir with smaller telephone companies. They claimed that the baby Bell would be able to cross-subsidize its rates and offer rates below actual cost. What is not speculation is that the City of San Diego dropped plans to construct its own telecommunications system, satisfied that the consortium would be able to provide them with sufficiently low rates.

FAVORABLE LEGISLATION

Favorable legislation can be used to promote telecommunications. Government promotion of communication media is as old as the republic itself. The U.S. Constitution stipulates that Congress has the power to establish post offices and post roads. While the constitution is not itself a declaration of telecommunications policy, it provides the framework for asserting that the government is responsible for providing the means of communication. Rural Free Delivery (RFD) is a further example of a government policy established to provide communication opportunities. RFD is particularly relevant to an examination of rural telecommunications policy in that it demonstrates the government's willingness to cross-subsidize the expense of rural communication systems by revenues from more cost-effective routes. Prior to RFD mail service was available to rural areas: it just required that residents come to a rural post office to pick up their mail. In the interest of creating a more equitable system, the government provided the same home delivery in rural areas that it had been providing in more densely populated areas. Postage costs were not increased for mail sent to rural routes. Rather, costs were based on averages across the system without taking into account the added cost of rural service.

Unfortunately, cross-subsidization can only operate in a system where the more lucrative routes cannot be taken away by competitors who offer service at a lower price, not having to subsidize the higher cost routes. Known as "cream skimming" in many industries, a company can offer lower cost service as long as it can choose which customers it serves. Whether mail service, telephone, cable television, or any other system which has its operation costs tied to mileage covered, it will always be most cost efficient to provide service to more densely populated areas. The Postal Act of 1845 was passed to prohibit competition (Sherman, 1980). There was a serious threat that private postal services would cream skim by providing service along the most lucrative routes, particularly the eastern seaboard, leaving the federal system to provide service to the less densely populated southern and western United States.

The "information superhighway" metaphor so popular just a few years ago is interesting in light of the postal road requirement of the Constitution. The analogy would be that Congress, which two centuries ago provided the roads on which private citizens sent messages, should now provide the 21st century equivalent.

Without a doubt, the federal government's acceptance of the concept of universal service has been a tremendous boon to rural telecommunications development. First proposed as a goal by AT&T, the federal government was quick to accept the notion. The company's 1910 annual report stated, "The telephone system should be universal . . . affording opportunity for any subscriber of any exchange to communicate with any other subscriber of any other exchange . . . some sort of connection with the telephone system should be within the reach of all" (Tunstall, 1985). AT&T's desire to extend its network to virtually every American home may have been good for the nation, but it was certainly in the best interest of AT&T. Their network's value was directly tied to its reach. As a result, extension of the network increased the network's value and AT&T's control. As a social goal, the United States saw telephone service for every American home as worthwhile. In 1934, Congress passed the Communications Act, which included the objective for the Federal Communications Commission (FCC) to "make available, so far as possible, to all the people of the United States, a rapid, efficient, nationwide and worldwide wire and radio communication service with adequate facilities at reasonable charges" (U.S. Code). Beginning in 1949, telephone service into rural areas was subsidized as a result of federal legislation (National Telecommunications and Information Administration, 1988).

Acceptance of the concept of universal service paved the way for approval of cross-subsidization. AT&T was able to reduce the cost of rural telephone service by using revenues received from more profitable urban customers. The system was nothing new, because the federal government had been financing rural free delivery to postal customers using the same sort of cross-subsidy for decades. This policy has remained virtually uncontested by Americans. It would be speculation to assume that phone customers approve of such a policy. It is more likely that the majority of Americans are not aware of this cross-subsidization. At least one policy analyst speculates that the rural subsidy continues because of the political power of rural interests in Congress (S. Harris, 1998, p. 20).

Of course cross-subsidy works best when there is no competition, otherwise a competitor can easily cream skim the customers who are receiving artificially high prices. Before divestiture, AT&T actually had three different cross-subsidies it could use to keep the cost of rural service artificially low. First, income from more densely populated areas could be used to subsidize less profitable rural areas, as mentioned above. Second, AT&T charged a higher cost for long distance service to keep local service lower priced. This was consistent with a universal service philosophy. The value to the network was in having everyone connected; thus keeping basic service costs low allowed the maximum number of subscribers. One of the first results of divestiture was that AT&T lowered long distance costs while the regional Bell operating companies, which now had responsibility for maintaining the local network, had to increase costs and add monthly access charges. Third, under the AT&T monopoly, businesses could be charged more for telephone service than residences. While it is arguable that this cost is actually tied to network demands (because businesses often use their telephones more often than residences), there was never a direct correlation. Small businesses that used only one telephone line have traditionally been charged almost double the residential rate. As long as one local exchange carrier existed, this system could work. With the increased competition from local exchange carriers made possible by the Telecommunications Act of 1996, a natural result will be a reduction in cost for business users while the subsidy for residential users dries up. For rural customers, this almost certainly means higher rates for basic service. Rural telecommunications systems do not have the density of urban systems, thus if their cost per mile is the same, their cost per user is higher. The estimated cost for monthly telephone service for one rural part of Washington is $442 (Levinson, 1998). Telecommunications companies in Pennsylvania and Washington have made requests through their state utilities commissions for higher rural residential service rates than those they offer other customers (Bonnett, 1997, p. 25).

A number of states have already begun the "regulatory reform" process where rates become more cost related. Massachusetts, Illinois, Michigan, and California all have instituted actions to move rates to a more cost-related basis (Duesterberg & Gordon, 1997, p. 42). These actions do not generate increased revenues but result in rate increases for some users and decreases for others, based on the actual costs of doing business. In most instances, rural rates for the same services can be expected to be higher than they are for more densely populated areas. The natural result of most of these rate-restructuring processes is to increase the cost of local service while decreasing the cost to consumers for long distance services. There is some disagreement over whether this will help or harm rural customers. At least some policy analysts believe that rural customers' total bills will actually be lower because rural customers make more long distance calls than urbanites. While the local access charge will be higher for rural customers, their long distance bill will be significantly lower than it has been, thus in the end their savings may be more than for urban customers (Keyworth, Eisenach, Leonard, & Colton, 1995, p. 9). This may be a convoluted way of examining the question, though. If the nature of rural life is such that it results in more long distance telephone usage, the fact that the per minute rate for long distance calls is lower may be of little consolation. A useful analogy might be to think of two people painting their houses of significantly different sizes. If the paint store has a sale, the one who owns the larger home saves more, but (s)he also still pays more. The bottom line under either pricing plan may be that rural residents pay more per month for telephone use than urban customers do.

Because of divestiture and increased telecommunications competition, funding for universal service previously provided by within-company cross-subsidization must now be provided by government mandate. The Telecommunications Act of 1996 attempted to further the principle of universal service in a competitive era by creating a universal service fund (USF). Telecommunications providers will pay into the fund. The USF expands upon the earlier concept of universal service, adding to its original goal by providing low cost telecommunications service to schools, libraries, and health care centers nationwide. While the goal is indeed a noble one, telecommunications companies have characterized this form of cross-subsidy, now mandated by the government rather than voluntarily offered by industry, as a hidden tax on telecommunications customers. Several telecommunications companies plan to increase their charges to customers to compensate for these payments. Knowing their customers would balk at a rate increase, AT&T has introduced plans to itemize the cost of the USF on every customer's bill. President Clinton strongly supports the USF as a means of providing low cost Internet access to schools. While the FCC supports the USF, the chairman does not believe customers need to pay more. Chairman Kennard has asserted that because of savings the companies are realizing from relaxed regulation, there is enough "surplus" to pay USF fees without increasing the cost of service to customers. Several members of Congress have already expressed their belief that the USF should be eliminated (Schiesel, 1998). The debate and modification of the USF is far from settled. The FCC continues to receive comments on the funding formula, cost models, and just about every other component of the USF, down to whether funding ought to be based on calendar years or fiscal years.

Pacific Bell might assert that funds for universal service could be provided without government involvement. The company has created a $50 million initiative to promote access in low-income, underserved California communities. Cynics might assert that the fund was created only to try to appease regulators and ease the way for the company's merger with SBC Communications. Regardless of the motivation, the end result is a partnership between the local exchange carrier (in this instance, one of the original Bell Operating Companies) and nine community coalitions representing more than 100 special interest groups. Community representatives have hailed the project as a "model for the nation" (*Business Wire*, 1998).

One other difficult issue with universal service is definitional: just what service and how universal? Decades ago, "service" was easier to define. Dial tone, or POTS (plain old telephone service), was what telephone companies meant by "service." Universal meant to every home. Now, however, POTS is not the issue but rather more advanced telecommunications services, PANS (pretty amazing new stuff), with greater bandwidths for moving data and video. While most proponents still argue that universal means to virtually every home, some are acknowledging that large bandwidth services may have to be provided at some local access point, such as schools and libraries. It may be too much to expect that universal service will mean high bandwidth service to every home. Rather, it may mean POTS to every home and high bandwidth service to every community. As many have asserted, "[r]edefining the concept of universal service for application to information and communication services in the Information Age is one of our most pressing and difficult policy problems" (Curtis & Schement, 1995, p. 59).

Section 706 of the Telecommunications Act of 1996 directs the FCC and every state regulatory commission to "encourage the deployment . . . of advanced telecommunications capability to all Americans" (U.S. Code). Congress set the goal, but left it to the regulators to determine the best means of achieving it. Nowhere in the Act is there a specific recommendation of how to provide "advanced telecommunications capability" (defined as high-speed, switched, broadband telecommunications capability). There is, however, a very specific threat. The FCC is charged with conducting regular research to determine whether advanced services are being deployed in a reasonably timely manner. If the FCC determines it is not, "it shall take immediate action to accelerate deployment of such capability by removing barriers to infrastructure investment and by promoting competition in the telecommunications marketplace." Section 706 is a congressional ultimatum to states to provide advanced telecommunications universally, or the FCC will step in and do it. It is important to note that Congress does not dictate how the states should make the services available, but it enumerates precisely what it expects the FCC must do to achieve the goal if the states fail (removing barriers to infrastructure investment and by promoting competition). While Congress provides no real incentive in the form of assistance, states have an understandable desire to keep the federal government out of their affairs and will work to comply with Section 706, lest the FCC step in.

Actions by federal agencies other than the FCC can affect the provision of telecommunications services. The Internal Revenue Service (IRS) has begun an investigation into the provision of Internet service by nonprofit organizations (Reiss, 1998, p. 6). Nonprofit networks often use a portion of user fees to provide free or subsidized service to low-income clients. The IRS questions whether these nonprofit providers should be entitled to tax-exempt status. The agency claims that an Internet service provider (ISP) is not entitled to tax-exempt status if it cannot demonstrate that it operates differently from commercial ISPs. Obviously, the loss of tax-exempt status would increase operating costs and cause the provider either to increase rates or provide fewer free connections.

An estimated 350 nonprofit ISPs could be affected by an IRS ruling. The nonprofit ISPs have privatized the concept of universal service for Internet access. Without creating a nationwide collection system, the nonprofit ISPs use their revenues to provide subsidized service without being required to do so. If the IRS rules the ISPs are not entitled to a tax exemption, thousands of low-income Internet users could be off-line. How ironic that a government-mandated universal service fund is acceptable policy for providing low-cost telecommunications but a voluntary-participation consortium may be regulated out of existence. Oregon Public Networking is one such provider that may not be able to continue subsidized service.

At the state and local levels, there are examples of favorable legislation to stimulate telecommunications development. Laws which make it possible for telecommunications providers to have access to property make it possible for telecommunications to expand their reach. Zoning laws that restrict the construction of antennas can impede it. Regulations need to change as the reality changes. Author Jane Jacobs tells of public health regulations in Toronto that prohibited operating outdoor cafés for decades after the horses and horseflies that necessitated the regulations were gone (Harris, 1998a, p. 19). Diligence is required to ferret out those regulations that impede development while enacting new rules to prevent abuses.

One other example of changing local telecommunications regulatory structure is rate regulation. While the FCC once handled most of the rate regulation, more and more of this responsibility has shifted to the state public utility commissions. The method states use for this regulation shows each individual state's priorities. The two dominant methods of rate regulation have been rate of return and price cap. Under rate of return regulation, the telecommunications provider is guaranteed a certain percentage return on its investment, whereas price cap regulation fixes the maximum a telecommunications provider can charge for a particular service. In theory at least, rate of return allows a telecommunications provider the opportunity to invest in system upgrades without being less profitable, because those expenses can be included in the cost of doing business. Rate of return regulated telecommunications providers can thereby provide their customers with the latest technological advances without a loss of profit. The negative aspect of rate of return regulation is that it provides no incentive for the telecommunications provider to operate as efficiently as possible. Because the profit margin will always be the same, rate of return regulation allows telecommunications providers to operate inefficiently and still be successful. Only the customer is affected in the form of higher rates.

Price cap regulation, on the other hand, provides tremendous efficiency incentives. If the state utility commission only sets a maximum price for a service, the telecommunications provider is theoretically free to make any profit margin it can. Any efficiencies it can create drop directly to the bottom line. The negative aspect to price cap regulation is that system upgrades, research and development and other advances that don't result in immediate revenues are less attractive to telecommunications providers.[1] The Telecommunications Act of 1996 is explicit in its preference for pricing regulation variants over rate of return regulation. The Act does not prohibit rate of return, but it favors price regulations because they are more consistent with a free market approach, something fundamental to the Telecommunications Act.

The tremendous variation among states' pricing regulation schemes is a testament to the differing opinions on what structure works best. Local telecommunications providers drive many of these decisions. Providers and regulatory commissions have worked together to try to devise systems that encourage development while still allowing profitability (Williams & Barnaby, 1992, p. 1). Cole (1995, p. 36) describes a number of reasons state policies differ:

- State legislators have differed in how much authority they have delegated to utility commissions
- There is tremendous variance in the states themselves; some are mostly urban, some rural
- Some states have hundreds of phone companies, others just a few
- Political pressure varies among the states
- Regulators and regulatory environments differ in each state
- Regulatory staff and expertise differ in each state
- Some states are more affected than others by the actions of neighboring states

The changes in telecommunications system ownership since passage of the Act demonstrate some of the differences in the states. If it did nothing else, the Telecommunications Act was expected to open up local exchanges to competition, dominated for decades by one dominant carrier in each community (for most communities, that was the local Bell operating company). Policy analysts see deregulation and increased competition as almost analogous when discussing the Act (see Shaw, 1998). It is clear that competitive environments are different in each of the states and some of that difference can be attributed to the regulatory environment, which makes competition more likely in some places than others. In Illinois, the state's largest local exchange company, Ameritech, lost more than 200,000 lines to competitors. In Iowa, the state's largest carrier, U.S. West, sold off more than 80,000 lines to competitors: more than 6 percent of the state's total lines. In Indiana, fewer than 200 lines have been transferred from incumbent local exchange carriers to new competitors (Sword, 1998). While Indiana has no urban area to match Chicago (a definite draw for competitors), its density is higher than Iowa's. The difference cannot be attributed simplistically to urban-rural differences.

GOVERNMENT INVOLVEMENT IN INFRASTRUCTURE DEVELOPMENT

Many outcomes are possible when government actually gets into the business of telecommunications. There are actually three different levels for state involvement in the construction and provision of telecommunications infrastructure. At one extreme are states that invest heavily in constructing their own networks. At the other extreme are states that prohibit government entities from providing telecommunications infrastructure. In the middle are the vast majority of states that have taken no statutory action either way, leaving the door open for communities to decide for themselves about their involvement in telecommunications infrastructure. Government involvement in infrastructure allows a state to construct systems which private-sector companies would never construct because they would never be cost effective. At one extreme, the State of Iowa constructed the Iowa Communications Network (ICN): a $500 million fiber-optic system to provide telecommunications services to public agencies across the state, particularly two-way video for distance learning for schools. Private telecommunications companies were not interested in providing the service, because many of the schools on the network would never use it enough to generate adequate revenues for the provider. What the state characterized as contributing to rural development may actually hinder it, as private telecommunications providers have lost their biggest customers in rural Iowa and are less likely to invest in system upgrades. In a 1997 survey of Iowa telephone companies, nearly one-third of all respondents cited the ICN as the one factor that could inhibit their companies' ability to continue to upgrade their systems and services offered to customers (more detail on the survey can be found in Chapter 3). It is impossible to know whether more private telecommunications investment would have occurred without the ICN, but it is certain that private systems would have traffic that they do not currently carry.

In contrast to Iowa, some government entities have not only constructed their own telecommunications systems, but have begun providing services to private telecommunications providers: a reversal of the traditional role. While financially successful, the scheme puts the government in the unusual position of selling its services to private companies. A couple of California communities have created revenue streams for themselves by providing telecommunications services for private companies needing to lease capacity. In Anaheim, the city leases nearly two-thirds of its fiber capacity to a private telecommunications provider. The city installed the system and retains about a third of the capacity for the city's public utility system, but earns revenue on the majority of its lines. In Burbank, the ratio of city-used lines to leased lines is reversed, but the principle remains the same. Burbank adds to the city coffers by leasing one-third of its fiber capacity to the local entertainment community (Byerly, 1998, p. 18). In contrast, the ICN mentioned above also has excess capacity, but state law prohibits it from leasing this excess capacity to others. Telecommunications providers in Iowa adamantly fought any efforts to expand the role of the fiber network further into services provided by private providers (Caristi, 1998, p. 617). Tremendous regulatory philosophical differences exist between states that are heavily involved in infrastructure development (such as Iowa) and states where such involvement is precluded by state law (such as Texas or Missouri).

Minnesota also constructed a statewide distance learning network. Although not extended to every Minnesota county, the network is similar to Iowa's in that it was created to serve the same function of two-way, interactive distance education. The Minnesota system, however, involved telephone companies from the outset. The network was formed when the state legislature created a task force that included telephone companies and educators (Adams, Lewis, & Stephens, 1991, p. 68).

With the passage of the Telecommunications Act of 1996, the regulatory landscape has changed. The most obvious change is the mergers that are possible because of relaxed scrutiny of conglomeration and cross-ownership. An unanticipated result, however, may be an increase in the number of municipalities forming their own cable companies in communities where cable television is already available commercially.

Cable television has always been unique in the way in which it is regulated. While telephone companies look to a state public service commission and the federal government for their operating rules, cable television companies must seek the approval of communities to obtain a local franchise. In granting approval, communities have been able to extract various levels of compensation from those bidding for cable franchises. Most communities extract a franchise fee, paid by customers. In addition, one or more access channels are required of the provider in most communities. In some instances communities have requested or were offered other incentives to award a franchise.

One area where the federal government preempts local authority is in rate regulation. The Telecommunications Act of 1996 provides only limited cable rate regulation. What little rate regulation does exist has a sunset provision, resulting in no rate regulation whatsoever after 1999. Already, a number of communities have passed local referenda authorizing their cities to begin construction of a city-owned cable telecommunications service. In almost every community where these referenda have passed, they have been the result of frustration with the local cable provider and the inability to receive regulatory relief. Prior to 1996, increasing rates and frustrations with service would have been cause for local action, but under a telecommunications regulatory structure that limits punitive action by local communities, frustrated customers must often choose between the existing provider and no service at all. Thus the move toward increasing municipal "overbuilds." The cable industry defines overbuilds as two wired systems providing cable television services to the same households. The town of Medina, Washington, with only 800 cable subscribers, is contemplating a huge investment in a municipal cable system "after fining the [cable] company over bad service, without prompting much of a reaction" (*Seattle Times*, 1998, p. B3).

From a policy perspective, government certainly has a more defensible position when it provides services in a community where the private sector has opted not to provide service. When a private company offers service, however, it is more difficult to assert that the government should intervene. This issue becomes even more pronounced when the service discussed is cable television because of the inherent implications for First Amendment issues concerning a government agency taking responsibility for determining which communication services are provided to the public (Ramey, 1993). Common carriers such as telephone systems, which accept all speakers and assess fees equally, are less problematic from a First Amendment perspective. But because operating a cable system involves choosing which communicators' messages are carried (i.e., which cable channels to include), the task of deciding who should be included is better left to nongovernment interests. Concerns have already been raised over a proposed municipal cable system in Muscatine, Iowa. Pressure has been brought to bear on the city to provide only "clean" programming over the system, months before the system was even in operation (*Iowa Cable News*, 1998).

While the impact of municipal overbuilds will be substantial enough on cable television companies, it carries even more possible impact with the inclusion of other telecommunications services. Certainly a city that provides fiber-optic lines to homes can do much more than simply carry HBO and ESPN. Because the Telecommunications Act also provides for competitive local exchange carriers, communities that go to the expense of installing fiber-optic cable can help finance the cost by providing other services, most notably telephone and Internet service. La Grange, Kentucky, was motivated to investigate a municipally-owned cable system because of frustration with their current provider, but may move into telephone and Internet service because of the economics (Ellis, 1998). Thus, local telephone companies and ISPs may face increased competition from municipally owned systems as a result of cities' frustration with their local cable television provider.

There is evidence to suggest that a municipality that operates only a cable television-based telecommunications system will not generate the necessary revenue to recoup the investment. Paragould, Arkansas, began a cable television system after frustration with a local provider. After seven years of operation, the city has yet to break even on the endeavor given the large capital expenditure, interest, and principal payments. Without their tax subsidies, they have a negative cash flow (Rizzuto & Wirth, 1998). In fact, the Cable Telecommunications Association (CATA) contends that not a single municipal overbuild system has ever been able to break even, much less show a profit. Tax subsidies and increased costs for municipal gas and electric utility customers have been used to keep systems solvent. In the case of Elbow Lake, Minnesota, the city had to sell the system of about 300 subscribers to a private vendor at a loss (CATA, 1998).

When state players are involved in constructing their own infrastructure, they are tremendously advantaged over private players. Cross-subsidies are always possible for any diversified company, but the potential for city governments is tremendous. Cable television system construction can be funded from virtually any city revenue source, including the franchise fee which cities charge cable companies for their rights of way. In Ohio, cable and telephone companies are supporting legislation that would attempt to "level the playing field" for municipally operated telecommunications systems. The bill would require a community to conduct public hearings about the use of public funds for such a purpose and require that a publicly owned system pay the same taxes as a private system pays (*PR Newswire*, 1998b).

There already is evidence that municipalities involved in providing some utilities will seek to add to the services that they provide, thereby increasing their revenue base. Shrewsbury, Massachusetts, has always provided electricity to its residents. In the 1980s, it added cable television to its services. In 1998, the city became one of the nation's first to provide long distance telephone services (Rosenberg, 1998). The general manager of Shrewsbury Light & Power has stated that the revenue from long distance service (estimated to be $200,000 annually) will be used to upgrade the city's cable television system.

At least one state has acted to preempt municipalities from constructing their own telecommunications system. The State of Texas passed the Texas Public Utility Regulatory Act of 1995. It prohibits the provision of telecommunications service by municipalities. This law was challenged and upheld by the FCC (1997). The action by the Commission was not an endorsement of the Texas law but rather a declaration that the Texas law did not conflict with the Telecommunications Act of 1996. In fact, the Commission went so far as to caution other states from such outright bans:

Despite our decision not to preempt, we encourage states to avoid enacting absolute prohibitions on municipal entry into telecommunications such as that found in [the Texas law]. Municipal entry can bring significant benefits by making additional facilities available for the provision of competitive services. At the same time, we recognize that entry by municipalities into telecommunications may raise issues regarding taxpayer protection from the economic risks of entry, as well as questions concerning possible regulatory bias when separate arms of a municipality act as both a regulator and a competitor. We believe, however, that these issues can be dealt with successfully through measures that are much less restrictive than an outright ban on entry, permitting consumers to reap the benefits of increased competition. (13FCC Rcd 3460)

It was within Texas's jurisdiction to determine for itself that municipal ownership of telecommunications systems was inappropriate. Despite the Commission's admonition, this decision opens the door for other states, which may elect to prohibit state involvement in telecommunications infrastructure.

In an earlier case, the South Carolina Supreme Court upheld a decision that prevented a municipality from constructing a cable television system (Orangeburg v Sheppard, 1994). The Court ruled that a cable system did not qualify as a utility under state law, and therefore a city was not authorized to construct or operate a system. Interestingly, the South Carolina law in question would have authorized the city to construct a telephone system. In an age where a single fiber carries both, Orangeburg might be within state law to construct a telecommunications system that carried both voice and video.

The State of Missouri recently banned municipalities from the telecommunications business. The prohibition is quite specific and different from the Texas legislation in two significant ways. First, the Missouri statute provides several exemptions for government-owned telecommunications systems, including medical and educational networks and Internet-type services (Missouri Statutes, 1997). This suggests that the state believes it may be necessary to provide these sorts of services, at least in some communities, which will not be available through commercial telecommunications providers. Second, the state has a sunset clause attached to the statute, so the law will expire in 2002. Of course Missouri may elect to renew the rule at that time, but in including the clause in the original regulation, the state acknowledged that the economic and regulatory conditions which existed when the law was passed in 1997 may be different from those which exist five years later.

The issue of whether municipalities (or states) should be in the business of constructing infrastructure is not easily decided. In a post-Telecommunications Act environment, competition is favored. Ideally private industries would compete with one another, but should competition come from a government entity? Our national preference for private enterprise-provided over government-provided services would suggest that government should not compete with private industry. But what if private industry won't provide the service or provides it at an unreasonable cost? Marketplace proponents assert that a telecommunications provider that overcharges for services will quickly draw a competitor who can undercut those prices and still generate a profit. Realistically speaking, the marketplace is not sensitive enough to every need. As one critic put it:

If the consumer is poor and the service is Internet access, the need may not be met by the marketplace—period. If the consumer is a small business and the service is a very high-speed data line, the need may not be met by the marketplace at a price the consumer can afford. The overall community well-being can suffer in the absence of service, or of affordable service. (Trainor, 1998, p. 55)

Rural communities may be forced to engage in telecommunications endeavors if they are unable to attract private providers. The danger is not that the government will lose money: government services have never been intended to be profit centers. The greater danger is that a community might provide a service that is of no interest to a commercial entity today, but that becomes financially attractive at some point in the future.

The ICN illustrates this, for when Iowa first considered a statewide fiber-optic network, it was seen as something that no private telecommunications company would provide. The state wanted the system to handle its own intrastate long distance communication and to provide an interactive two-way video system for distance education. In the 1980s, no company believed there would be a revenue base to support fiber to every rural Iowa county, several with populations under 10,000. The state continues to expand the system, which has been applauded by education professionals (*PR Newswire*, 1998a). With the explosion of the Internet, traffic along the fiber lines dramatically increased. Iowa hospitals increased their use of telemedicine. State National Guard programs began making extensive use of the system for remote site training. Businesses clamored for access to Internet and video-conferencing services that the ICN could provide. Of course, because law prohibited the system from carrying commercial traffic, it could not take on business customers. On the other hand, private telecommunications companies lost major prospective clients in the schools, libraries, hospitals, and state government that use the ICN, making profitability less likely for any new system trying to meet the needs of commercial users. Because those ICN users are able to use the state system rather than private telecommunications carriers, private providers do not have the higher levels of traffic they would have enjoyed if there were no ICN. Pressure continues to mount for sale of the system and many suspect the state will divest itself of the system within the next five years (Roos, 1998, p. 1).

SUMMARY

While the Telecommunications Act of 1996 created a great deal of uncertainty, it created an equal amount of opportunity. As Sidak and Spulber (1997, p. 101) demonstrate, states have always had the authority to create contractual arrangements with utilities. The new regulatory environment, however, provides increased possibilities, including rate structuring.[2] States now have the ability to experiment with different models in attempting to promote rural telecommunications development. Whether states choose to construct their own infrastructures, provide grants for others to construct them, or pass legislation that encourages private investment is too complex for a simple formula to determine. Different mixes of state involvement have already been implemented and innovation need not stop with the already defined models. Community leaders ought to be involved with telecommunications providers and state public utility commissions in devising systems which best serve their states. This recommendation may not necessarily be new (Cutler & Sawhney, 1991, p. 236), but it is even more relevant in the post-Telecommunications Act world in which we live.

Government entities at all levels must decide whether, and how much, to invest in telecommunications infrastructure. Government investment is appropriate to construct systems and provide services that would not otherwise be available. Recognizing that demand and technology change so rapidly, every state-constructed system ought to be developed with a privatization plan. The exact timeline may be impossible to predict, but there ought to be an underlying assumption that eventually any telecommunications infrastructure constructed by the state will be privatized. This ought to encourage support by the telecommunications industry, which will recognize state action not as competition for customers, but as stimulus for increasing the reach of advanced telecommunications systems. States will construct systems that might not be seen by private investors as profitable. When (or rather if) the systems later become profitable, the state can sell the systems. If the systems never become profitable, then the government provides a telecommunications service that would otherwise not be provided. Government should be in the business of providing services available to residents that otherwise would not be, especially when the services are seen as essential in the 21st century.

The fact that we now operate under a telecommunications regulatory system that will be increasingly varied among states is a benefit. The history of cable television regulation provides a good example of the result of varied approaches. Communities have been able to negotiate on an ad hoc basis with cable television providers. The result has been a wide disparity between the levels of service and satisfaction. While there may be an inherent advantage to larger communities in their negotiations, smaller communities are in no way without negotiation strength.

Rural communities must assert their need for advanced telecommunications services, whether through an incumbent telecommunications provider, by inviting a competing telecommunications provider into their market, or by demanding state involvement. State involvement might be simply eliminating anticompetitive regulatory structures, a greater level of involvement by providing low-cost loans or grants to construct systems, or it might be the dramatic action of constructing the needed infrastructure. In each instance, communities must declare their need clearly and convincingly. In this age of varied regulatory environments, the adage about the squeaky wheel getting the grease is germane.

NOTES

1. This explanation is rather simplistic. In fact, state utility commissions can regulate using both price cap and rate of return, applied either to different services or to different sized telecommunications providers. What's more, rate of return can be based on capital, equity, investment, or some combination. The formulas and applications are far beyond the scope of this text. For a general understanding, see Thompson, (1991).

2. In spite of the Supreme Court decision in AT&T v Iowa Utilities Board (119 S. Ct. 721, 1999), states still have tremendous latitude in determining rate structures and regulatory schemes. While the AT&T decision was a strong endorsement of the FCC's authority over that of state regulatory boards, it did not eliminate the traditional role of state regulators in determining pricing structures.

3

TELEPHONE COMPANIES: PROVIDING ALL THE RIGHT CONNECTIONS FOR VIABLE RURAL COMMUNITIES

Peter F. Korsching and Sami El-Ghamrini

INTRODUCTION

Maintaining vital rural communities requires the input of many different actors and resources combined in a complex formula and executed through a more-or-less planned course of action. Success in this endeavor is as much art as science. Yet fifty or so years of research and practice in rural and community development have yielded principles useful for assisting communities in their quest for viability. Some of these principles relate to factors largely beyond the control of the local community such as proximity to an urban center, an interstate highway, a government installation, an institution of higher learning, a natural attraction, and national and global social and economic forces. Other factors are more localized, such as the local physical infrastructure of roads, transportation and utilities, and institutions such as schools and churches. Then there are the intangibles, such as the quality and vision of local leaders. Without leaders' insight into local problems and solutions, their linkages to needed resources, and their ability to mobilize the local population, little development can occur.

In today's society with an economy driven largely by information and communication, telecommunications are an important component of rural community viability and require a supporting local state-of-the-art infrastructure. A telecommunications infrastructure will not guarantee community viability, but without such an infrastructure local viability, growth, and development are exceedingly difficult. Much as initially the railroads, later all-weather roads, and more recently the interstate highway system were gateways to growth and development for rural communities, so being connected in the information system is the current gateway.

Those who view telecommunications as being invaluable in rural development offer several supportive arguments (Gillette, 1996; Dillman, 1991b; Hudson &

Parker, 1990; National Telecommunications and Information Administration, 1995; Parker, Hudson, Dillman, & Roscoe, 1989; Parker, Hudson, Dillman, Strover, & Williams, 1992; Read & Youtie, 1996; Salant, Carley, & Dillman, 1996; Schuler, 1996; Office of Technology Assessment, 1991). First, through these technologies the speed of transmission of information and data is universally fast, thus reducing historic pressures on firms to locate in urban places. Conversely, they introduce a powerful incentive for urban firms and public agencies to relocate to rural areas because of generally lower land and wage costs, more relaxed environmental regulations, more abundant natural amenities, and stronger pro-business climates. Telecommunications technologies also are promoted as increasing the efficiency and competitiveness of rural businesses. They allow telecommuting, where employees are able to work at locations, often rural, that are distant from central offices, and they encourage the "lone eagle" phenomenon in which rural entrepreneurs start businesses, often in their own homes, that serve distant markets. Finally, through innovative applications such as distance learning and telemedicine, these technologies help stay the erosion of social institutions in small communities.

Others are more dubious of the projected benefits of telecommunications (e.g., Dillman & Beck, 1988; Dillman, Beck, & Allen, 1989; Glasmeier & Howland, 1995; National Association of Development Organizations, 1994; Tweeten, 1987; Wilson, 1992). Critics contend that telecommunications allow urban firms to exercise greater domination over their local subsidiaries, subsume activities formerly conducted at the local level, and enhance their access to rural consumers. Another concern is that rural areas typically are not well equipped to exploit these new technologies in terms of their local technology infrastructures, skills to successfully use the technologies, and cultural values supportive of adoption.

For rural communities the question is not whether telecommunications technologies will influence local well-being. The telecommunications revolution is forcing extensive changes in all communities, rural and urban. The question is, will rural communities be competitive players in the telecommunications-based development game? To be competitive players it is essential that rural communities have a state-of-the-art telecommunications infrastructure.[1]

Development of such an infrastructure and its use for the community's benefit hinge on the visions and actions of local leaders. Few rural communities have considered organizing themselves to promote telecommunications-based development. Examples of public and private organizations in rural communities using relatively innovative and sophisticated telecommunications technologies to some capacity in their operations are not uncommon (Leistritz, Allen, Johnson, Olsen, & Sell, 1997). Yet few uses of these technologies go much beyond the specific internal operational needs of the organizations, and potential community benefits are lost. With some notable exceptions, managers or administrators of organizations, including publicly funded organizations with missions to provide information and education, have not considered the benefits that might accrue to their organizations and their community through the sharing of technology and technical knowledge and skills (Gregg, Abbott, & Korsching, 1996; Pellerin, Gregg, Ramsey,

Abbott, & Korsching, 1996). For example, in Iowa most libraries and all local offices of the Cooperative Extension Service are connected to wide-area networks. They could become community information centers by allowing public access to the networks rather than limiting use to organizational personnel (see Chapters 7 and 11).

Most rural communities that have current telecommunications technologies along with innovative utilization for optimum local benefits also have a group of local leaders that include telephone company managerial personnel (Read & Youtie, 1996; Parker, Hudson, Dillman, Strover, & Williams, 1992; Wilson, 1992). Involvement of telephone company managerial personnel adds additional knowledge and urgency to information technology development and use. Sawhney, Ehrlich, Hwang, Phillips, and Sung (1991) and Cutler and Sawhney (1991) suggest that a catalyst for telecommunications-based community development is the participation of local telephone company personnel in community leadership. Innovative telephone companies they studied generally were characterized by having at least one creative manager.

There is an additional factor influencing innovative behavior of telephone companies in rural communities. It is the degree to which the company is an integral part of and contributes to the social capital of the community. Social capital "inheres in the structure of relations between and among actors" (Coleman, 1988, p. S98). Its "features of social organization, such as networks, norms and trust . . . facilitate coordination and cooperation for mutual benefits" (Putnam, 1993, p. 35). Social capital is a public good that includes generalized reciprocity among residents of a community.

That is, the assurance community members have that their altruistic actions will be rewarded at some point ensures their willingness to contribute to others' welfare. People are thus less likely to opt out of civic responsibilities and social attachments, thereby creating more certainty and stability, as well as becoming models for future cooperation. (Wall, Ferrazzi, & Schryer, 1998, p. 311)

In this chapter we examine the role of Iowa telephone companies in rural community development. Specifically, we examine the current state of their technological plants and the services they provide, their plans for upgrading their physical plants, and factors that managers perceive as possible impediments to upgrading. We also examine telephone company managers' perceptions of the role of telephone companies in local development and their perceptions of state and federal telecommunications policies.

Data are from two sources. In 1994 managers of Iowa's telephone companies participated in a survey to determine the past and continuing roles of Iowa's 153 telephone companies in local development. Of 144 requests for interviews, 134 were completed, a response rate of 93 percent. Managers affiliated with multiple telephone companies were interviewed for one company only. In 1997 a follow-up survey of the telephone company managers was conducted. One hundred thirty-six managers responded to this shorter interview schedule. Both surveys were conducted through telephone interviewing.

TELEPHONE COMPANY CHARACTERISTICS

About one-half of Iowa's telephone companies are cooperatives or mutuals, the rest are privately or stockholder owned. Most companies are small, local operations. Ninety percent of companies in the survey have 11 or fewer full-time employees. Only the three large, absentee-owned corporations (U.S. West, GTE, and Frontier) have more than 90 full-time employees. The companies are fairly current in the provision of basic services. Nearly all of the companies (131) had their entire customer base on private lines, and the remaining three companies had 99 percent private lines. About 90 percent (122) of the companies provided equal access to long distance carriers serving the state, contrary to the Sawhney et al. (1991) generalization that independent telephone companies often have not been able to provide equal access and still maintain a reasonable profit margin (the Federal Communications Commission has mandated all companies provide equal access). Ninety-five companies (71%), mainly the small independent companies, have linkages, arrangements, or agreements, with other companies outside their service areas, such as other independent telephone companies and utility companies. The primary benefit gained through these linkages is saving money by sharing facilities, equipment, and personnel.

Managerial experience and technical expertise are both high for these companies and should serve them well. Eighty-six percent of the respondents to the survey were general managers and/or presidents, averaging 25 years of experience in the industry, 20 years in the present company, and 13 years in the current position. Fifty-eight percent of the respondents had at least some college education, including a few who had completed a two-year technical degree in telecommunications. Additionally, 77 percent have some other formal training in telecommunications such as short courses, seminars, and other types of technical training. Again, Sawhney et al. (1991) suggest that managers of rural telephone companies primarily have on-the-job training and little formal education, but Iowa's telephone company managers tend not to reflect that generalization.

TELEPHONE COMPANY TECHNOLOGIES AND SERVICES

Much of the research on telecommunications and rural development seems to confirm that telecommunications is a necessary but not sufficient condition for economic development. "Benefits to rural areas are especially visible given the distance barriers that rural businesses and citizen must overcome" (Sawhney et al., 1991, p. 160). Therefore, an initial prerequisite for rural communities to achieve and/or maintain viability is the development of a telecommunications infrastructure on par with urban areas. Figure 3.1 provides information on the percentage of telephone companies offering services in 1997 that are considered innovative, yet quickly becoming indispensable to current business, government, and organiza-

Figure 3.1
Percentage of Telephone Companies Currently Providing Nine Innovative Services or
Planning to Add Them within the Next Five Years

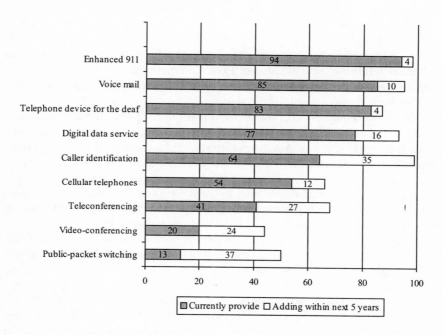

tional operations. Also included for those companies that did not offer the service
is the percentage of companies planning to offer the service in the next five years.
The nine services are teleconferencing, voice mail, cellular telephone, enhanced
911, telephone device for the deaf, video-conferencing, caller identification, digital
data service, and public packet switching. The 1994 survey also inquired about
three-way calling, dedicated lines, call waiting, and call forwarding, but because all
or nearly all companies offered each of these services in 1994 they were not repeated in the 1997 survey.

Three of the services, enhanced 911, voice mail, and telephone device for the
deaf, were offered by nearly all telephone companies, and according to stated
plans, coverage of these services will be nearly universal within five years. It should
be noted that a state program funded by the Iowa telephone industry provides a
telephone device for the deaf for eligible customers. Two other services, digital data
service and caller identification, are currently offered by somewhat fewer companies, three-fourths and two-thirds, respectively, but within five years nearly all
companies plan to offer caller identification and over 90 percent plan to offer digital data service. Within five years of the survey about two-thirds of the telephone
companies will offer teleconferencing and cellular telephones. Although only
about half of the local carriers currently provide cellular service, most of the state
has cellular service through a variety of other providers. Video-conferencing and

public packet switching, the most innovative and perhaps most technologically sophisticated of the services, are offered by the fewest companies, 20 percent and 13 percent, respectively. About 50 percent of the telephone companies will or plan to offer video-conferencing and public packet switching by 2002.

Researchers may discount projections made in anonymous questionnaires as being overly optimistic, but we feel projections of when services will be added for those companies that did not offer them are fairly objective and realistic. We had asked managers the same question in 1994. Comparing the 1994 projections with services provided only 3 years later in 1997, the proportional increases were commensurate with respondents' projections.

We also were interested in the physical plant of the telephone companies because several of the innovative services require special hardware such as fiber optic lines and digital switches. Almost two-thirds of the companies reported having some fiber lines. One hundred thirty of the 134 companies reported having digital switches in the 1994 survey. The remaining four companies planned to replace analog switching with digital switching by 1998. Seventy-nine percent of the telephone companies had Signaling System 7 for routing calls (SS7) in 1997, and an additional 18 percent planned to add SS7 in the next five years.

In summary, Iowans have access to basic telephone service, or universal service. Universal service is what was formerly known as plain old telephone service (POTS) plus additional features such as dual-tone multi-frequency signaling (DTMF), access to emergency service, access to interexchange services, and access to directory assistance. Furthermore, most telephone companies offer some advanced features such as voice mail, enhanced 911, and telephone device for the deaf. Fewer companies offer more innovative features that specifically enhance the commercial and industrial sectors, particularly public packet switching, videoconferencing, and teleconferencing. Also, the surveys asked the telephone companies only if they offered the services, not how extensive coverage of the services was in their service areas. Lack of availability and limited coverage may provide barriers to the development and expansion of certain industrial sectors.

Results in these studies of services provided by telephone companies are consistent with a study of telecommunications technologies and services used by residents and businesses of 20 communities in six Midwestern states, Iowa, Kansas, Minnesota, Nebraska, and the Dakotas (Leistritz et al., 1997; Johnson, Allen, Olson, & Leistritz, n.d.; Allen, Johnson, Leistritz, Olsen, Sell, & Spilker, 1995). The technologies with higher levels of usage were the older, less innovative technologies such as FAX machines, answering machines or services, and personal computers. More innovative technologies such as teleconferencing, e-mail, and electronic or satellite data transfer were used by 30 percent or less of businesses and by less than 20 percent of residents. Although the research does not address the availability of the various services to the respondents, results did indicate that the respondents' evaluation of the local telephone company and its services helped explain telecommunications use.

Upgrading the Local System

Telecommunications technologies, services that are created through those technologies, and the services that are demanded by customers are rapidly and continually changing. To keep pace with customer needs and demands telephone companies must be in the position to upgrade their physical plants as new technologies become available and new services become feasible. The timeliness and scope of the upgrades will depend on several factors including customer demand for services, competition from other companies, the telephone company's resource availability, motivation of the telephone company's manager, and public policies governing the industry.

We asked the managers of the telephone companies in both the 1994 and 1997 surveys what could hinder their companies' ability to continue to upgrade their systems and the services offered to customers. Although wording of the two questions was not exactly the same (in 1994 there was no limit to the number of responses whereas in 1997 we asked what one factor could hinder upgrading) the questions are sufficiently close to allow comparison. The questions were open-ended and the responses were collapsed into major categories. Table 3.1 contains the response categories and the percentage of responses for each.

Table 3.1 indicates some major differences between 1994 and 1997 in what telephone managers perceived as factors hindering their ability to upgrade their plants. The policy environment within which telephone companies operate had changed radically between the two surveys, with the federal government's Telecommunications Act of 1996. Conversations with administrative personnel of Iowa's two major telephone organizations, the Iowa Telecommunications Association and the Rural Iowa Independent Telephone Association, revealed that the two major issues facing Iowa telephone companies are competition and deregulation. Moreover, competition is intensified by deregulation of the industry as

Table 3.1
Factors Hindering Telephone Companies' Ability to Upgrade System and Services

	Percent Response	
	1994	1997
Competition from state (ICN)	16	34
Competition (especially from cable, cellular, and other companies in local loop)	11	25
Government regulations	33	13
Availability of funds	15	12
Insufficient customer demand for profit	15	11
Access revenue	0	5
High cost of technology	10	0
Total	100%	100%

spelled out in the Telecommunications Act (Jerome, 1998; Pletcher, 1998). Over half of the respondents to the 1997 survey mentioned some manner of competition as a factor hindering the ability to upgrade. The competition issue has dimensions arising from the Telecommunications Act, which are salient to telephone companies in all states, but it also has dimensions unique to Iowa. The dimensions unique to Iowa pertain to the Iowa Communications Network (ICN).

Hindrance to Upgrading: The Iowa Communications Network

The Iowa Communications Network is a state owned and operated fiber optic system with at least one point of presence in each of Iowa's 99 counties. It was constructed primarily for educational purposes with legal users limited to educational institutions (K–12 schools, public and private colleges and universities), libraries, medical facilities, and state and federal agencies. Construction began in 1991, and by 1994 the system's backbone was completed with 105 sites online. Currently, over 400 sites are online with more planned in the near future.

A primary motivation for constructing the ICN was to bring advanced telecommunications to rural areas of the state. Many of the advanced services the ICN provides were not available to rural areas nor would they have been available in the foreseeable future without state government initiative. To further facilitate use in rural areas, the fee structure is not distance sensitive, that is, the costs for any specific use is the same regardless of the distance the signal must travel. The Iowa Department of Economic Development views the ICN as a tool for economic development, particularly in rural areas. Through the ICN rural communities have services and amenities, such as video classrooms for distance education, formerly available only to urban residents.

Construction and operation of the ICN has not been without controversy (Caristi, 1998). Telephone companies perceive the ICN as state-supported competition. And because it is state supported, in terms of both the initial construction costs and the continuing operation subsidies (user fees are below cost and do not pay for the system's operation), the telephone companies perceive the competition as unfair. One telephone company manager responded to the questions on factors that hinder the ability to upgrade the system, "The ICN is state-subsidized and I can't compete with that. I may [be] competing with my own tax dollars."

As controversial as construction and operation of the ICN is, so is its future disposition. There are strong disagreements in the state legislature on actions the state should take. Disgruntled critics of the system have pressured for the sale of part or all of the system. In a recent criticism this question was asked—if the ICN is so great why have other states not followed Iowa's example and implemented similar state owned and operated fiber optic systems (Yepsen, 1999)? Proponents see the ICN as a unique resource that should be further developed and maintained. Telephone company managers, however, tend to be in relatively close agreement with each other on the ICN's future disposition. In the 1997 survey we made reference to the ongoing debate on whether the state should continue to own and operate the ICN network or whether it should be sold or leased to private companies. We then asked the managers to select one of the major options in the debate listed in Table 3.2. The respondents also had the prerogative to mention some other option or respond "don't know."

Only 5 percent of telephone company managers preferred the status quo of a state owned, operated, and subsidized ICN. The 91 percent who selected or suggested options wanted to see a change in the ICN's ownership and/or operation. By far the most popular option, selected by over two-thirds of the managers, was to sell the network to private companies. Interestingly, none of the managers wanted state ownership with some of the capacity leased to private companies. It appears that at this point the telephone companies are sufficiently disenchanted with the state's venture into what they consider to be their domain that they are not willing to engage in any collaborative efforts with the state. Overall, the responses reflect a desire on the part of the telephone companies for level competition by requiring users to pay fair costs, by privatizing the network, or by strictly limiting the use of the network to educational purposes.

In the long run, public ownership of one of Iowa's major telecommunications networks may work to the detriment of the very communities the system is designed to help, and indeed, may have some negative impacts for the state. A primary motivation for construction of the network was to provide advanced telecommunications services for educational purposes at a reasonable cost to areas of the state that otherwise would not have access to such services. Private telephone companies were not willing to make the investment, at least not in the foreseeable future, because returns on infrastructure investment in sparsely populated remote rural communities are slow. The ICN provides advanced telecommunications in-

Table 3.2
Telephone Company Managers' Preferred Options for the Future of the Iowa Communications Network (ICN)

Option	Number	Percent
State owns and operates the ICN network and pays some or all of the costs for making additional connections at schools and other eligible sites.	7	5
State owns and operates the ICN network, but new users are required to pay the full costs of their connections.	14	10
State owns the ICN network, but leases some of the line capacity to private companies that, in turn, lease lines to users.	0	0
State sells the network to private companies.	93	68
Other: State owns and operates, and provides services only to schools and libraries.	10	8
General comments for changes.	7	5
Don't know.	5	4
Total	136	100

frastructure to remote schools, colleges, libraries, hospitals, and state and federal government offices. For these institutions the creation, storage, and dissemination of knowledge and information is critical and greatly enhanced by ease of communications, and facilitates their ability to serve their respective clienteles. Unfortunately, in many rural communities these institutions also tend to be the largest going concerns and therefore the largest users of telecommunications. In other words, the largest source of potential revenues is effectively eliminated as a customer to the local telephone service provider (Caristi, 1998). Thus, whatever small incentive there may have been initially for the telephone company to eventually expand its services is now further reduced. Through the ICN's operating policy one of the primary strategies for telecommunications infrastructure development in rural areas, the aggregation of sufficient traffic to make the investment profitable, is excluded (Cutler & Sawhney, 1991). As Parker (1998) states, "If school and government traffic is segregated on a separate restricted-use network, it is unlikely that there will be enough other potential traffic to justify the investment needed for the advanced services that can stimulate economic development" (p. 629).

Therein lies the rub. Under current legislation the private sector, for which the economic development benefits of the ICN might be very large, is not an eligible user. Furthermore, given the perceptions of the telephone company managers that through the ICN the state is unfairly competing with them and tapping their largest potential revenue sources, the telephone companies are not likely to expand advanced services into unserved areas. "Not only does restricted use policy block access to the entities that can create jobs and raise wages, but the 'creaming off' of a major source of potential traffic makes it unlikely that commercial telecom competitors will step in to fill the commercial needs" (Parker, 1998, p. 629). This problem has not gone unrecognized. In a study on the role of the telecommunications industry in Iowa's economic development, the consulting firm Arthur D. Little, Inc. (1992) recommended that to derive full potential for economic development from the ICN the state should remove any legislative restrictions (including eligible users) that might impede its progress. "We recommend Iowa officials reconsider the constraints placed on ICN by the statutory requirements on use of the network, payment for it and implementation or construction" (Arthur D. Little, Inc., 1992, p. III-34). Of course, telephone companies would strongly oppose such action. Those defending the ICN state that construction of the system initially was open for bid to the private sector, but no potential bidder was willing to extend the system to every part of the state (*Government Technology*, 1997). Industry representatives, on the other hand, counter that a number of bidders responded to the initial request for proposals, but they were not allowed to integrate existing facilities into construction of the system.

Hindrance to Upgrading: The 1996 Federal Telecommunications Act

Telephone companies' concerns also originate from the deregulation aspects of the 1996 Federal Telecommunications Act. Whereas public utilities such as telephone companies serving the local community, the local exchange carriers, historically have been regulated monopolies, the Telecommunications Act opens the local exchange to competition. The deregulation of competition provides options to remote communities for building the needed infrastructure through private and/or public resources when the local exchange carrier consistently refuses. To achieve connectivity, North Adams, a rural community in a remote location in the Berkshire Mountains of Massachusetts, has been developing plans to construct major telephone trunk lines. Telephone companies say that with the sparse population in northwest Massachusetts there is insufficient demand. In response, North Adams has organized a task force of local leaders and officials, Berkshire Connect, and will draw upon public and private funds to lay a fiber optic cable (Goldberg, 1999). Municipalities across the United States that feel their needs have been ignored by the local carriers are either exploring or already have built local fiber optic telecommunications systems that provide quality telephone, cable, and Internet service (Newcomb, 1997).

A major concern of the existing telephone companies is that competitive local exchange carriers will focus their efforts on securing primarily those areas within a local exchange from which the most profit may be derived and exclude the less profitable areas. For example, the City of Hawarden in northwestern Iowa was given approval by the Iowa Utilities Board (IUB) to establish its own telephone company as a competitive local exchange carrier within a larger exchange maintained by Heartland Telecommunications (*Des Moines Register*, 1999). Although the approximately 2,500 residents of Hawarden may have had a legitimate grievance in terms of the quality of service provided by the local exchange carrier (see Chapter 4), by creating its own local exchange from the most densely populated area of the existing local exchange the city of Hawarden was engaging in "cherry picking" or cream skimming, selecting only the most profitable areas and leaving the less profitable areas for the existing local carrier. Cherry picking, or cream skimming, obviously is detrimental to the existing telephone company, but it also is detrimental to the quality of telephone service in the areas the competitive local exchange carrier excludes from competition. If the competitive carrier is able to siphon off a large share of the profits, decline in revenues for the existing carrier will affect the company's ability to maintain, much less upgrade technologies and services to what most likely will be the larger part, geographically if not demographically, of that local exchange. The result may be the establishment of a more-or-less permanent underserved population, or what Wresch (1996) terms, "information exiles."

In the 1997 survey we asked telephone company managers their perceptions of the likelihood of entry by a competitive local exchange carrier in any of their exchanges within the next 3 years. Their level of trepidation about the possibility of facing competition is reflected in Figure 3.2. Only 12 percent felt it would not at all be likely that they would be facing competition in their own exchanges, and about two-fifths thought it very likely or almost certain. On the other hand, those who are concerned about competition probably need to be concerned more with companies that currently are not local carriers, such as cable, cellular, and other companies in the local service area. About one-third of the company managers do not

Figure 3.2
Likelihood in the Next Three Years of Seeing (a) Entry by Competitive Local Exchange
Carrier in Any of the Respondent Company's Exchanges, or (b) the Respondent
Company Becoming a Competitive Local Exchange Carrier in Exchanges Not Currently
Served

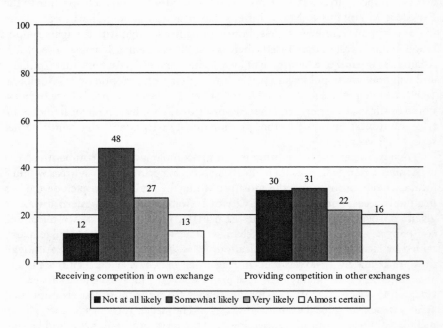

see it very likely that they will engage in competition in exchanges they are not serving, and an additional third state it is only somewhat likely.

Over the past year-and-a-half approximately 40 competitive local exchange carriers have been approved by the IUB, and as of this writing, four are actually operating (Jerome, 1998). There are some regulatory hurdles that potential competitive
local exchange carriers must overcome before establishing operations in rural exchanges. Small rural telephone companies have a rural exemption, that is, they are
exempt from opening their networks to the sales of services, network elements,
and interconnection to other providers. The exemption can be lost if another company desires to operate in the rural telephone company area of operation and the
IUB determines that the request is not unduly economically burdensome to the
customers and is technically feasible. Of particular concern to local exchange carriers is competition from "wireless" competitors such as cellular service providers,
primarily because they are exempted from the regulations (Case Study).

CASE STUDY: WIRELESS COMPETITORS IN THE LOCAL EXCHANGE

Consolidated Telephone Cooperative is a rural telephone company located in southwest North Dakota with about 8,500 customers and services that include mobile tele-

phones, cable and satellite television, and Internet service (Gruley, 1999). Following the Telecommunications Act of 1996, Consolidated expanded its geographic service area by establishing a competitive local exchange in Dickinson, a town of about 17,000 population served by U. S. West. To establish the competitive local exchange, Consolidated applied for and received approval from the state's public utility commission and then negotiated an agreement with U. S. West for interconnection with and use of their telephone network.

Consolidated itself is now under siege from Western Wireless Corporation, a company that offers cellular telephone services. In efforts to expand its customer base Western Wireless is utilizing "fixed wireless" technology, a system of a wireless telephone that remains stationary in the home or business (Gruley, 1999). Western Wireless offers its basic service at about $1 less than Consolidated, but the real incentive for customers to use their system is savings in long distance charges. Because it is a wireless system, customers do not pay the access charges for long distance service that they pay through Consolidated. Some Western Wireless customers are using the service only for long distance calls and have maintained service through Consolidated for local calls (Gruley, 1999). The upshot of this is that Western Wireless customers have substantial savings on long distance telephone service and Consolidated is losing revenue from loss of access charges. But much more than access charges are at stake. Western Wireless has requested regulators in North Dakota and 12 other western states to declare it eligible for universal service funds that are paid to rural telephone companies to offset costs of providing telephone service to remote rural areas. The request on the action is still pending (Wilhelmson, 1999).

To set up its system, Western Wireless leased 2,000 telephone numbers from Consolidated, but did not tell Consolidated that the numbers would be used for fixed rather than mobile phones. Western Wireless also did not seek authorization to establish a competitive local exchange through North Dakota's public utilities commission nor did it negotiate an agreement with Consolidated (Wilhelmson, 1999). Currently, wireless carriers are exempted from state regulation because the use of wireless as a direct competitor of traditional telephone service was not foreseen and therefore not included in the regulations. In other words, Western Wireless is sidestepping not only regulations and procedures on establishing a competitive local exchange, but also Consolidated's rural exemption status. The case currently is in court and being closely watched by several groups because of its possible far-reaching effects on extant local service providers, potential wireless competitors, and the telephone-using public.

As factors perceived to hinder the ability to upgrade the system and services, competition and deregulation taken together increased from 60 percent to 72 percent between 1994 and 1997 (Table 3.1). On the other hand, concerns about the high cost of technology dropped from 10 percent to zero over the same 3 years. Considering the price of advanced telecommunications technologies, for example a digital switch costs in the range of $300,000–500,000 and has a life expectancy of about eight years, we expected the cost of technology to be of greater concern than the results indicated. The obverse of the high cost of technology, available financial resources to pay for the technology, also decreased as a concern. Taken together, in 1994 nearly one-third of the managers responded that high cost or lack of financing were factors hindering the ability to upgrade, and this dropped to less than one-fourth in 1997, although this still is a sizeable number of companies concerned about financing.

Table 3.1 also indicates that 5 percent of the telephone company managers stated access revenue was a factor hindering upgrades in 1997, a factor not mentioned in 1994. Access (or switching) revenues come from fees paid by long distance carriers to the local exchange company for use of its network in connecting long distance calls. For most Iowa telephone companies 60–70 percent of revenues are derived from access or switching of long distance calls, and these revenues help keep down local telephone rates which are provided below cost (Jerome, 1998; Pletcher, 1998). Thus we return to the issue of competition. If the existing local exchange carrier faces competition and a major loss of access revenues from another carrier in the high service density area of an exchange, insufficient financial assets might not only be an impediment to upgrading but also may result in lowering the quality of services in the remainder of the exchange or in other exchanges to compensate for the lost revenue.

In summary, competition and its potential impact on revenues are the major concerns expressed by Iowa's telephone company managers in relation to their ability to upgrade their systems. Nevertheless, the data also show that many companies already are providing some innovative services and have plans for further improvement in the near future. Perhaps the competition they fear also is an incentive to upgrade their systems to achieve a competitive edge. And upgraded systems, in turn, will have an impact on the viability of the rural communities they serve.

TELEPHONE COMPANIES AND LOCAL DEVELOPMENT

As participants in the local economy, telephone companies should not only provide a full spectrum of innovative services to their clientele, but should also have a strong interest and be involved in the development activities of the local community. It is in their interest to maximize the business potential of communities they serve thus increasing their revenues through community development and growth. Telephone companies will be more innovative when the leaders have perceptions that innovation is important, and these perceptions are reinforced through their involvement in local development activities (Sawhney et al., 1991; Amabile, 1988).

Telephone company managers were asked to rate the importance of several types of infrastructure to local economic development in the 1994 survey. Managers rated the importance of water/sewer/waste disposal, transportation, telecommunications, retail and service sector, recreation, health care, and education from 1, not important, to 10, very important. Not surprisingly, telecommunications was rated very high, about 9, surpassed only by education, approximately 9.1. Recreation was lowest at 6.8, and the others ranged from 7.5 to 8.5.[2] So telephone company managers see telecommunications as being important for economic development.

This general question was further pursued in the 1997 survey with more specific questions on the importance of telecommunications to various economic development issues. We also asked about the importance of telecommunications to four community institutions and amenities that are important to, if not directly related to, sustainable economic development. The results are presented in Table 3.3. Generally, telecommunications rates high in importance to all the economic development issues except quality of working conditions, volume of retail trade, and local business creation. Interestingly, the issue to which telephone company managers attribute the highest importance of telecommunications is recruitment of new industries and businesses. Business and industry recruitment, also known as "smokestack chasing," has been shown to be the least beneficial of community economic development strategies. But the managers do see the importance of telecommunications to the other strategies of community economic development, especially to a strategy deemed particularly important in rural areas, innovative applications in home-based businesses. Overall, managers seem to appreciate the importance of telecommunications for local economic development.

Telephone company managers also appreciate the importance of telecommunications for local institutions and amenities, in particular, educational and medical services. It is surprising, though, that as many as 4 percent of the managers felt telecommunications were not important for educational services.

To assess telephone company involvement in local economic development we asked managers if their companies were involved in any of several economic devel-

Table 3.3
Telephone Company Managers' Perceptions of the Importance of Telecommunications for Economic Development and Local Institutions

	Very Important	Somewhat Important	Not Important
Industry and business recruitment	87	12	1
Creation of home-based businesses	84	15	1
Business retention and expansion	82	17	1
Profitability of businesses	82	17	1
Educational services	81	15	4
Medical services	79	19	2
Job creation	78	19	3
Future growth of the local economy	76	23	1
Good-paying jobs	74	23	2
Library services	64	31	4
Local business creation	63	36	1
Volume of retail trade	60	34	6
Quality of working conditions	53	42	5
Entertainment and recreation	40	51	9

Note: Some rows may not equal 100% because of "Don't know" responses.

opment activities. Table 3.4 indicates that over two-thirds of the managers stated their companies are involved in some community economic development activities. Interestingly, helping with development of public services such as education, health, and welfare is among the top three activities in which the managers mentioned participating despite rating these lower in importance. Managers must realize the importance of such services for economic development and local viability. On the other hand less than two-thirds of the managers stated they consult on innovative applications of telecommunications to new businesses or existing businesses (62%), or provide support to local entrepreneurs starting new businesses and working on retention and expansion programs (60% each). Those who do not consult on innovative applications of telecommunications may be managers of companies that lack state-of-the-art infrastructure and therefore cannot provide those services. Similarly, managers not working to assist entrepreneurs and existing firms may also be associated with companies that do not have advanced technologies. On the other hand, entrepreneurship often is not perceived to be as important for economic development as the more traditional recruitment of business or industry, and many local economic development programs are geared more to recruitment of new firms than to support of existing firms (Loveridge, 1996)

FACTORS AFFECTING TELEPHONE COMPANY SERVICE PROVISION

From the two surveys we see that telephone company managers for the most part appreciate the importance of telecommunications for community economic development and many state their companies participate in local economic development. But there are differences among managers in perceived importance and participation. There also are differences in the number of innovative services the

Table 3.4
Involvement of Telephone Companies in Community Economic Development Activities

Activity	Percent Involved
Working with local or county economic development organizations	70
Assisting with recruitment of business or industry	70
Helping with development of public services (education, health, welfare)	70
Consulting on innovative applications of telecommunications to new or existing businesses	62
Working on retention and expansion programs for existing businesses	60
Providing support to local entrepreneurs starting new businesses	60
Working on development of local infrastructure other than telecommunications (roads, water, etc.)	46
Job training programs	26

companies provide their customers. We now examine factors that may be related to the differences among Iowa telephone companies in the services they provide. We include several telephone company manager characteristics, some characteristics of the telephone company's organizational structure, environmental conditions, factors the managers consider important to providing advanced services, and the community's social capital.

Manager characteristics include age, years of education, years in the current position, in the same company, and in the telecommunications industry, whether or not they had any formal telecommunications training such as vocational or technical school, and their rating of the importance of telecommunications for continuing local development.

Research on the structure of organizations such as telephone companies has shown that size of the organization is consistently related to the organization's innovativeness (Strang & Soule, 1998). Kimberley (1981) identifies four dimensions of organizational size: physical capacity, available personnel, clients served, and available resources. For the telephone companies these dimensions are measured as linkages with other companies outside the service area, such as joint plowing and laying of cable, pole attachment agreements for cable, consortiums of maintenance personnel, the number of employees, the number of telephone access lines (customer connections), and financing from several sources including the Rural Utilities Service, the Rural Telephone Bank, the Rural Telephone Finance Cooperative (RTFC), and the local bank.

Primary environmental elements for telephone companies, as we have seen, are federal and state regulations. Telephone company managers were asked in separate questions if there were any state or federal regulations which made it difficult for their companies to upgrade their systems, yes or no.

Factors that the managers' might consider important to providing advanced services are skills and training of personnel, cost and return on technology investment, and government laws and regulations. Each of these factors is a scale composed of two items that the respondents rated on a four-point continuum from not at all important to very important.

Community social capital is approached through the concept of entrepreneurial social infrastructure (ESI). "ESI may be viewed as a particular format for directing or converting social capital into organizational forms that encourage collective action" (Flora, Sharp, Newlon, & Flora, 1997, p. 627). One of the important components of ESI for community economic development identified by Flora et al. is the mobilization of resources, that is, collective and individual investment for the common good.

ESI mobilization of resources is measured by response to the question "Has (company) been involved in any of the following activities? (a) Providing support to local entrepreneurs starting new businesses; (b) Assisting with the recruitment of businesses or industries to locate in the local area; (c) Working on the development of local infrastructure other than telecommunications (such as roads, water and sewer lines, rail access, etc.); (d) Helping with the development of public services such as education, health and welfare services; (e) Job training programs; (f)

Consulting on innovative applications of telecommunications to new or existing businesses; (g) Working on the retention and expansion programs for existing businesses or industries; (h) Working with local or county economic development organizations; (i) Any other area of local economic development?" A summated scale of telephone company activities in local economic development was computed that ranged from 0–9 with a mean of 5 and a Cronbach's alpha reliability coefficient of .76.

Telephone company telecommunications innovativeness is measured with a scale that sums the level of use over the nine innovative services in Figure 3.1. Respondents were asked "Does your company offer (service)?" If the response was no, then the respondent was asked "Will you be adding (service) within the next five years?" Possible responses were "No," "No, unsure," and "No, but will add it (service) in the next five years." The responses from the two questions were then combined and coded as: Yes the service is offered = 4; No, but it will be added in the next five years = 3; No, and uncertain as to whether it will be offered in the next five years = 2; and No, it will not be offered = 1. The resulting innovativeness scale ranged from 12–36, with a mean of 21 and a Cronbach's alpha reliability coefficient of .71.

Table 3.5 contains correlations for those variables significantly related to telephone company innovativeness at the .05 level of significance. Relationships are all in the weak to moderate range except for involvement in economic development activities, which is strongly related to telephone company innovativeness ($r = .53$). Of the telephone company structural characteristics the number of employees and number of access lines are moderately correlated with telephone company innovativeness ($r = .26$ and $.27$ respectively). Linkages with other companies is positively but weakly related to innovativeness ($r = .17$). Linkages with other companies also is positively related to two of the measures of company size, number of employees, and number of access lines (correlations not shown in Table 3.5). Larger companies apparently expand utility of their own resources through linkages with other companies.

The only telephone company manager characteristic related to company innovativeness is the manager's perception of the importance of telecommunications to local economic development ($r = .23$). The importance of leadership in local telecommunications development, especially leadership from the telephone company, is strongly supported in the literature (Read & Youtie, 1996; Schmandt, Williams, Wilson, & Strover, 1991). Inhibiting state regulations have a positive relationship with innovativeness ($r = .18$) which was not expected. Managers' perceived importance of personnel skills and training is positively related to innovativeness ($r = .21$).

To further assess the relative importance of the variables we examined for explaining telephone company innovativeness, we used stepwise multiple regression with the seven variables in Table 3.5 significantly correlated with innovativeness. Only two variables, involvement in economic development activities and number of employees, entered the regression and explained 30 percent of the variation in innovativeness. Involvement in economic development activities explained the

Table 3.5
Correlations for Variables Related to Telephone Company Innovativeness at .05 or Higher Level of Significance

Variable	Correlation Coefficient
Manager characteristics	
Perception of importance of telecommunications	.23
Telephone company characteristics	
Linkages with other companies	.17
Number of employees	.26
Number of access lines	.27
Environmental elements	
Inhibiting state regulations	.18
Factors manager considers important	
Personnel skills and training	.21
Mobilization of resources	
Involvement in economic development activities	.53

majority of this variation (23%). Thus we see that those telephone company managers who understand the symbiotic relationship between the welfare of the company and the welfare of the community as indicated by their involvement in local development activities will invest resources to upgrade the company's telecommunications infrastructure for the common good of the community.

CONCLUSION

Few authorities on rural community economic development would question the importance of telecommunications for rural community viability. The chapters in this volume on individual community sectors discuss telecommunications' importance for libraries, hospitals, newspapers, businesses, farmers, economic developers, local government, and the Cooperative Extension Service. In each sector there is broad variation in the extent of use and in the specific types of uses of telecommunications, but overall, the increasing importance of telecommunications to each sector, and through the sector to the larger community, is clearly evident. State-of-the-art telecommunications infrastructure increasingly is a necessity for viable rural communities.

Rural Iowa is similar to the rural areas of other states in that many rural communities do not have the state-of-the-art telecommunications infrastructure (Goldberg, 1999). Although universal telephone service is close to being universal and several enhanced telephone services are available to most customers, for many communities the technology to provide the truly innovative services is lacking. Less than half of all existing telephone companies planned to be offering videoconferencing and public packet switching by 2002. Furthermore, in our surveys of the other community sectors the quality of local telephone service generally did not rate very high. For those sectors that were asked the question, the largest percentage of respondents rated local telephone service adequate or lower on a rating from poor to excellent with adequate being the midpoint. Respondents did perceive telephone services improving in the future. This corresponded with many managers' statements that within the next five years their companies would be adding services not currently offered.

How many companies upgrade their systems and how quickly those upgrades are accomplished will be influenced by the environment in which the companies operate and the availability of financial resources. The major environmental concerns of telephone companies currently are deregulation and competition. The two issues are linked, because deregulation provides each company opportunities to compete for customers of other companies but also exposes each company to competition from other companies. The effects of competing with a state owned and operated system, such as the Iowa Communications Network (ICN), are probably deleterious for telephone companies, and may also be deleterious for isolated rural areas unless the ICN is opened to use by private as well as public entities. Competition from the private sector, on the other hand, may be beneficial for both telephone companies and communities. When faced with an uncertain environment, one strategy used by organizations to adapt to that environment and ensure survival is the adoption of innovative technologies to improve efficiency and the delivery and quality of services. By adopting technologies and providing services that rural communities desire and require, existing telephone companies will be less exposed to direct competition, including competition from municipalities organizing their own telephone companies.

Just under one-fourth of the managers were concerned about financing for upgrading. Availability of financial resources, of course, is not unrelated to competition. If a competing local service provider uses predatory pricing through cherry picking, available financial resources such as the critical access charges may decline. Rural telephone companies do have a number of sources for funding improvements and expansion of services. U.S. Department of Agriculture's Rural Utilities Service (formerly Rural Electrification Administration) is a major source of funds for many rural telephone companies. Other sources of financing include the Cooperative Finance Corporation, the Rural Telephone Finance Cooperative, the Rural Telephone Bank, the National Telephone Finance Corporation, and the industry's universal service fund.

Perhaps most important is recognition of the symbiotic relationship between the local telephone service provider and the community, including the telephone company's contribution to and participation in the community's social capital (Brunner, 1998). This places responsibility on both the community and the telephone company. One of the most consistent findings in the research on successful telecommunications infrastructure development and use in community development is the critical role of local leadership. Telephone company personnel often are involved in local development activities, and indeed, are included among the group of local development leaders (Read & Youtie, 1996). Thus, they become the catalysts for action on telecommunications. But in some communities, as our research indicates, telephone companies are not involved in local development. Lack of participation in rural community development activities is a particular concern with large, absentee-owned telephone corporations (see Chapter 12). In communities where telephone company personnel are not involved, community leaders should make special efforts to include telephone company managerial personnel in their planning and development groups.

On the other side of the equation, many rural communities have leaders who do not understand the potentials of telecommunications or who do not have the vision for its innovative application (Bleakley, 1996; Dillman, 1991b). In the 1994 survey we asked telephone company managers if local business owners and managers recognized innovative applications of new telecommunications technologies, and if they had the skills and motivation to use them. In a range of strongly disagree (1) to strongly agree (5), telephone company managers perceived business owners and managers to be relatively strong on skills and motivation to use new technologies (4.0), but somewhat weaker on recognizing applications (3.7). Business owners and managers must be helped to recognize and implement competitive strategies using telecommunications in their enterprise operations. Examples of organization with a major focus on technology training include the Center for Rural Development in Somerset, Kentucky, and the Northeast Wyoming Economic Development Coalition serving five counties in Northeast Wyoming (McMahon, 1998).

A deeper concern arises from a survey of Iowa local economic development professionals (see Chapter 12). The respondents were asked to rate the use of telecommunications technologies by eleven different sectors of their communities, from nonexistent (1) to high (4). The three lowest, other than churches, were sectors from which one would expect the core of leadership for local development: economic development (2.0), city government (1.9), and the chamber of commerce (1.7). Such communities, with leaders and officials lacking commitment to telecommunications, may require external assistance from community development specialists with programs for planning, organizing and implementing development that incorporates telecommunications. Telephone company managerial personnel can have an important leadership role in these communities.

The physical infrastructure necessary for rural communities to participate in the telecommunications- and information-based economy is conventionally provided by the local telephone company. The sophistication and modernity of the infrastructure depend on the innovativeness of the telephone company, and in turn, upon the interests and participation of the telephone company managers in local development. Therefore, managers of telephone companies occupy a critical position in decisions and actions affecting the well being of rural communities. Rural community leaders need to understand that their communities' welfare is strongly linked to the innovativeness of local telecommunications service providers, and they need to be conversant with strategies for involving the providers in local development activities.

NOTES

1. Although we tend to consider state-of-the-art telecommunications technologies as embodying fiber-optic lines, digital switching, and wireless systems, copper lines actually can handle some fairly sophisticated services such as compressed video.

2. Although it was not included in our survey, other recent Iowa surveys consistently have found that housing ranks high in needed infrastructure, including the survey of local economic development professionals (Chapter 12).

4

WHEN PUBLIC LEADERSHIP OUTPERFORMS PRIVATE LEADERSHIP: THE CASE OF PUBLIC TELECOMMUNICATIONS UTILITIES

Montgomery Van Wart, Dianne Rahm, and Scott Sanders

INTRODUCTION

Government economic development efforts are undertaken for many reasons and through a host of mechanisms. For example, government economic development is undertaken when the private sector has insufficient funding, inadequate expertise, excessive risk, or when there is a critical need for public control of common resources to ensure widespread access. Mechanisms include tax reductions or abatement, private-public agreements, quasi-governmental corporations, and publicly controlled agencies. Examples of such economic development activities by government have included, at various times, the U.S. Post Office, airports, railroads, sports stadiums, highways, economic development districts, tax abatements to lure lucrative stores and businesses to cities or states, tax reductions for residential improvements, financial support for the invention or implementation of new technologies (especially conversion from military and NASA research), public works projects intended to provide economic development (e.g., the Tennessee Valley Authority), tax supported projects to invent new products or solutions to help an industry (often expensive basic research), and a variety of economic stability mechanisms (e.g., Social Security, the Federal Reserve Board, the Federal Home Loan Board, numerous federally subsidized insurance programs, etc.).

One economic development initiative that is not often recognized, however, is the government opting to deliver a product or service as a consequence of private sector failure to do so. When governments take on these sorts of activities, they often create organizations very similar to private sector organizations. Indeed, they are called "public enterprises," a name that denotes their hybrid character. Today in rural areas there is a growing trend to create new public enterprises for telecommunications services. The new information age allows rural citizens to "connect" with the explosion of information services and to increase their community's competitiveness by offering options such as telecommuting and rural siting of firms dependent on a telecommunications capacity. But without first-rate telecommunications infrastructure and services—the right connections—such communities are unable to offer the information-based opportunities that many firms and citizens demand. This chapter examines the degree to which local governments have the right connections and the importance they assign to them. It also examines the creation of a new type of public enterprise—rural telecommunications utilities—that is a response to deficits in telecommunications provision. The development of these public enterprises has been an economic development strategy adopted by rural communities in the wake of private sector abandonment.

Public enterprises today play a small but significant role within the U.S. economy as they have throughout our history. In fact, older public enterprises such as railroads, electric utilities, water utilities, and even airports are under increasing pressure in the United States and around the world, where they have been more extensively used (Yergin & Stanislaw, 1998). Although American tradition reveals a consistent belief in the private sector as the driving economic force for the country (Hughes, 1991), this does not negate "the deep-seated preference of the American people to employ the powers of government to influence the economy *at strategic points* needing special attention" (Goodrich, 1967, xxxiv [italics added]). Government influence can occur through the creation of public enterprises as well as through other means, including public agency programs, anti-trust actions, regulatory regimes, and public subsidies. Within the United States, the general purpose of public enterprises has been to supplement the capitalist system rather than to displace it, as in many other countries (Ryan, 1997). Economically and practically, this has meant that "public enterprises have followed the pattern of industrial development in the country" (Kwoka, 1996, p. 8), with public enterprises being created in new industries and privatization of public enterprises often occurring in more mature industries (Goodrich, 1967; Wilcox & Sheperd, 1975).

Historically, public enterprises began in basic infrastructure (e.g., waterworks) and postal service in the late 1700s and shifted to water and land transportation (e.g., canals, turnpikes and later railroads) in the first half of the 1800s. They then shifted focus to electric and gas utilities at the turn of the century and next moved to select public works, farm promotion, and economic protection agencies in the Great Depression era. After World War II public enterprises shifted to a variety of international assistance, social welfare, and economic development projects. Because today a major cutting edge of the economy is advanced telecommunications, it is little wonder that one of the most important debates about the *creation* of public enterprises is related to telecommunications utilities.

Much discussion has focused on the older public enterprises such as municipal electric and water utilities, which are increasingly being considered for reduction, heightened competition, or outright privatization. Reduction would mean increased dependence on self-sustaining revenues. Heightened competition means that the private sector is being encouraged to compete with public agencies similar to the way garbage collection is now frequently bid out to many organizations rather than exclusively to a city department. Outright privatization means that the enterprise is sold entirely to the private sector with no direct public control.

However, new economic initiatives continue to receive government assistance or direction through various types of subsidies such as sports arenas in urban areas (Swindell & Rosentraub, 1998) and other select new public enterprises (Wettenhall, 1998). The creation and promotion of public enterprises are important economic development policy tools. Even on strictly economic terms, economists have occasionally found that public enterprises "were significantly more efficient than their privately owned counterparts" (Hausman & Neufeld, 1991, p. 414).

After briefly outlining public enterprise history, this chapter reviews high-quality telecommunications systems and the challenges faced by many rural areas. Drawing on case and survey study data from rural Iowa communities, factors necessary for high-quality telecommunications systems success involving local government (as well as those leading to failure) are discussed. These trends, utilizing local government as the central actor in telecommunications innovation, strongly document the need for the community development framework outlined in Chapter 1; that is, success tends to depend on a publicly oriented, community-based improvement process that is primarily influenced by a variety of local actors in concert with local government. Trends showing the importance of the new (public enterprise) telecommunications utilities indicate how the different types of community capital—physical, financial, human, and social—are enhanced in those communities which elect to provide their own advanced telecommunications services.

DEFINITION AND HISTORY OF PUBLIC ENTERPRISES

Despite a fierce streak of individualism and a strong belief in Smithian liberal economics, Americans have always been willing to selectively consider accomplishing the public good through public enterprises and other government-directed means. Analysts and auditors agree that there is no consistent definition or name for such public enterprises (Rainey, 1997; Government Accounting Office, 1989). One of the foremost scholars in this area says succinctly that a public enterprise is "a business owned and operated by government" (Wettenhall, 1998) before discussing the richly complex topic. The general sense of public enterprises is that they are publicly owned or initiated and that they are largely or wholly self-supporting through income other than direct appropriations. Public enterprises exist at all levels of government, sometimes called government corporations, government-sponsored enterprises, but sometimes they are executive agencies or departments as well. At the federal level, the Government Accounting Office (GAO) classified 45 organizations as public enterprises in its 1988 study: 10 banking related, 3 education related, 9 farm related, 6 housing related, 8 industrial related, 3 investment related and 6 others. Examples include the Consolidated Rail Corporation, Corporation for Public Broadcasting, Federal Deposit Insurance Corporation, Gallaudet University, Legal Services Corporation, National Park Foundation, Rural Telephone Bank, Tennessee Valley Authority, and the United

States Postal Service. Although Social Security and Medicare are not currently administered as public enterprises, they have many appropriate characteristics and could be run as such with sufficient political will. At the state level, traditional examples include select transportation projects and port authorities and, in recent history, the state lotteries. State universities, community colleges, and state and county hospitals have some public enterprise characteristics but are, in general, highly dependent on appropriations. At the local level, the most common examples of public enterprises are water, electric, and gas utilities, although other local enterprises exist such as sewer districts, airports, and housing authorities. Municipal and municipal-like utilities constitute the largest class of public enterprises; for example, there are 1,971 public water utilities and rural cooperatives in Iowa alone. Yet in terms of the overall economy, public enterprises have generally been less preferred as an economic tool of government intervention than government regulation (as Caristi demonstrates in Chapter 2).

Federal creation of new public enterprises has come in waves, a view consistent with Schlesinger's observation that there are cycles of support for governmental solutions (1987). In the Federalist period, national solutions were typical for problems ranging from opening up the west through the creation of the Cumberland Road and the Erie Canal, to a substantial postal service (which made up the largest component of the federal work force). Only the railroads received much support after the era of national projects ended around 1820. In the meantime, states and to some degree local government were encouraged to initiate infrastructure projects suited to their specific needs. A period of robust public enterprise expansion at the local level began in the late 1800s and early 1900s after the Progressive's reforms had affected local government (Hellman, 1972; Schap, 1986). (In the case of Iowa, during the period shortly after the turn of the century not only were many public electric utilities created, but almost all private water utilities in Iowa were bought out by municipalities, which were booming at the time). The New Deal brought an era of large public enterprise creation, with activities as diverse as the Tennessee Valley Authority, federal insurance programs, and regional banks. Although over 20 public enterprises have been created at the federal level since World War II, they have either tended to be smaller (e.g., the Saint Lawrence Seaway Development Corporation), involve stock ownership (e.g., the Federal Home Loan Mortgage Corporation), or are enterprises created to handle public bailout situations (e.g., the government railroad corporations). The creation of housing, redevelopment, airport, and gambling enterprises has been common at the state and local levels since the 1960s.

The willingness to use public enterprises in the United States is overwhelmingly practical rather than ideological, as has often been the case in Europe and elsewhere (Anderson, 1966; Goodrich, 1967; Redford & Hagan, 1965). Kwoka (1996) notes that there are three major rationales for public enterprises in the United States. First, public provision is a "possible solution to the problem of natural monopoly." Second, public provision "is appropriate for services with public goods properties." And third, public provision "may be appropriate where capital requirements are extremely large or fraught with risk" but there is a critical public interest involved (pp. 7–8). At the local level, all three characteristics are evident. Public enterprises have often acted as guards against excessive charges by private monopolies. They do this by creating an environment in which new competition may emerge (known as "threat of entry" competition), providing price comparisons (known as "yardstick" competition) and as a means of hastening needed service (Emmons, 1993; Schap, 1986).[1] In addition, the services offered have tended to be those affecting the community at large, often services that are not easily divisible, such as power and water utilities. Finally, the cost, maintenance, and risk involved are often extensive, as is the case with water systems, sewer districts, or public housing authorities, respectively.

THE IMPORTANCE OF RURAL TELECOMMUNICATIONS AND CURRENT ORGANIZATIONAL USAGE: AN IOWA STUDY

Generally speaking, telecommunications infrastructure is necessary for citizen access to information and services for both private and public sector organizational efficiency and for community economic development (Read & Youtie, 1996). To investigate the government's general involvement and entrepreneurial activities with regard to addressing advanced telecommunications needs, two separate methods were used. In the first research phase, in the spring of 1997, a survey of municipal clerks and administrators from 275 of the largest towns and cities in Iowa was conducted by telephone. The population of respondents was selected from the *Iowa League of Cities 1996–1998 Directory*. Of those contacted, 265 responses were received yielding a response rate of 96 percent. Through the telephone survey we established a broad information base on the current and planned use of telecommunications technologies in Iowa's local governments.

The second phase of research focused on rural communities known for their exceptional telecommunications capacity, as well as those communities where there is poor telecommunications provision. Twenty-five in-depth interviews were conducted with local government officials and telecommunications industry experts in Iowa.

The series of cases is drawn from Iowa, which typifies the norm for local utility creation. Iowa has both true urban and rural areas. Over 94 percent of its 950 cities and towns have fewer than 8,000 inhabitants. Iowa has unusually clear-cut examples of public telecommunications provision at both the state and local level. The state has a variety of "backbone" distribution systems for its "wired" telecommunications, allowing for a wide range of private and public service provision patterns. Finally, Iowa's generally favorable climate toward local utilities provides more examples and experience than other states.

In the survey of Iowa local government officials, the data reveal that 89 percent of local government administrators report telecommunications technologies as important or very important to education services, 86 percent to library services, 81 percent to medical services, 79 percent to the future growth of the local economy, 70 percent to the recruitment of new industries and businesses, 67 percent to maintaining good paying jobs, 66 percent to the creation of home-based businesses, 60 percent to local business creation, 58 percent to job creation, 55 percent to the quality of working conditions, and finally, 46 percent to the volume of retail trade.

This general level of awareness of the value of telecommunications technology is underscored by particular emphasis on its consequences for economic development. Nearly half of all city administrators indicate that their local government is greatly involved in their community's economic development efforts, 42 percent report that their city's telecommunications technology base has been used in the past to promote the community's economic development efforts, and 77 percent of those who have not used it for economic development in the past indicate that it will likely be used in such promotional efforts in the future.

Looking to local governments as organizational users themselves, many benefits are perceived to accrue from advanced telecommunications technologies. As Figure 4.1 shows, a high percentage of city administrators and clerks report as important or very important the features of saving time (89%), providing immediate access to information (88%), providing better customer service (85%), improving accuracy (80%), and providing better communication with people and organizations outside of the local organization (80%), as opposed to 59 percent within the organization and saving money (79%). As interactive uses of the Internet become more common, such as the downloading of forms, accessing customer-specific information, that is billing records and filing and paying bills, complaints and licenses, these percentages will likely increase. While large cities such as New York (NY) (Poulos, 1997) and Phoenix (AZ) are becoming proficient in this area, much of the public sector, including most small cities and towns, are lagging for many reasons. Kenneth Thornton (1997), general manager of IBM Global Government Industry, speculates that government is well behind industry for three reasons. "First, policy leadership must be stronger in leading electronic initiatives. . . . Only seven percent of policy leaders across North America said they understood technology well enough to be comfortable making policy decisions. Second, historical barriers all too often keep government agencies form cooperating with each other. Third, governments are seemingly caught between a never-ending list of citizen desires and pressures to reduce taxes. New funding vehicles must be found" (p. 52). Although this study did not examine the prospects of "cyberdemocracy" (public participation in the processes of governance via Internet-based technologies), it is clear that such participation becomes increasingly possible for voting, plebiscites, and cyber town meetings in the future, as the Internet becomes more widespread (Grossman, 1995; Hacker, 1996).

Figure 4.2 shows the access to and use of advanced telecommunications technologies by size of municipality. For example, 46 percent of local government administrators in Iowa's smallest towns stated they had Internet access, although access might be at the community library rather than on their desktop.[2] This percentage progressively increases to 86 percent for the largest cities. When asked if their organization uses an e-mail system, only 14 percent of smaller towns responded affirmatively, but larger towns responded affirmatively 22 percent; that percentage increases to 37 percent for smaller cities and leaps to 76 percent for the

Figure 4.1
Proportion of Government Officials Rating as "Very Important" or "Important" Seven
Benefits of Telecommunications Technologies for Municipalities

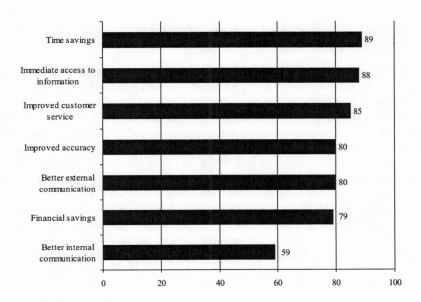

Figure 4.2
Access to and Use of Advanced Telecommunications Technologies by Size of
Municipality

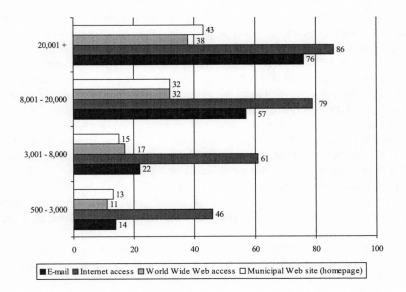

largest cities. Similar patterns are reported for a homepage and for World Wide Web access. In all cases, the level of usage increases substantially as the size of the city increases. This general trend confirms our intuitive speculations.

CURRENT CHALLENGES TO THE PROVISION OF STATE-OF-THE-ART SERVICE IN RURAL AREAS

Despite telecommunications' acknowledged importance, there are special problems for rural areas such as the high cost of installing and maintaining service for relatively small numbers of users (Sussman, 1997). In the past, some of these problems have been solved by using regulatory regimes such as price averaging and cross-subsidies. Today however, with increasing levels of deregulation, new public decisions about how to handle these structural problems are required. One response has been the set-aside of special federal funds for needy cases. Another response has been to require rural local users to pay more of the true cost of service, rather than be heavily subsidized by urban and corporate users as in the past. While complete lack of access is uncommon, it does occur. For example, lack of a critical mass in some very small communities results in not having cable television systems.[3] Even though in large cities basic service provision may not be an issue, the extent of service is. The City of Des Moines is factoring in cable television bandwidth as a part of its franchise negotiations with the local cable television provider, TCI. According to Assistant City Manager Rick Clark, who wants the franchisee to upgrade fiber optic service to the door rather than only the distribution backbone system, "our concern is that if we don't shoot far enough, we're going to end up with a system that doesn't have enough capacity to allow the citizenry to compete in the world" (Eckhoff, 1997, p. 1A). Very common today in rural areas is the lack of capacity in telephone systems for additional lines or direct Internet access. These systems generally have copper wire backbones and the expense of system upgrades does not currently match the revenue potential of the new Internet markets. Bob Haug (1997), a public electric industry advocate, notes that "unfortunately, some current service providers seem to have written off Iowa's rural communities when it comes to investment in new technology" (p. 9A). This makes non-dial-up access to the Internet as a feature of local telephone service unavailable in many areas. A related problem is services that are planned but delivery is delayed for years (Emmons, 1993; Schap, 1986). Well-wired communities (such as those with extensive fiber optic systems) have a distinct competitive advantage in being able to offer immediate service "to the curb." To the curb service in telecommunications today means that there is substantial bandwidth (which facilitates speed in high demand uses) available directly to households and local businesses (not just local "hubs"), either through coaxial cable that is configured for multiple uses (rather than just cable television as in the past), or increasingly, through fiber optic cable itself.

Although inferior service is a major problem in many places, it is keen in rural areas. In worst case scenarios, getting service personnel to install new approved lines takes months and routine service calls must be ordered well in advance. It is increasingly common in the larger telephone companies to have large regional centers for service, so that substantial areas of one state are covered by the service center in another state. Further, telephone upgrades are often simply not technically available in rural areas. Cable television providers are often under attack for using too much of their available bandwidth[4] for paid advertising channels and too little for popular channels. "Dissatisfaction with cable TV and, in some cases, telephone service seems to be at an all-time high" (Haug, 1997, p. 9A). Although such frustration is hardly limited to rural areas, urban areas have far more alternatives (Eckhoff, 1997). Key to Internet service is speed, but sharing a limited number of lines sometimes means extremely slow speeds, which has been documented to lead to lower Internet enjoyment and usage.[5]

The high rates charged sometimes result from a lack of competition, which was partially capped in the past by regulation. State boards regulate telephone rates. The federal government regulates cable television rates. In the case of cable television, annual rate increases nationally were still around 6.9 percent for 1997. This may be because the deregulated market has not brought significant competition except for 81 communities nationally, according to a new Federal Communications Commission (FCC) study (*Des Moines Register*, 1998). In the year following the 1996 Telecommunications Act deregulation, the cable television industry only experienced a 2 percent loss in market share (from 89 to 87%) to competitors such as direct broadcast television, although it experienced rapid expansion in the size of the market overall (FCC Hearing, 1997).[6] In the case of the Internet, lack of competition has a deleterious indirect effect for users. Due to lack of competition, most large, rural telephone systems refuse or delay system upgrades that lead to more lines and bandwidth. Therefore local users often cannot get the additional lines needed for computers, or the telephone lines cannot handle the bandwidth needs of the Internet. Of course anyone can get America Online, Prodigy, and other Internet providers with long distance charges when local access is not available, but these charges can be exorbitant. Competition in the local telephone market has also been minimal to date, despite the high expectations of the Telecommunications Act of 1996.

The final problem is a lack of local expertise. In fact, some regional telephone providers are withdrawing their local experts to urban service hubs to cut costs. The lack of local training and expertise discourages experimentation, problem solving, and innovation in a timely manner. Although special universal service funds are being established under the 1996 Telecommunications Act for rural schools, libraries, and health care facilities, local entities must apply for them. "Insofar as the technical expertise required to apply for universal service funds is a scarce and unevenly distributed resource, the FCC rules may very well discriminate against those rural communities that are in greatest need of support" (Garcia & Gorenflo, 1997, p. 2).

These trends are born out in the survey data of Iowa local government administrators when considering telecommunications adequacy. When respondents were asked whether their telecommunications systems were adequate, 33.2 percent of those in small towns rated them poor to below average. As Figure 4.3 shows, the

Figure 4.3
Percentage of Municipalities by Population Size Rating the Quality of Available
Telecommunications Services within Their Community as "Below Average" or "Poor"

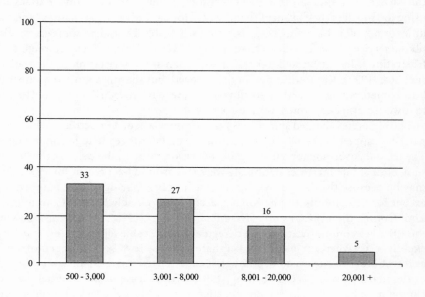

percentage rating their systems inferior drops to 27.1 percent for larger towns
(3,001–8,000), plummets to 15.8 percent for small cities (8,001–20,000), and
dwindles to 4.8 percent for large cities (20,001+). Clearly many rural Iowa towns
feel that their telecommunications services are frustratingly substandard.

    These types of problems are more likely in rural areas because of inherent eco-
nomic biases. Free markets tend to work best when there is a sufficient population
base to provide competition and a service or product can be easily divided among
providers. To counter the lack of competition in rural areas and to protect them
against under-servicing, universal service was implemented, providing for service
minimums and cost caps. Claire Milne (1997) explains that "universal service is for
users. The intention of universal service policies is to fulfill the needs of people
whom the free market might fail to serve. . . . These 'marginal' markets fall into
three main categories: residents of remote rural areas, people with low-incomes
and disabled people" (p. 1). However, it is currently thought by many to have reduced competition and even led, according the chief economist of the FCC, to a
"culture of entitlement to broad subsidies that encrusts our telecommunications
policy" (Farrell, 1997, p. 2). Weakening universal service provision could lead to a
further deterioration of service in some areas. According to industry observer
Blake Harris (1997), "American telecom policy for most of the last century—in
large measure unchanged by the 1996 Telecom Act—has been shaped by the political pressure for universal service, price-averaged, below-cost residential service"

(p. 124). He goes on to assert that "the national average price for a residential home is about $15 a month for line and dial tone. The cost, averaged nationally, to provide this service is about $25 a month" (p. 124). Basic service, especially rural service, was subsidized by long distance rates, special features, business rates, toll fees, and so on. Because many rural areas are experiencing absolute population decreases, the economics of Internet service provision is currently difficult, that is, increased demands placed on the telephone system for more telephone numbers and Internet access often exceed current capacity yet are insufficient in the short term to justify extremely expensive system upgrades. To overcome these inherent market and structural challenges in rural areas and small cities, more aggressive strategies are often necessary—some private, some public.

PUBLIC PROVISION

Five key factors lead to success or failure in providing appropriate advanced telecommunications technologies in rural areas in either the private or public sectors: technological sophistication, expertise, commitment, financial resources, and marketing (Table 4.1). Is there wide access to services, low cost, high quality, and integration of systems (technological sophistication)? Is their expertise available within the community or in easy access of the community? Is there a commitment for involvement with local businesses and residents, a partnership with the community, and a local champion to make sure that the services are provided? Are there the financial resources available to provide the initial technological infrastructure, either from the existing telecommunications providers (in the private sector) or from an external source such as an electric utility (in the public sector)? Finally, is there good promotion to educate people about the technology and to stimulate competition when it exists? Usually the private sector is responsible for the provision of telecommunications technologies and services to rural communities; however, there are cases where the private sector does not provide the appropriate telecommunications technology in either a timely or moderately priced fashion. The question then becomes, what can communities do when private, nonlocal telecommunications services fail to comply with local demands for systems upgrades, improved quality, and service? Some communities opt to provide telecommunications through public enterprises.

The trend toward increased public provision in Iowa is most clearly demonstrated by the pattern of recent elections forming city telecommunications utilities. Out of thirty-two referenda for public provision of telecommunications, only two were defeated,[7] while the majority were affirmed with more than 80 percent of the vote. This trend of public provision of telecommunications is increasing despite the fact that it is often opposed with vigor by private providers. In the City of Spencer (population 11,000) the incumbent cable television provider outspent the citizen's group that backed the referendum by 130 to 1. Despite this, the referendum passed by a 91 percent margin. According to Haug (1997), there are at least four reasons for cities' actions: "dissatisfaction with current service, the hope of economic development, a desire to improve local educational opportunities and

Table 4.1
High-Quality Rural Telecommunications: Factors Leading to Success and Failure

Factors Leading to Success
1. Technological sophistication
   - wide access to services
   - low cost for services
   - high-quality services
   - integration of systems
2. Expertise
   - in-house expertise available
3. Commitment
   - local involvement of residential and/or business
   - informal partnership with community
   - strong internal champion public; lack of private leadership or capacity
4. Financial resources
   - private utility is locally owned; public utility supports initial investment
5. Marketing
   - good promotion when competition
   - good education when monopoly

Factors Leading to Failure
1. Technological sophistication
   - restricted or no access to services
   - high cost for services
   - low-quality service
   - weak integration of system
2. Expertise
   - expertise not available
3. Commitment
   - lack local involvement of residential and/or business sectors
   - no local champion
   - no real partnership with community public; leadership position delegated to unwilling public administration
4. Financial resources
   - corporate ownership without local investment public; no local utility to provide substantial start-up funds and bond preferences
5. Marketing
   - poor promotion when competition
   - poor education when monopoly

preparation by city electricity and gas utilities for competition in those industries" (p. 9A). The function of the telecommunications utility varies in these communities from cable television solely to all high-end telecommunications services, but all of these communities are clearly trying to build their public capacity should they need to expand their services. The City of Hawarden provides a primary example.

Hawarden is a city of 2,439 about 30 miles north of Sioux City in northwestern Iowa. With 63 percent of the electorate voting in its 1994 referendum, Hawarden's citizens voted 96 percent in favor of an all-purpose city communication utility (inclusive of telephone service). Their legal ability to administer the telephone operations was supported by recent legislation and various regulatory board rulings, and although legally challenged, was upheld by the Iowa Supreme Court.[8]

The stated purpose of the municipal communication utility was to make Hawarden a telecommunications leader and center rather than a rural backwater town. Its municipal utility constructed a fiber optic system in town, completely bypassing the outdated copper wire system that Heartland Telephone Company (formerly U.S. West) maintains.[9] At this writing, Hawarden is the only fully integrated city-owned communication utility in the state with telephone, cable television, Internet service, and high-speed data transfer. With the legal capacity to administer telephone operations having overcome its last hurdle in the Iowa Supreme Court, other cities are expected to follow suit.[10] Hawarden provides services at below market averages. Its quality in such areas as channel access (39 channels) has been rated "acceptable" to "good," while in other areas, such as Internet speed (1 megabyte per sec), it has been rated "excellent." Hawarden is able to provide this service largely by bypassing the telephone system as an Internet provider and instead relying on its proximity to a high-capacity fiber distribution system called Pioneer Holdings, which adjoins to a power system.

Hawarden's expertise is still modest and is largely provided by the city electric utility and supplemented with contracted services. The public "leadership" was provided jointly by the mayor and the public works director, who were willing to take risks and be aggressive in pursuing a vision of the city as a telecommunications leader. The city will increase its expertise by adding a telecommunications position. The perception of poor service that existed prior to the public take-over made public acceptance of a city-run operation relatively easy. Also key to the public service expansion was an alliance with Pioneer Holdings. In partnerships with North Iowa Power Cooperative, Long Lines, and MCI, Pioneer Holdings leases space on the fiber loop backbone in the northwest part of Iowa.

Hawarden's finances are stronger than many rural communities because of its ownership of the electric utility. Communities in Iowa with their own electric utilities tend to be better situated to provide telecommunications utilities because revenue surpluses and bond capacity can be directed to critical start-up expenses. Additionally, municipalities with electric utilities already possess citywide rights-of-way, expertise in delivery such as peak-demand systems, staff expertise regarding cable installation and maintenance and requisite equipment. Nationally, 77 percent of all municipally owned cable systems are in cities with an electric utility.

In Iowa the trend is even more pronounced: 100 percent of municipally owned cable television systems are paired with the electric utility. Of course, once established such telecommunications utilities can be self-supporting or can provide a surplus, depending on community goals. Hawarden has been aggressive in marketing its internal capacity (personnel, financial resources, commitment, etc.) to its community. This resulted in the original nearly unanimous vote for municipal telecommunications expansion and continuing widespread community praise and support, despite legal and technical challenges.

Hawarden is not the only aggressive public sector telecommunications provider in the state, however. The City of Madrid (pop. 2,400), only twenty miles from Des Moines, represents another rural example. Madrid could not convince its local telephone company (GTE) to lay additional cable and provide hook-up for local Internet access because it was deemed too small a market. Like Hawarden, the city-owned utility was able to connect with the Pioneer Holdings system by paying the modest hook-up fees. The City of Harlan (pop. 5,150) telecommunications utility offers enough capacity to local Internet users (10 megabits per second) to fit most commercial applications as well. Furthermore, Internet access does not require a second phone line in Harlan. The City of Harlan, like a number of other local cable television utilities, offers a service that is slightly more expensive than the private provider but has a number of upgrades (such as additional channels). This has provided local customers with a choice of service and with rates that are below the regional average and below the rates charged when a monopoly existed. The City of Cedar Falls (population: 34,000) represents a more urban example. Cedar Falls' fiber optic system provides Internet bandwidth of 4 megabytes per second and high-speed transfer of 10 megabytes at very competitive prices. Perhaps because Cedar Falls still has an alternate private cable television provider and has a larger audience immediately adjacent in a major city (Waterloo, 66,500), its marketing materials are glossy and extensive, an uncommon feature in Iowa's fragmented and largely noncompetitive market.

Nationally, the public sector has been far more active recently in the creation of public utilities than in the preceding decades. As a feature of the Salt River Project, Phoenix (AZ), Palo Alto (CA), Anaheim (CA), Colorado Springs (CO), Holland (MI), and Springfield (OR) all currently lease public fiber optic lines for private sector use or are planning to do so in the near future. A number of other communities around the country—including Gainesville (FL), Marietta (GA), Wadsworth (OH), and Pacific County (WA)—are actively expanding their telecommunications capacity. In addition, many large communities, such as Frankfurt (KY), Braintree (MA), Lincoln (NE), Chattanooga (TN), and Tacoma (WA), are investigating telecommunications utility opportunities. On the other hand, the imminent deregulation of the power industry has not yet affected municipal electric and gas industries, but is likely to do so in the next decade (*Public Power Weekly*, 1998).

Public implementation of telecommunications services at the local level is too new to see examples of failure per se; the only failures at this point are cases in which there is a failure to act despite the need and opportunities to do so. Yet many local jurisdictions that have inferior telecommunications services may assess their situation only to find no realistic options. Unless rural communities in Iowa or other states can connect to one of the telecommunications backbone networks at a reasonable cost, which is often not possible because of isolation, the issue may be moot. Telecommunications expertise in small communities is generally lacking. Local leaders may be part-time or fully occupied with other practical matters. Most importantly, small communities often must pay for the upgrades, connection costs, and consulting fees and to do so they must be financially capable of bearing those expenses. For smaller communities this capacity is sometimes linked to the presence of an electric utility, although a few communities are investigating the use of other means such as their public cable television franchise.

FUTURE TRENDS

Future trends are not clearly discernible because of a number of unresolved "wildcards" in the telecommunications industry in Iowa and across the nation. First, numerous legal issues have yet to be resolved. Second is the uncertain development and disposition of the backbone distribution systems. Third is the evolving innovation in satellite and wireless technology.

While the basic principles of the Telecommunications Act of 1996 are generally known and the FCC is vigorously promoting market competition, many of the effects of the Act are still unclear and are being argued in various court battles. This will have a direct bearing on the operation of local telecommunications utilities. For example, in an U.S. Court of Appeals decision involving Abilene, Texas, municipalities were ruled "entities" and allowed to compete under the same rules as private providers under Section 253 of the Telecommunications Act of 1996. In another decision, the FCC was required to allow state regulators to take the lead in implementing the new deregulation rules (AT&T v. Iowa Utilities Board, 1999).

In terms of competition enhancement, Direct Broadcast Systems (DBS) and other telecommunications service providers still complain of extensive vertical integration in cable television industry programming, mutual stock membership, and in-kind trading leading to disguised price differentials (FCC Hearings, 1997). Also, although basic universal service continues to be regulated somewhat in the 21st century, it is unclear whether such regulation will be particularly effective in holding prices down in areas with no competition. Telephone service will be regulated at the state level to ensure that affordability of basic service is maintained, but at costs significantly higher than current charges. Especially needy users can apply to a fund created to subsidize special circumstances in high-cost (primarily rural) areas. In the case of cable television, cable companies are currently allowed to increase prices to cover upgraded services or higher costs. Consumer groups have complained that this has led to excessive basic packages inflated with numerous frills that are too expensive to be considered basic.

At the state level, a number of key issues are being worked out regarding right-of-way issues, the responsibilities and prerogatives of original telecommunications providers, and the legality of municipally owned telephone (and other tele-communications) services. It remains to be seen whether local private sector telecommunications providers (telcos) will be allowed to serve larger areas from which they are currently barred in Iowa. Such expanded service could stimulate competition. It also remains to be seen to what degree many of the publicly owned fiber optic systems can lease space (bandwidth) or directly service fee-paying clients other than government offices, educational systems, and hospitals. For example, Iowa has the largest publicly owned fiber system (operating in all 99 counties) in the country, costing nearly one-half billion dollars. Although the Iowa Communications Network (ICN) was designed for distance education (and technically available only to K–12 schools, state government, state universities, the National Guard, and a few other special users), "more than 80 percent of the traffic is something *other than* distance education" (Caristi, 1997, [italics added]). Currently neither local government (except Des Moines) nor the private sector is legally allowed direct access and privatization is a serious policy consideration.

A second wildcard is the development of the backbone distribution systems. Two models of competition are possible: competition based on services has been the primary U.S. and European Union model in telecommunications in which former monopolies have been forced to open up their systems under deregulated fair practice provisions. Another model, encouraged in the United Kingdom, promotes the development of alternate infrastructure systems altogether. This is generally a more robust form of competition and requires less regulatory oversight (Taschdjian, 1997). Thus the creation and success of various backbone distribution systems is critical in determining the nature of the competitive playing field. In Iowa the most important question is how the publicly owned, statewide ICN mentioned immediately above is developed. Public access to this system, whether through a change in the legislation about eligible users or through privatization, would immensely change the competitive landscape. The privately owned telephone system is certainly more comprehensive, but its inadequate capacity for increased usage in many areas seems unlikely to be upgraded rapidly without significantly better revenue streams. Using the example cited earlier, while the costs of upgrading the Madrid telephone-Internet system were less than $10,000, the revenue from a couple hundred potential Internet users simply did not, in the opinion of GTE, warrant the telephone company's expense and time. Iowa also has a private telco fiber system, the Iowa Network Services (INS), but it is limited to the specific areas of its service provision. And the Pioneer Holdings system, a 400-mile fiber optic loop, is located only in the northwestern part of the state. Currently this patchwork of systems allows for options in some places but also has extensive lapses of competition in rural areas. Legislative changes would probably be necessary for rapid improvement of the competitive environment which would in turn provide better telecommunications service options.

The third wildcard is the technological advancement associated with satellite and wireless transmissions, as well as other possible technological innovations. The potential of wireless telecommunications backbones is currently only partially utilized, but national policy has begun to open up the wireless spectrum by selling large amounts of wireless spectrum to the private sector. Although it is highly unlikely that this will substantially affect telecommunications systems in Iowa in the short-run, the long-term possibilities are likely to be major as more spectrum is available and is used more effectively. Another example is the current capacity to increase copper wire capability, by a factor of 10 to 100, depending on the type and condition of the wire. Utilization of this technology would revitalize much of the copper wire system going into homes. However, this innovation, which has been known for a number of years, has yet to stir the interests of the major telephone companies to perfect the technology and install it universally. The financial question has been the major disincentive. What does the telephone company get out of the expense of the system upgrade? The answer is very little because the new income streams go to companies such as America Online and Prodigy.

CONCLUSIONS

The telecommunications area is changing radically as is the public sector role. At least three trends seem clear.

- First, because telecommunications and information technologies are increasingly important commodities, rural areas are not exempt from this pressure to connect. In fact, the need to connect for economic development, telecommuting, citizen involvement, and governmental efficiency and effectiveness may be even more important for rural than urban communities; the latter tend to have other private sector options. For all the discussion about leveling the playing field by connecting rural communities to high-end services such as the Internet and high-speed data transfer, such connections are frequently limited or lacking.

- Second, in a deregulated environment, competition is increasingly important as a lever for improvement. Because rural service in the past was often indirectly subsidized through universal service regulations and because today strong incentives for infrastructure upgrades are generally only in major markets,[11] there is little real competition in the rural telecommunications market. Some competition may be achieved by simply continuing to deregulate the industry. Yet to achieve systems with full capacity parity, rural areas will either have to pay more, see the continuance of strong universal service provisions in legislation, receive help through publicly supported systems (such as a privatized ICN system in Iowa), and/or find creative local options through strong local—*either* private or public—leadership.

- Third, because rural areas are often underserved in an unregulated market and strong monopolies may continue to dominate in the short term, public enterprise solutions may more often be a viable and necessary alternative. Such units can provide entrepreneurship at the local level where those communities with the need and the capacity can respond to their own local needs and shape their own future. This is not a new concept. In 1932 Franklin Delano Roosevelt said that "where a community . . . is not satisfied with the service rendered by the private utility, it has the undeniable right as one of the functions of government to set up . . . its own governmentally owned and operated service" (quoted in Emmons, 1993). Public enterprise has renewed importance in an era in which economic development is critical for communities and it makes sense in an age that commonly places economic development on the governmental agenda. Further, these cases of public enterprise even provide some competition at the macroeconomic level, where large regional providers know that erosion of their customer base is a genuine possibility. These cases also encourage them to be more proactive in installing system upgrades that are of marginal immediate financial utility. The irony, of course, is clear: although we are in a "less-government" era and some public utility industries will experience increased competition and perhaps some decline, there is a significant class of cases where local government public enterprise creation is *increasingly* critical for economic development. In this class of cases, local telecommunications utilities do not function as micromonopolies, but provide yardstick comparisons, threat-of-entry competition, and local determination alternatives[12] critical to a well-functioning capitalist market.

Whether or not the trend to create new public telecommunications enterprises becomes a major national trend following the Iowa example, or whether the restrictive environments in some states predominate, remains to be seen. Certainly there is a strong historical precedent for the creation of local utilities in leading economic areas, just as there has been modest pressure to privatize in more mature public enterprise industries. Although Iowa's trend has been primarily among small, rural communities,[13] some of the cities actively expanding their public enterprise telecommunications capacity nationally have been moderately large. Further, the common types of services provided by these new public enterprise telecommunications utilities are not yet entirely clear. Possibilities range from cable television, Internet, high-speed data transfer, or telephone service to all three major services. Of course the continuing integration of computer and communications systems makes these services enormously important and communities have a huge stake in assuring access to high-quality, low-cost services, regardless of whether they are provided privately or by a public enterprise.

NOTES

1. For a contrary view, see Vennard, 1968. Edwin Vennard expresses the general views of the power industry and at the time was the managing director of Edison Electric Institute.

2. Because of the extensive statewide, state-run system that connects libraries, the access data for smaller communities is probably overstated. E-mail, World Wide Web data, and maintenance of a homepage probably give a more accurate picture of the actual organization usage rates for the Internet.

3. In the first research phase telecommunications issues were largely confined to telephone, Internet, video-conferencing, and high-speed data transfer systems. For the case studies, cable television was also included because it often is part of the "wired" mix of telecommunications services that can be considered and which can effect integrated systems in the future.

4. The term "bandwidth" is the industry standard in discussions of telecommunications demand and usage. Bandwidth is used for both telephone and television standards.

5. For example, one of the rural local residents directly interviewed in this study stated that although his community had local Internet access, he could never get through to place an order. He speculated that if purchasing the service was that difficult, that the service itself must be dreadful, and he opted to pay the toll charge to connect to the university's system.

6. Deregulation of cable television first occurred in 1982, but was re-regulated in 1992 and subsequently deregulated again in 1996 with other telecommunications sectors.

7. Nationally, the success rate also seems high, according to anecdotal reporting. However, two examples of voters declining telecommunications outside Iowa are Coldwater, Michigan, where the existing cable television system vigorously fought the initiative, and Moorhead, Minnesota, which was opting for municipally provided telephone service.

8. Governor Branstad signed House File 596 authorizing the Iowa Utilities Board to issue certificates of public convenience and necessity to municipal telecommunications utilities in 1997.

9. The cost was considerable for Hawarden, approximately $4.6 million, but was actually less than the $6 million bid placed by the city to buy the old system that was turned down.

10. The private telephone company association is expected to challenge the recent ruling in federal court. Towns that have already been authorized to add telephone service are Coon Rapids and Laurens, Iowa. Others looking into it include Muscatine, Carroll, and Spencer.

11. A good example of the pressure for system upgrades comes from the largest cable television provider in Iowa, TCI. Proprietary cable television systems are notoriously difficult to dislodge from their city franchises no matter how dissatisfied the customers may be. TCI has spent over $20 million in system upgrades in its Des Moines market alone to successfully fight off enthusiasm for switching companies or for a municipal telecommunications utility.

12. It is often considered appropriate to have "default" providers of critical services for the public good, so that universal access to services is available to all individuals. Proponents of municipal enterprises argue that they have the right and responsibility to be the default providers of services when the private sector fails to provide any or adequate critical services.

13. The cities of Cedar Falls (pop. 34,000) and Muscatine (pop. 22,881) are the largest to date in Iowa.

Part II

Using Telecommunications in the Rural Community

5

RURAL BUSINESS AND TELECOMMUNICATIONS TECHNOLOGIES

G. Premkumar

INTRODUCTION

The new millennium is expected to herald a society that is going to be critically dependent on information and the technologies to process that information (Burstein & Kline, 1995). Researchers predict that our society will employ more people in information-related industries than in previous decades, and that access to information and skills in using the technologies will become predominant determinants of success for individuals, organizations, communities, and nations. Hence, it is critical for researchers in rural development to evaluate the impact of these information technologies on rural America. The objective of this study is to examine various communications needs of small businesses, the communications technologies that are required to meet those needs, and the factors that facilitate the adoption of these technologies.

Researchers have questioned whether certain categories of business (e.g., small business) or certain regions (e.g., urban or rural) are adversely affected by these new information technologies (Parker, Hudson, Dillman, Strover, & Williams, 1992). As we move to an information society, access to information and its efficient processing will become vital to the local economy (Parker, Hudson, Dillman, & Roscoe, 1989). The changes in information and communications technologies provide both opportunities and threats to small businesses located in rural communities (Allen, Johnson, & Leistritz, 1993; Rowley & Porterfield, 1993). The growth of the Internet, ubiquitous computing, and instantaneous communication from anywhere to anyplace have made geographic locations and distances less relevant, especially in the service industry. Markets that were closed due to geographic constraints have become accessible and global dissemination of information from the most remote corners of the world has become feasible. Rural businesses can compete with their urban counterparts in the same market and may have a competitive edge in terms of lower labor costs and overhead. This may motivate urban businesses to take advantage of the cheap labor resources and potential markets in rural communities by relocating some of their operations in these communities.

Modern communications technologies also create threats to growth and survival of rural businesses. History has shown that availability of basic infrastructure, such as rivers, railroads, and interstate road systems facilitates local economic development. The economic survival and growth of some rural communities has been significantly affected by lack of access to the interstate road system. The modern economy is significantly influenced by the availability of information, and therefore economic growth would be greatly facilitated by the availability of a good information technologies and telecommunications infrastructure (Yarbrough, 1990). The infrastructure would include availability of high bandwidth telecommunications lines, local carriers to provide a variety of digital services using the high bandwidth capacity for businesses and end-consumers, availability of information technology vendors and service providers, and local expertise to exploit the technologies. Unfortunately, telecommunications infrastructure seems to follow a similar pattern as the interstate road system. Market forces that dominate the creation of telecommunications infrastructure may inhibit extension of universal access to all areas in the country, because demand in urban areas is greater than in rural areas. Hence, rural businesses are caught in a vicious cycle—lack of communications infrastructure reduces the demand for communications services, which further constrains future investment in the infrastructure. This may result in rural businesses being further alienated from mainstream economic activity. An additional source of threat is that these new technologies provide the opportunities for businesses to bypass the rural areas in the United States and relocate in developing countries where the labor and overhead costs are even lower. Many software and information related firms have relocated their data entry and software development in developing countries, taking advantage of the lower labor costs and live satellite and other online links to their counterparts in the United States.

COMMUNICATION NEEDS AND CHANNELS

The need for communication is a key factor that initiates the adoption and diffusion of modern computer and communications technologies in most businesses. The latest trend in integrated supply chain management has highlighted the importance of communications technologies and interorganizational systems for businesses. Supply chain management is concerned with the integration of all activities associated with the flow of goods and services from the raw material supplier to the end user (Handfield & Nichols, 1998). Information flows along the supply chain become the most critical component to ensure smooth flow of goods and services. Communications technologies integrate the various systems that generate, process, and store the information along the supply chain and ensure effective supply chain management. All members in the supply chain, regardless of whether they are small or large operations require modern communications technologies. The communication chain is only as strong as the weakest link, and therefore, large businesses are interested in ensuring that the communications links with small firms are as good as their links with large firms. For example, large retailers such as Wal-Mart provide training and support to their small suppliers to enable them to electronically link up to their network. These innovative business practices are forcing small businesses to adopt the latest communications technologies or run the risk of economic isolation if they lack these technologies.

The supply chain consists of a variety of members or trading partners and can be described in multiple levels depending on the depth of detail required. The major trading partners in a single level supply chain is shown in Figure 5.1. Essentially the chain involves the flow of goods and services in one direction, cash in the opposite direction, and information flow in both directions. The primary objective of supply chain management is to improve the efficiency of the chain by removing any constraints to free flow. Any bottleneck in the flow of information will result in excess inventory or unavailability in retail outlets, customer dissatisfaction, and increased costs for products and services. In the current business environment, firms, whether small or large, must be electronically integrated with other members in the chain to actively participate in a supply chain.

The active trading partners for an organization are its customers, suppliers, logistics firms (carriers) who ensure the flow of goods and services in the supply chain, and financial institutions such as banks who are responsible for movement

Figure 5.1
Business to Business Interorganizational Commerce—Extranets

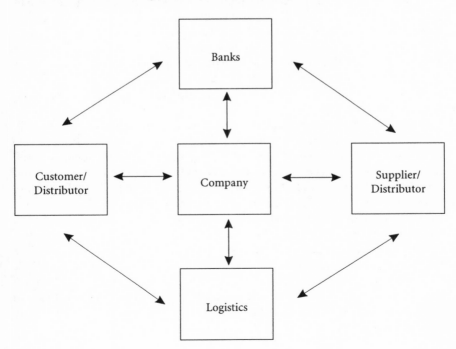

of cash along the chain. The typical interactions between a supplier and customer would include request for a quote, the quotation, purchase order (P.O.), P.O. acknowledgment, invoices, shipping notice, payments, and so on. The interactions with the bank would include payments and loans, and the interaction with the logistics firms would include bill of lading, shipping notice, invoices, payments, and so forth. It is estimated that more than 50 business transaction communications happen among the trading partners before an order is completed.

The type of communication among the trading partners is dependent on a variety of factors including industry type, product or service characteristics, location of partners, access to technology and information, and information quality. The communication characteristics vary along many dimensions (Zmud, Lind, & Young, 1990). Some of them are:

- Richness of medium—a medium that is less ambiguous, uses multiple cues, provides immediate feedback, and has ability to personalize the message (Daft & Lengel, 1986)
- Communication mode—Immediate or delayed communication
- Information type—structured, semistructured, or unstructured
- Processing—need for computer processing of communicated information
- Location—geographic location of communicating parties

A variety of communications channels can be used for communication. The common ones are: face-to-face, telephone, facsimile (FAX), U.S. mail, e-mail, electronic data interchange (EDI), online data access, and video-conferencing. Each channel has its advantages and disadvantages. The use of a particular communications channel is dependent on the communication task characteristics. For example, a face-to-face communication would be appropriate for a rich unstructured communication (e.g., negotiating a sales contract), while EDI would be a better choice for highly structured communication that needs further processing (e.g., sales invoice).

While larger organizations actively embraced modern communications technologies, smaller businesses, especially in rural communities, have been slow in adopting these technologies (Raymond & Bergeron, 1996). To better understand their adoption our research team collected data from 78 firms selected from five rural Iowa communities. The businesses were selected from the local directory of businesses and by references from other businesses. Face-to-face interviews, aided by a structured questionnaire, were used to collect the data. Sample characteristics are shown in Table 5.1. The respondents were from a variety of industries with a greater emphasis on retail sales, financial services, and real estate. The respondents also had a wide range of experience in the use of computers. While 26 percent of the respondents had less than 5 years experience, 36 percent of them had more than 10 years experience. The firms also varied in size with about 46 percent of them having less than 5 employees and 26 percent having more than 25 employees.

Table 5.1
Sample Characteristics

	Number of Firms	Percentage of Firms
Industry		
Manufacturing	11	14
Retail sales and wholesale trade	28	36
Service	14	18
Finance, insurance, or real estate	19	24
Other	6	8
Company Size		
Less than 5 employees	6	46
6 to 10 employee	11	14
11 to 15 employees	4	5
16 to 20 employees	3	4
21 to 25 employees	2	3
More than 25 employees	20	26
Annual Sales Revenue		
Less than $1 million	26	33
$1 to $50 million	21	27
More than $50 million	3	4
Don't know	28	36
Computer Experience of Company		
Less than 5 years	21	27
6 to 10 years	28	36
More than 10 years	28	36

Table 5.2 illustrates the use of communications technologies by the respondents. Ninety-four percent of the firms use FAX technology. Direct online data access is used by 72 percent of the firms and EDI by 32 percent of the firms. As the sophistication level of the technology increases there are fewer users of the technology. Based on general estimates of investments in these technologies, it seems that the number of users is also inversely related to the costs of these technologies. Because two or more participants are required for use of communications technologies, the utility of the technologies increase as the number of users increase and is most useful when there is universal access. A critical mass of users is initially required to create a threshold level of usefulness that will motivate additional businesses to adopt the technologies (Markus, 1987). Table 5.2 indicates that FAX has almost reached universal access among Iowa businesses to become a common medium for business communications compared to EDI or other technologies.

Table 5.2
Percentage of Communications Technologies and Information System Applications

Communications Technologies	Percentage of Population
FAX	93
Modem for data transfer	80
Online data access	72
E-mail	41
EDI	32
Internet	28
Satellite	19
Video-conferencing	6
Information System Applications	
Word processing	96
Database	86
Spreadsheets	85
Accounting systems	96
Customer billing systems	85
Payroll systems	80
Sales order entry & point of sale system	67
Production / inventory	65
Personnel	55
Purchasing	53
Engineering—CAD	30

FAX technology is easy to use, relatively inexpensive, and accepted as a formal document for most business transactions. However, FAX is inefficient because the information undergoes multiple transformation from paper to electronic format leading to data reentry at both ends. Use of a modem for data transfer is a relatively simple technology that is useful for infrequent communications. Direct *online access* is a more efficient way to transmit the information because it avoids the conversion. However, trading partners may be reluctant to provide direct computer access to each other due to security concerns, and technical compatibility with a wide variety of systems used by the trading partners makes its implementation a complex task. Also, online access is normally restricted to only a few transaction communications rather than a whole range of business communications. Normally online-access is popular with franchisees/subsidiaries and trading partners with whom long-term business relationships based on trust have been developed. *E-mail* provides the broadest scope for communication and is closest in analogy to the paper mail system. However, e-mail is normally restricted to text-based messages and attachments, and the information has to be transformed to be used in other internal transaction applications. For example, a sales order received by e-mail must be opened by a person, the information retrieved and reentered into the order processing system. *EDI* overcomes this drawback because it lets transaction information be directly transferred from sender's computer application to receiver's computer application without manual intervention. However, it is a more complex technology that requires extensive cooperation and trust between the two trading partners.

Internet access is the latest communications technology that provides access as well as disseminates information worldwide and has the potential to provide all the benefits of earlier technologies. The open standards, common protocols, and extensive availability make it a global communications medium for all business communication. New software that integrates direct access, EDI, and applications with the Internet infrastructure will make it a preferred medium of communication. Satellite communications is more complex and requires more investment and expertise. Video-conferencing is a newer technology that requires significant investment and is used to provide face-to-face audio-video communication to the participants. It is popular in distance learning, providing training and education to remote rural towns.

Figure 5.2 shows the communications media currently being used by the organizations for common business transaction communication with customers and suppliers. Except for sales invoice, which is currently sent through the U.S. Mail and FedEx service, telephone and FAX are the most popular media. However, other online electronic media are slowly becoming popular. For instance, online data access and EDI are increasingly being used for purchase orders. An important factor determining the use of communications technologies is the location of suppliers and customers. These two partners account for a major share of business information communication in most organizations. The location of suppliers and customers are shown in Figure 5.3. It can be seen that for most respondents in our survey a large proportion of their suppliers are located outside the community, but a significant proportion of customers are located within the community. This necessitates greater use of electronic communication to interact with suppliers but much less use of communications technologies to interact with customers. Technologies such as EDI are normally initiated by customers who would like their supplier to be electronically integrated with them. Because the customer base for most of the respondents are primarily local they probably did not have much external pressure to adopt these technologies. If the customer base is external to the community, the firms might feel a greater compulsion to use more electronic communications technologies.

An important advantage in using electronic communications technologies such as EDI and online data access is the flexibility in processing the communicated information in internal systems without manual intervention. This, however, requires that the firm has integrated internal systems that can seamlessly take the information for further processing. In many small firms the level of computerization may be limited and therefore they may not be able to benefit from these tech-

Figure 5.2
Percentage of Use of Six Communications Media by Local Business for Four Business Transactions—Invoicing, Ordering, Shipping, and Purchasing

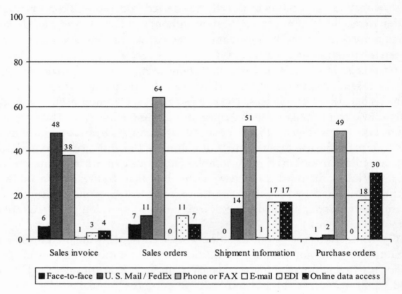

Figure 5.3
Percentage of Suppliers and Customers Located Outside the Community

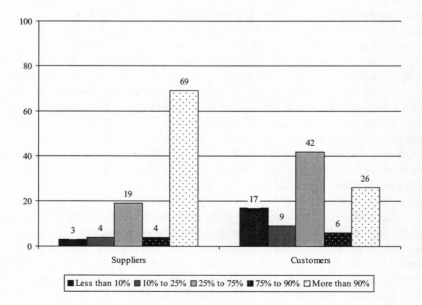

nologies. In fact, in some firms, it may be more of a burden to set up the necessary infrastructure to enable them to provide electronic access to trading partners. Hence, an assessment of the level of computerization of their internal systems would provide an indication of the benefits from adopting newer communications technologies.

The information system applications used by the respondents are shown in Table 5.2. The number of firms using the three basic computer packages (word processing, spreadsheet, and database) is very high. Among the business applications accounting systems are the most popular (96%), followed by customer billing systems (85%) and payroll systems (80%). We do not see extensive automation in purchasing and ordering systems, two important systems for electronically integrating with trading partners. The lack of automation of the ordering and purchasing systems may be a disincentive to try new communications technologies, such as EDI. Perhaps because the firms are small or lack the expertise, they do not perceive the need to use these technologies to automate their operations.

FACTORS INFLUENCING THE ADOPTION OF COMMUNICATIONS TECHNOLOGIES

While communication need is one of the factors to influence the adoption of communications innovations, a variety of other factors also influence their adoption. The traditional innovation adoption and diffusion literature examines a wide variety of innovations in different contexts and provides interesting insights on adoption of information technologies. Rogers (1995) states that the decision to adopt and use an innovation unfolds in stages. The first stage is *awareness*, in which the firm or firm's personnel seek information about the innovation. In the second stage, *persuasion*, they are persuaded by information about the technologies from others, such as vendors, to adopt the innovation. Then the firm's personnel go through a *decision* stage to decide whether to invest in the innovation, and an *implementation* stage in which the innovation is implemented in the organization. Finally, in the *confirmation* stage, actual outcomes are measured and compared with initial expectations. Various characteristics of environment, organization, and the individual organization members influence whether the innovation is adopted. The characteristics of the innovation also play a significant role in influencing the adoption decision. In a population of firms, the diffusion process starts slowly, takes off rapidly after an initial period, and eventually levels off.

The innovation adoption literature identifies five major categories of characteristics influencing adoption of information technologies in organizations (Kwon & Zmud, 1987). The categories of characteristics are (a) product/innovation, (b) organizational (or structural), (c) environmental, (d) task, and (e) individual characteristics. The most significant characteristics under each category are:

- Product/Innovation—relative advantage, complexity, compatibility, and cost
- Organizational—product champion, top management support, information technology expertise, organizational size, and organizational resources

- Environmental—external pressure, competitive pressure, vertical linkages, and external support
- Task—structure, autonomy, and uncertainty
- Individual—education, age, experience, and psychological traits

The innovation characteristics most frequently studied are relative advantage, complexity, communicability, divisibility, cost, profitability, compatibility, social approval, trialability, and observability (Tornatzky & Klein, 1982). Of these, high relative advantage, low cost, high compatibility, and low complexity are found to be consistently related to adoption. *Relative advantage* of an innovation is the degree to which the innovation is perceived as better than the idea it supersedes (Rogers, 1995). A rational adoption decision in an organization would involve evaluating the advantages of the new and the old technologies. The new communications technologies provide many benefits to the adopters in terms of reduced turnaround time, better customer service, reduced costs, and timely information availability for decision making. *Cost effectiveness* is the degree to which benefits from the adoption of an innovation are commensurate with the costs. Although hardware costs have come down, the software development and computer operation costs have not had a commensurate decline. For small businesses the cost of hardware and software remains a major deterrent to adoption (Premkumar, Ramamurthy, & Nilakanta, 1994). *Complexity* is the degree of difficulty associated with understanding and learning to use an innovation The complexity of the technologies creates greater uncertainty for successful implementation and therefore increases the risk in the adoption decision. Firms initially were reluctant to adopt EDI because the technology was perceived to be very complex (Premkumar et al., 1994). An innovation's *compatibility* is the degree to which it is perceived as being consistent with the existing values, past experiences and needs of the potential adopter (Kwon & Zmud, 1987). The use of computers and modern communications technologies can bring significant changes to the work practices of businesses, and resistance to change is a normal organizational reaction. It is important, especially for small businesses, that the changes are compatible with the views of the owner/manager. For example, the owner/manager needs to feel comfortable using these technologies and should not feel that automation may reduce his or her control over daily operations. Using proprietary technologies in interorganizational systems could lock up a firm with a specific customer or supplier. This could go against the values of some owners/managers who may be fiercely independent and not wish to be influenced by one or two large suppliers or customers.

Some of the organizational factors that influence the adoption of information technologies are product champion, top management support, information technologies expertise, size of the business, and slack resources (time, funding, technical skills). *Product champions* create awareness of the innovation and its benefits, market the innovation within the organization, mobilize support for the innovation, and shepherd its implementation (Beath, 1991). *Top management support* is critical for creating a supportive climate and providing adequate resources for adoption of new technologies (Palvia, Means, & Jackson, 1994). Their support is critical for communications technologies because the use of these technologies requires the cooperation of trading partners. Close interaction is required among trading partners' senior management to create close electronic business partnerships, examine legal and other issues, and facilitate the adoption of these technologies. In small businesses the information technology *expertise* in the organization and an understanding of the potential of information technology by the owner/manager play a role in facilitating the adoption of new technologies (Thong, Yap, & Raman, 1996). Firms that do not have the necessary information technology expertise may be unaware of new technologies or may not want to risk adopting these technologies. Larger *size* organizations are found to have greater slack in resources and are therefore able to experiment with new innovations. Even within the small business category larger businesses are able to take risks with new technologies. They are also able to more easily mobilize adequate financial resources required for implementing innovations. Availability of adequate technical *resources* is another critical factor for many small businesses. Small firms, especially in rural areas, will have a difficult task of hiring people with technical skills to implement communications technologies (Palvia et al., 1994).

The external environment also plays a significant role in the adoption of new technologies, particularly communications technologies. Communications technologies are interactive technologies whose utility to a business increases only if used by trading partners as well (Premkumar & Ramamurthy, 1995; Gatignon & Robertson, 1989). Hence, there is increasing *external pressure* exerted by trading partners (customers and suppliers) to adopt these technologies. For example, many firms adopted EDI due to demand from customers. Small businesses are more vulnerable to customer pressure. They are more likely to be economically dependent on bigger customers for their survival, as illustrated by the experience of small firms adopting EDI to satisfy the demands of large firms such as Wal-Mart or General Motors. Strong *vertical linkages* of franchisees and subsidiaries to their parent organizations facilitate the transfer of innovative technologies. The larger parent organization can use its size advantage to experiment with innovative technologies and then transfer the innovations to the smaller units. Firms may also be required by their franchisers or parent units to have certain information technologies for communication purposes. *External support* from vendors is critical for small firms without adequate information technology (IT) expertise to try out these new technologies (Thong et al., 1996). Organizations are more willing to risk trying new technologies if they feel there is adequate vendor or third party support for the technologies.

Organizations adopt innovations and use them for specific tasks. Information technology research has evaluated the roles of various task characteristics, such as task structure, autonomy, and uncertainty on innovation adoption (Saunders & Courtney, 1985). A *structured* task is found to be more amenable for IT adoption compared to an unstructured task, as it is easier to automate a structured task. Greater *autonomy* in the performance of the task is found to facilitate innovation adoption. Task *uncertainty* tends to create greater information search that motivates the use of IT for the task. However, the uncertainty may also be an inhibitor to adoption due to lack of task structure. The use of an innovation in multiple task environments makes it difficult for researchers to clearly identify task factors that influence the adoption decision.

Individual characteristics of the adopters or decision makers, such as education, age, experience, and psychological traits have been found to influence adoption of innovation (Palvia et al., 1994). In large organizations individual characteristics have less influence because most technology adoption decisions are made on a collective basis and are not optional decisions for individual users. Characteristics of the decision maker(s), however, may influence the adoption decision.

ANALYSIS OF THE FACTORS RELATED TO ADOPTION OF COMMUNICATIONS TECHNOLOGIES

While the factors listed in the earlier section have been studied in the context of adoption of different innovations, there have been few studies related to adoption of communications technologies by rural businesses. This study will examine which of the many factors discussed earlier has an impact on adoption of communications technologies. Because the potential factors are so numerous, a subset of the factors that have been found to be significant in earlier studies was considered for this study. The first three sets of independent variables—characteristics of the innovation, organization, and environment—were considered for the analysis. Task and individual factors were not considered because they were expected to have a much lower impact on an organizational level adoption decision. The innovation characteristics considered were relative advantage, cost, complexity, and compatibility. Organizational characteristics included were top management support, IT expertise, and size of the business. Environmental characteristics included were competitive pressure, external pressure, and external support.

The independent variables were measured using multi-item Likert-type indicators based on operationalization in earlier research studies. The dependent variable, adoption of information technology, was a dichotomous variable based on adoption of any one of the four communications technologies: e-mail, data transfer using modem, online data access, and EDI. The items measuring the variables were included in the survey of 78 small businesses described earlier.

Table 5.3 provides the results of t-tests that measure the difference in mean values between two groups—adopters and nonadopters—for innovation, organization, and environment characteristics. *Relative advantage* is the most significant factor among the innovation characteristics. Firms adopt technology only if they perceive a need for the technology to overcome a perceived performance gap or exploit a business opportunity. During the interviews some of the benefits mentioned by the adopters were reduced turnaround time, increased transaction speed, access to current information, and reduction in data entry errors. Although some nonadopters were not aware of the benefits of the technology, those who were aware did not find any need for these technologies in their businesses. For example, a small firm with only five employees and a local customer base may not

Table 5.3
Differences between Adopters and Nonadopters

Variable	Nonadopters Mean (S. D.)		Adopters Mean (S. D.)		t-Value	Significance Level
Innovation						
Relative advantage	2.99	(0.94)	3.90	(0.59)	4.34	0.0001*
Cost	3.30	(0.76)	2.90	(0.79)	-1.46	0.149
Complexity	2.70	(0.82)	2.16	(0.84)	-1.88	0.063
Compatibility	2.63	(1.02)	1.84	(0.83)	-2.81	0.006*
Organization						
Top management support	2.97	(0.63)	3.69	(0.92)	2.46	0.016*
Product champion	3.37	(1.06)	3.89	(1.01)	1.35	0.181
IT expertise	7.44	(2.90)	11.24	(5.43)	2.44	0.017*
Size	12.83	(23.48)	135.92	(470.60)	0.90	0.371
Environment						
Competitive pressure	2.40	(1.09)	3.60	(1.03)	3.53	0.001*
External pressure	1.95	(0.85)	2.60	(1.08)	1.87	0.065
External support	2.46	(0.63)	2.32	(0.89)	0.50	0.619

Note: * indicates significant variables

find much need for communications technologies such as e-mail or EDI. The firm may use the computer in a limited way for accounting purposes. The other innovation characteristic that was also found to be important is *compatibility*. Because communications technologies involve communication between two partners as well as integration with internal information systems, compatibility of the various technologies is a very important factor. It is hoped that the move towards open standards in the Internet will alleviate this problem.

Among the organizational characteristics, *Top management support* and *IT expertise* were found to be significant. Top management's vision and commitment to the innovation is essential, especially in small businesses, to get adequate resources and support to implement the innovation. Their support is even more important for communications technologies because it involves interaction with trading partners and creating business agreements for using the technology. The use of these technologies could significantly change the way business is conducted within the organization as well as externally with its trading partners. Top management's commitment is required to overcome the resistance to change that is normal in such situations. In small firms it is likely that the owner/manager is the top management, and if he or she is not convinced of the technology's utility it is very unlikely to be adopted. In many small firms lack of IT expertise also is a significant deterrent to adoption of new communications technologies. Training existing personnel in new technologies is one strategy to overcome this problem. Another strategy would be to outsource the services to a third-party vendor. However, the availability of IT service vendors in rural areas is very limited. We did notice that having vertical linkages with parent units helped many firms to overcome the IT expertise shortfall. We found that firms that are subsidiaries/franchisees often have better communications technologies. A majority of them obtained their information systems support from the parent organization.

Size, as measured by number of employees, was not significantly related to adoption of communications technologies. Although there was a substantial difference in the mean values between adopters and nonadopters, the large variances resulted in low t-values. However, when we classified the respondents into three categories of size based on number of employees (small < 5; medium 6–20; and large > 20), we found that larger firms adopted these technologies more often than smaller firms. Larger firms have the resources to invest in technologies and the organizational slack to experiment with them. They may have in-house information systems support service to create the awareness, and initiate and facilitate the adoption of these technologies. The operations of smaller firms may be simple enough to be managed without these technologies and therefore have no need for them.

Among the environmental characteristics, there was a significant difference between adopters and nonadopters on competitive pressure. The digital revolution combined with the availability of network infrastructure has made it technically feasible and a socially acceptable business practice to use electronic means for business communication. Hence, it has become a strategic necessity for firms to have these technologies. A closely related factor, significant only at 0.06 level, was external pressure. Because the benefits of communications technologies are only realized if there is a critical mass of adopters, many firms are requesting and, sometimes, requiring their trading partners to be linked electronically with them to realize the full benefits of the technology. This is particularly true in the case of EDI. In the retail industry, for example, large chain-stores insist their suppliers use EDI.

BUSINESS AND COMMUNITY—VERTICAL AND HORIZONTAL LINKAGES

Community researchers have identified two types of relationships, or linkages, that are important to the welfare of the community (Warren, 1978). Vertical linkages refer to relationships a firm has with organizations and agencies external to the community that sustain its growth by ensuring infusion of technology, information, expertise, and other resources. Horizontal linkages refer to relationships the firm has with local institutions/agencies that facilitate its existence and growth. While vertical linkages are more useful for the introduction and initiation of new technologies into a firm, horizontal linkages ensure internal diffusion and sustained use of the technology within the firm. A factor that emerged in many of the interviews, especially with firms that were franchisees, subsidiaries, or divisions or larger companies, was that the flow of technologies came from their parent unit through vertical linkages. Very often, the corporate unit acquired the expertise and developed the necessary systems and then implemented the system in the local unit. Often these were implemented as direct online access systems. The author found during the interviews, that while vertical linkages facilitated innovation adoption they did not necessarily lead to awareness or diffusion of technologies within the company. For example, in a telemarketing firm, none of the employees, including the manager, were aware that their computers had modems that could link up with computers in the parent organization, even though at the end of the day they were uploading all their data to the corporate office using modems. Their knowledge was limited to clicking a button on the screen to send the data without realizing that they were using online data access for file transfer. In another instance, a local subsidiary of a national retail chain used online data access through point-of-sale terminals to access price and inventory information from a corporate database for their daily sales operations. The employees in the store were unaware that they were communicating with a remote computer in the corporate office. In these situations, the local people were never involved in the implementation and never thought of newer methods of exploiting the technologies to make them more productive and competitive.

Rogers (1995) identifies three different levels of knowledge about innovation: what the innovation is, how the innovation works, and why it works. The first level provides the "awareness knowledge" on the innovation, the basic information requirement for making an adoption decision. The next level, "how-to knowledge," is essential to use an innovation effectively. The amount of how-to knowledge required increases as the complexity of the innovation increases. Finally, "principles knowledge" provides information on the functioning principles and basic theories on the innovation. Normally, this knowledge is not essential for using an innovation but would be required if the adopter had to troubleshoot or improve the existing innovation.

It is important for a firm to determine the level of knowledge required for the communications technologies used by its employees. Are computer technologies as simple as telephones to use and therefore the firm needs to provide only how-to knowledge, or do users need more knowledge to enable them to find innovative uses? This would have a bearing on the training and education provided by a firm to its employees. It seems the vertical linkages provide the how-to knowledge, but in certain situations more detailed principles knowledge would help the firm to expand use of the technologies.

Community development scholars emphasize the importance of horizontal linkages and networks for maintaining viable communities. We did not find evidence of horizontal linkages as mechanisms for spreading benefits of the technologies within the community. Most businesses indicated that they received very little technical or other support from the community to use communications technologies. This could be attributed to the limited scope of the survey covering only five communities. Alternatively, this could be due to the very nature of the sector when compared to other sectors of the economy such as education and government. Businesses operate in competitive environments and very often information on various aspects of their operations are of a proprietary nature that they would be unwilling to share with other businesses in the community. They may also be constrained by confidentiality agreements signed with their parent units. In some firms, the level of expertise on the technologies may not be sufficient to be of much use to the community. Nor does use of these advanced technologies by one firm necessarily trigger the development of local physical infrastructure in the community, although there are exceptions. For example, one firm's use of high bandwidth application may force the telecommunications provider to create the necessary infrastructure so that other firms can piggyback on the facility. In Kearney, Nebraska, the need for high bandwidth infrastructure by Cabella's, a mail-order firm, created the infrastructure for the entire community. Other similar instances of telemarketing firms improving the telephone infrastructure in the local community have been reported (Abbott, 1997; Parker et al., 1992).

While our study did not find evidence of horizontal linkages facilitating adoption there are case studies of successful community development using telecommunications technologies (Abbott, 1997). For example, ACENet, a nonprofit community development organization in Ohio, created incubator facilities with necessary telecommunications infrastructure to attract small businesses. The organization actively seeks funding from external agencies and other community networks to facilitate economic development in its area. In Nevada, Missouri, the local economic development commission used buildings that were vacated by a hospital to create a telework center to provide distance education facilities and telecommunications infrastructure for small businesses. These examples highlight proactive communities using telecommunications to attract businesses for economic development. Case studies generally do not indicate that a specific local business initiated action for economic development or was responsible for bringing in new telecommunications infrastructure, although the business leaders may have been members of the community action teams that created these initiatives.

This suggests that diffusion of information technologies within the business sector or from business to other sectors will not automatically occur. Systematic community efforts are needed to ensure that the knowledge is disseminated through channels such as continuing education classes, seminars, and incentives for service providers to share that information.

CONCLUSIONS

New information technologies have opened opportunities for small businesses in rural communities as well as exposed them to additional risks. This study evaluated the various communication needs of small rural businesses and the communications technologies that are required to meet those needs. Using the diffusion of innovation model, the study examined five broad categories of factors that affect the adoption of communications technologies: environment, organization, innovation, task, and individual. Subsequently, the impact of a subset of those factors, belonging to three categories (environment, organization, and innovation) on adoption of four communications technologies (e-mail, data transfer with modems, online data access, and EDI) was examined. The results are based on a limited sample of firms in five rural communities and must be viewed in that context. Nevertheless, the study provides useful insights into the use of communications technologies by rural businesses and suggests further areas of investigation.

The results indicate that facsimile is the most prevalent communications technology among rural businesses, followed by limited usage of modem and online access. The utility of these communications technologies depends on the level of integration with internal information technology applications. Most firms have computerized their accounting and billing systems but have not expanded into other functional areas, thereby minimizing the utility of these technologies. The results indicate that relative advantage, compatibility, top management support, IT expertise, and competitive pressure were important determinants for the adoption of communications technologies. The study also assessed the role of horizontal and vertical linkages in adoption and found that vertical linkages were predominantly more beneficial in getting the latest technology. Many businesses received their technologies from parent organizations. We also found that businesses do not typically share their knowledge with the rest of the community, indicating weak horizontal linkages. Based on analysis of survey data and interactions with businesses in rural communities, a few guidelines to facilitate the adoption of modern communications technologies are provided.

Because awareness of the technologies and their benefits (relative advantage) were significant factors influencing adoption, systematic efforts are needed to raise the level of awareness, beginning with the owner/top management, whose vision and support is critical for introduction of new technology. Also, because communications technologies reduce location specificity of markets, firms need to evaluate strategies to exploit that potential and increase the geographic scope of their business. They need to realize that the business environment is no longer "business as usual," serving only their community. The free flow of information is creating well-informed customers, willing to use online services for a variety of needs, and local firms need to compete harder for their business. For example, the dramatic growth in online banking services is creating significant competition for local banks, and online shopping is affecting profits of local business.

Because businesses typically do not share information about their technologies, it is important for community organizations such as chambers of commerce, and state and local government agencies to take the lead in educating small business proprietors in latest technologies and best practices as relevant to their industries. This will improve their technology awareness as well as reduce their perception of technology complexity.

Growth of the Internet creates opportunities for new businesses in electronic commerce. Local organizations can take the lead in creating incubator facilities to jumpstart some of these businesses. As we move into an information-based economy, these businesses are expected to have significant potential for growth.

Most respondents indicated that the local expertise in information technologies was minimal. Because adoption of these technologies is dependent on the availability of skilled personnel it is important to create partnerships with local community colleges and educational institutions to generate a workforce with appropriate skills. As the society moves towards an information-based society, it is critical for communities primarily dependent on agriculture to realize the economic trends and shifts in workforce employment and to develop necessary skills within their communities.

NOTES

A detailed analysis of the survey data is available in Premkumar and Roberts (1999).

The author wishes to acknowledge the help provided by Margaret Roberts in the collection of data. The author also wishes to acknowledge the support provided by the Rural Development Initiative at Iowa State University for completing this project.

6

THE COMMUNITY NEWSPAPER
IN AN ONLINE SOCIETY

Eric A. Abbott and Walter E. Niebauer, Jr.

INTRODUCTION

Newspapers serve two crucial roles in the community. The first is to provide advertising to support local businesses. This economic role is shaped by the service area of local businesses. It is the need to reach local consumers with advertising messages that leads to the opportunity for creation of local newspapers. Any change in businesses or the areas they serve affects community newspapers. The second role is to provide community information. All communities need some mechanism for coordination and social control. In very small communities, this information is communicated interpersonally. But as communities grow, newspapers have taken on some of these functions. Interpersonal channels in small towns still are crucial and carry information that is either too important to wait for the newspaper or too trivial to be printed. Local newspapers don't run investigative reports and for the most part don't criticize the community. They present the community's image to the world.

An understanding of these roles will help in predicting how new telecommunications technologies such as the Internet and World Wide Web might affect newspapers and community communication patterns. In this chapter, we begin by examining the roles of newspapers in rural communities and then utilize data from a survey of Iowa weekly and daily newspapers to analyze both the current adoption of new telecommunications technologies by newspapers and the extent to which these technologies are seen as transforming rural communities.

The economic trade areas that serve rural communities form a local, regional, state, and even national hierarchy. At the local level, consumers buy many of the basics needed to live—groceries, drugs, hardware, and gasoline. For more specialized products and services, consumers may travel to regional or even state centers, or order nationally or globally. Although a rural community may have an emergency response medical unit and even a doctor, specialized services are found in regional or state centers. Local clothing stores can be found, but selection is wider at regional or state malls. At the national (or now even the global level), catalogs and the Internet make it possible to shop anywhere for almost anything.

As rural communities and the newspapers that serve them evolved, shopping and trade patterns also changed. Before 1900, local trade centers served areas that consumers could ride to on a horse or walk to over paths. When Rural Free Delivery (RFD) brought postal service to rural areas beginning in 1896, roads were improved so that mail could be delivered and towns began to grow around the post office. Towns not selected for post offices, railroad stops, or road intersections declined and disappeared. Yarbrough (1997, p. 17) found that now there are virtually no towns of less than 1,000 that support a weekly newspaper in rural areas.

Iowa provides a typical example of changes in the newspaper industry. In 1911, there were 912 weekly newspapers in Iowa's 450 communities. In 1998, there were 340 total (weekly and daily) newspapers. Roughly 20 percent of Iowa's towns now lack any local newspaper and these towns are virtually all under 1,000 in population. Many are now served by newspapers from nearby towns that include some local news from each community they serve but, more importantly, provide advertisers with a mechanism for matching their trade areas to the newspapers' circulation.

As communities grew, a critical mass of advertising demand resulted in the creation of local newspapers. Their circulation areas matched almost exactly the trade areas of community businesses. Advertisers supply about 75 percent of the revenues of newspapers, with sales supplying the remaining 25 percent. Because the physical distribution of the newspaper requires about 22 percent of revenues, one might conclude that the newspaper is a product that, owing to subsidization by advertisers, is free to readers, who pay only for delivery (Maynard, 1995). Especially important are classified ads, which in rural areas are read as much as news (Schramm & Ludwig, 1951). These ads often make up a third of total advertising revenue.

At the regional level, daily newspapers serve larger towns and cities but also reach out to rural citizens who might be lured to the regional center for their specialized needs. Rural free delivery of mail provided an opportunity for regional and state daily newspapers to reach rural citizens for the first time. (Weeklies didn't benefit as much, because subscribers often made it to town once a week and picked up their paper then.) Dailies made vigorous efforts to sell subscriptions to farmers and rural community residents who, because of improved roads and the advent of the automobile, now could not only read about regional towns, but could occasionally shop there. The massive road paving projects of the 1920s extended the distances rural residents could travel to the state capital and nearby major cities.

In 1913, when parcel post was added to Post Office services, rural residents found they no longer needed to rely on the Wells Fargo wagon or local merchants. Catalog ordering expanded; consumers could now buy from distant businesses. Thus, economically, various levels of newspapers have evolved, serving overlapping areas. As trade areas shifted, newspapers shifted to match them.

The information role served by newspapers is similarly shaped by the trade areas that are served. Schramm and Ludwig (1951), pioneers in research concerning weekly newspapers, described "the extraordinary hold which weekly newspapers have on their readers and the important part these newspapers play in socializing their communities" (p. 301). In their analysis of weekly newspaper readership, they found that the weekly paper has severe competition from radio and dailies when it comes to sports, political news, and opinion, "but it has very little competition in recording the everyday events of its own community" (p. 313). Byerly (1961) found that community newspapers serve as a unifying force for the community and that they stimulate thinking about local problems and projects, create interest in government and elections, and promote local welfare and projects. Stempel (1991) found that newspapers continue to lead other media by a wide margin as the source of information about local communities (what the mayor of your city is doing, what's happening in your local school system, etc.).

Tichenor, Donohue, and Olien (1980) undertook an analysis of the communication role of the community press and found that "among the social controls that maintain the norms, values and processes of a community, those that regulate the generation and distribution of information are some of the most pervasive. Newspapers are highly visible mechanisms of this type and, as such, their functions necessarily fit into a pattern that varies predictably according to size and type of community" (pp. 102–103). Newspapers exert two types of community control, they concluded. The first type, *feedback control*, occurs when the press serves to alert the community about some important activity or situation. The second type, *distributive control*, concerns routine information provided about community residents, events, or elites. Tichenor et al. (1980) found that when a community is small and homogenous, the press tends to carry mainly distributive information and to strive for consensus and tranquility. In larger more pluralistic communities, more feedback information is carried and the press sees its role more as a watchdog for the community.

One key development affecting both the economic and communication roles of the newspaper in the community has been what sociologist Roland Warren (1963) referred to as "The Great Change" in communities. Yarbrough summarized this change as "our transition from locality based communities that are relatively self-sufficient, relatively bounded, which you can study and understand as a single unit, over to the type of system that we have today—which is very much an open system in which the horizontal ties and linkages within the community have been greatly weakened and the vertical ties, the kinds of organizations that link to the outside—the bank, the newspaper and virtually every other institution within the community—have strengthened" (1997).

One major concern about the arrival of online technologies is that they might further enhance vertical linkages and promote communities of interest over communities of locality. John Dewey (1916) explained that "people live in a community, [by] virtue of things which they have in common; and communication is the way in which they come to possess things in common" (p. 5). Kielbowicz (1988) notes that a community may be organized along geographic lines (locality) or lines of interest. In either case, the community derives its cohesion from shared information, symbols, ideology, and experience (Kielbowicz, 1987, p. 169). The original ARPANET (the military predecessor of the Internet) designers predicted in 1968 that future online interactive communities would rest on common interests rather than on shared geographic space (Dicken-Garcia, 1998, p. 23).

Warren (1963) noted that printed newspapers have always played both vertical and horizontal communication roles in communities. They discuss local issues and concerns that bind people together (horizontal), but they also report on developments outside the community and include material generated elsewhere (vertical). However, as Yarbrough has pointed out, the trend has been strongly toward vertical linkages and online technologies seem poised to further strengthen these linkages (Yarbrough, 1997).

NEWSPAPERS AND TECHNOLOGICAL CHANGE

The new telecommunications technologies now affecting newspapers and other institutions in rural communities are important because they have the potential to alter the economic and communication functions of newspapers. Currently, for example, merchants purchase advertising from newspapers that then circulate the information to consumers. If merchants could communicate directly to consumers—say via e-mail or a Web page—they might need the services of newspapers less. This would transform the economic interdependence of newspapers and merchants, but it would also affect the communication function the newspaper plays, because it is the advertising revenue that makes it possible for the newspaper to play its communication role. Past technological changes have had major impacts on both the economic and communication role newspapers play. One, from 1830-1890, saw the arrival of faster printing presses, cheaper papermaking processes, mechanical typesetting, and photoengraving. This coincided with the rise of mass circulation newspapers that carried national news to communities. Local newspapers, which had reprinted clippings from big city newspapers, could no longer compete in this area. To survive, they began to reinvent themselves to provide local news and they also increasingly emphasized local advertising (Yarbrough, 1990).

A second revolution, from 1960 to the present, has resulted from computerization of the writing, editing, typesetting, and composition processes and the wider use of photo-offset printing. Computerization has made it much easier for newspapers to utilize information provided from outside the community (although "ready-print" newspapers—printed on one side with news and advertising at central plants and then shipped to local newspapers that printed the other side—were used for years as a means of inserting outside information into local newspapers. Such information usually was not news, but rather timeless features and entertainment articles). Computerized information, because it can be transmitted electronically, also means that local community publishers now can make their copy available to a global community. The development of the Internet and other online applications utilizing computers and modems greatly enhances both receiving and transmitting possibilities for newspapers.

To explore the implications of computers and online technologies for newspapers and their communities, we first need to examine the current state of adoption and use of these technologies in rural communities. This is especially important because relatively little research has been done on rural or community newspapers. Much of the data published thus far has been about large daily newspapers and an assumption is often made that smaller or rural newspapers either are or will soon follow the same trends. Garrison's 1997 study of the uses of online technologies by newspapers studied only U.S. dailies with circulations of more than 20,000. Virtually all previous major studies of newspaper computerization and online use have focused on daily newspapers only (Johnson, 1995; Soffin, Heeter, & Deiter, 1987). Garrison's study, as he himself noted, left out two-thirds of U.S. dailies and all of the weekly newspapers, which constitute the great majority of all newspapers published. Harper (1996) called for studies of small newspaper markets, reasoning that because they require fewer resources to start up, they may see a profit before larger papers. It is important to remember that we are still in the relatively early stages of the online information revolution. But we should not conclude that weeklies will necessarily replicate the experience of big-city newspapers.

FOUR KEY FACTORS

This chapter systematically examines four key factors relating to the ability of rural newspapers to utilize telecommunications technologies to increase rural community viability: (1) the technological base of rural newspapers; (2) the current levels of online sending and receiving of information; (3) specific use of the Internet, e-mail, and World Wide Web to gather and disseminate information; and (4) perceptions of the importance of online technologies such as the Internet for the future of rural communities. Data to examine each of the four factors were obtained from a 1997 mail survey of all 340 daily and weekly newspapers in the State of Iowa. Usable questionnaires were received from 192 (57%) of the 340 newspapers.

The technological base of rural newspapers refers to the computerization of both business and production aspects of the newspaper's operation. The more sophisticated the equipment, and the more applications (software and hardware) for which it is used, the easier the transition from print to online communications. A vision of a rural electronic online community newspaper providing local news, photographs, data and links requires a minimum level of computers, software, and training. Rural newspapers that lack this minimal level of technological sophistication cannot handle photographs digitally, design "electronic" pages, or manipulate text.

Many newspapers, especially dailies, have used computers to receive news from wire services for more than two decades. The greater the ranges of use of online technologies to send and receive news, advertising, and photos, the greater the ability to make the transition to online use of the Internet and World Wide Web. Rural newspapers with significant online capabilities are in a better position to develop new community information hubs and find new ways to move information around a community.

Specific use of the Internet, e-mail, and the World Wide Web to gather and disseminate information pertains to what the newspapers are now doing as well as what newspapers expect to be doing in the near future. Rural communities in general lack training and equipment to create Web sites, links, databases, and other building blocks of an online information service. Newspapers that have this capability may be in a leadership position to foster this development.

Technical capabilities alone will not usher in an era of online use by rural newspapers. There must also be a perception that such use will be important to the future of the newspaper and the economic development of the community it serves. Although there is a general technological utopianism (Iacono & Kling, 1996) about computers and online technologies among futurists and large newspapers, it is unclear whether smaller newspapers share their view. A long-term commitment will be necessary for these technologies to be transformative.

THE TECHNOLOGICAL BASE

Prior to 1960, the newspaper was in a precomputer era in which text was prepared by casting it, line by line, with linotypes. The content of past issues was saved in paper form as whole copies or clippings that were sorted into categories in the newspaper's morgue. The ability of one paper to share its content with another was restricted to exchanging a copy via the U.S. mail and to utilizing a telephone-line teletype system. A typical newspaper had a very large investment in typesetting machines, presses, and lead melted to form type.

Beginning about 1960 (Garrison, 1983; Yarbrough, 1990), computers became powerful enough to interest newspapers. Initially, the main use of computers was for financial and record-keeping purposes, but this quickly spread to the production area. Weekly newspapers, because they tended to have old presses and other equipment, and because their networking needs were smaller, were some of the innovators in computerization. Larger newspapers required more sophisticated networked systems that took longer to develop. Once developed, however, larger newspapers found that their scale of benefits from adoption was higher than that for smaller newspapers.

The advent of photo-offset printing made it possible to prepare type for newspapers without having to cast it in lead first. Old linotypes and presses could be eliminated along with their high-salaried operators. Lower-paid keyboard operators could type in the news on specialized typewriterlike machines that created punched tape that was fed into a computer compositor. Instead of each weekly newspaper having to own its press, faster offset presses was grouped, with one press serving the needs of three or four newspapers. Thus, the early focus on newspaper computerization was not on its use by reporters or editors. Emphasis was strictly on the "backshop."

By the late 1970s and early 1980s, improvements in computers had led to microcomputers (personal computers or PCs) that were capable of handling word processing, spreadsheets, and other tasks. This shifted the emphasis of newspaper publishers toward the possibilities of utilizing computers to create and handle the news itself. Once the text of the newspaper could be saved on a computer, the space-intensive morgues that stored clippings and past issues could be reduced in size by storing them digitally. Furthermore, unlike microfilm—which also saves space—digitally stored material could be searched electronically. Once an electronic morgue was created, it could be used by journalists to search for past information, but it could also be made available to subscribers or even other newspapers as a remote database.

Other developments resulting from computerization of newspapers and advances in computing hardware and software have included digitization and pagination. Digitization of photographs and other artwork has revolutionized photography at major newspapers. Digital cameras have made rapid transmission of news photographs possible. Manipulation and storage of photos on computers have made it possible to eliminate costly chemical photo labs at most major daily newspapers. Full pagination refers to the ability of newspapers to lay out all elements of their publications electronically, controlling the location and characteristics of photos, text, headlines, and advertising.

By 1994, virtually all newspapers—both weeklies and dailies—were computerized in some way, but a sizable number were still not moving into more advanced uses of information technologies such as online information access, electronic storage, and pagination (Davenport, Fico, & Weinstock, 1996).

To assess the extent to which rural newspapers have the technological base necessary to take advantage of the Internet and World Wide Web, Iowa newspapers were asked about the current computerization of both basic production and business practices. Production topics included the extent to which digitization of page design, word processing, photo manipulation, and advertising has occurred. Business uses of computers included use of computers for general billing, billing classified advertising, computerized ordering and selling, financial record keeping and budgeting, maintaining subscriber lists, and work scheduling.

Because previous studies have shown large differences in adoption and use of computers and online technologies due to circulation size, we divided both the weeklies and dailies into two groups each at the midpoint of their circulation. The 28 daily papers were divided at 6,662. All papers at or below this size were considered small dailies. All those above this size were classified as large dailies. Similarly, for the 164 weeklies (and semi-weeklies), small was defined as at or below 1,807 circulation, and large was 1,808 or above.

There are important differences among the four newspaper categories in terms of their technological base. Results in Table 6.1 show that small weeklies have significantly fewer of the production capabilities than the other three categories. While small weeklies have only about half of the 11 production capabilities, large dailies average 9.7 out of 11. Similarly, small weeklies use less than half of the computerized business practices, while large dailies use nearly all of them. In practical terms, what this means is that weekly newspapers are much less digitized at the present time and therefore less able to utilize the Internet and World Wide Web, while large dailies are almost all ready for the transition. There are a number of reasons why small weekly newspapers lag behind the larger ones in adoption. One

Table 6.1
Use of Base Computerized Production and Business Technologies, 1997 Iowa
Newspaper Study, N = 192

	11 Production Items (mean score)	6 Business Items (mean score)
Small weeklies (up to 1,807)	5.8	2.6
Large weeklies (1,808+)	7.4	3.1
Small dailies (up to 6,662)	9.4	3.6
Large dailies (6,663+)	9.7	5.1
One-way ANOVA test	$F = 20.7$	$F = 12.3$
	$p < .000$	$p < .000$

Note: For each of eleven production items, respondents were asked to check which ones were "used for the production of your newspaper." The items included use of computers for desktop publishing, word processing, photo and illustration manipulation, entering text for classified advertising, designing and building display ads, and scanning photographs and other material. One point was assigned for each positive answer. A one-way analysis-of-variance with a Scheffé test showed that each group is significantly different from the others in terms of production capabilities, with the exception of small dailies compared with large dailies.
For business practices, a similar approach was used, with one point awarded for each of the six business practices. An analysis-of-variance test showed large dailies are significantly different from both small and large weeklies.

weekly Iowa editor interviewed during the research said that at his newspaper, the high cost of new software was one important factor. He said he often waits for second- or third-hand software that is much less expensive. A second factor has to do with the capabilities of his staff. Until they have the training to use the new software or hardware effectively, he won't adopt. A third factor concerns the available pool of trained labor in the community. He is reluctant to adopt new software that only one person in the community might have any idea how to operate, because if that person moved away, he would be put in a difficult position.

Current Sending and Receiving of News, Advertising, and Photos

An essential building block for an online community is a large number of information providers and consumers that can exchange information electronically along with the ability of newspapers to collect, store, and manipulate this information. Newspapers with these capabilities can more easily provide an Internet or World Wide Web presence. While many businesses have only recently begun to utilize online technologies, many newspapers have done so for several decades.

By the late 1970s, many major dailies had begun to use computers to reduce costs in one major area—the receipt of wire news. Prior to computers, teletype machines clanking away at 66 words per minute were found in daily newsrooms. Editors would read through mounds of printed output, select the stories desired for the local newspaper, and then send the copy to the backshop where it would be rekeyboarded by a machine operator. In the 1970s, Datastream and Datanews services began providing an alternative—replace the teletype with a computer that could receive news transmitted at 1,200 words per minute over leased lines, microwave, or satellite (Garrison, 1983). When Datastream began providing news wire copy online, it increased transmission speed and greatly reduced costs of use because news stories did not have to be rekeyboarded. In the early 1980s, the majority of large daily newspapers were receiving wire news in this way, but local news such as press releases, organization announcements, legal notices, and so on was still being rekeyboarded at the office.

Next, computers with modems and word-processing software made it possible to receive press releases online. They arrived more quickly and again saved money by avoiding rekeyboarding. In 1982, only 31 percent of Iowa's dailies were even interested in receiving news releases and other institutional news electronically. By 1984, 67 percent of dailies were interested (Abbott, 1986, p. 1). Iowa State University Extension home and family news releases were sent via computer modem and telephone line to United Press International and Associated Press wire services beginning in 1983 and 1984, respectively. One reason for the increase in interest in electronic delivery was that newspapers themselves were converting internally to computer-based systems with online capabilities. A second reason was that the technologies for electronic transmission of information were improving. While transmission speeds of 300 baud were common in the early 1980s, 1,200 and 2,400 baud systems were in use by the latter part of the decade. At these speeds, newspapers could realistically begin to design systems to rapidly receive and process information electronically from outside the newsroom.

Weekly newspapers were slower to adopt online computer capabilities. One important reason for this is that much of the content for weekly newspapers comes from the local area. Few of these papers were connected to wire services—the major impetus big dailies had for adopting this new technology. Legal notices, an important source of revenue for weekly newspapers, could not be supplied in computerized form because local governmental units providing them had not yet adopted computers. Furthermore, many small dailies and weeklies were operating in rural areas where phone services were not yet capable of handling computerized transmissions.

Some researchers believed that online capabilities would be especially useful to rural newspapers. In addition to receipt of wire news and press releases, online systems would enable reporters and editors to tap information databases—census figures, statistical material, government reports, and so on that previously had not been easy to access. However, early studies found little use of this capability by these newspapers (Kerr & Niebauer, 1986). In a survey of all daily newspapers in Michigan in 1984, Soffin et al. (1987) found that only two were using any specialized online computer services and they were the two major newspapers serving the Detroit area. Of those not using online services, three-fourths said they were too expensive, half said they had no need of them, and 15 percent said they didn't know how to use them. Forty-two percent of these papers had a modem—the prerequisite for online communication. The authors concluded: "If one looked only at the use of databases by Michigan daily newspapers, one could conclude that on-line databases have very little impact on or importance to the newspaper industry" (Soffin et al., 1987, p. 9).

By 1994–95, Garrison (1997) found that there had been a rapid increase in the use of online computing at the largest newspapers in the United States. In a survey of all large newspapers with a circulation of more than 20,000 publishing a Sunday edition, Garrison found that by 1995 the percentage of reporters, editors, and re-searchers that used one or more online databases was 64 percent, up from 57 per-cent in 1994. At large dailies (defined in his study as more than 52,800 circulation), 90 percent were using online databases, compared to 40 percent for newspapers below 52,800 in circulation. More importantly, at large dailies, more than half of respondents said they use their online connections daily or more frequently, com-pared to 8 percent at the smaller papers. In fact, 44 percent of the smaller dailies re-ported that although they have online systems, they have *never* used them. Garrison also found that among users of online services, America Online and the Internet had increased dramatically in use (AOL went from 17% use in 1994 to 38% use in 1995; the Internet went from 25% in 1994 to 45% in 1995). Both of these services permitted e-mail, a use that was frequently mentioned by users. More traditional services, such as Lexis/Nexis, CompuServe, and others did not show increases in this period. In Michigan, Davenport et al. (1996) found that 37 percent of all daily newspapers in the state were using online databases by 1994 and 29 percent said they were using the Internet.

In Iowa, there are large differences in the use of nine online applications be-tween small weeklies and large dailies (see Table 6.2). Small weeklies are signifi-cantly less experienced in the use of a variety of online technologies and this could be expected to cause them to be slower in utilizing the Internet. Meanwhile, large dailies are already using an average of six out of the nine online forms.

Use and Expected Use of the Internet, E-mail, and the World Wide Web

Before 1994, very few citizens, businesses, and other information providers in local communities would have been able to provide or receive information in elec-tronic form. MOSAIC, the first truly accepted Web-based interface, was only adopted widely in 1994. Despite media hype about future use of the Internet, fewer than 10 percent of citizens were connected in that year. More recently, a Nielsen Media Research survey in June 1998 reported that one-third of Americans over the age of 16 use the Internet. As this technology becomes more widely adopted, inter-est is growing in its possible uses for community announcements, local bulletin boards, and commerce. Thus, community newspapers are currently considering how this technology might either help or hurt them in the future. Based on an analysis of adoption of the Internet in Blacksburg, Virginia, analyst Phil Buehler (1997) estimates that when adoption of the Internet exceeds 25 percent, a take-off in terms of local community use begins to occur, with churches and civic and busi-ness organizations all beginning to go online. With current use rates at 33 percent

Table 6.2
Uses of Online Technologies by Iowa Newspapers, 1997 Iowa Newspaper Study, N = 192

	Use of Nine Online Technologies (mean score)
Small weeklies (up to 1,807)	.94
Large weeklies (1,808+)	2.09
Small dailies (up to 6,662)	4.14
Large dailies (6,663+)	6.07
One-way ANOVA test	$F = 32.69$
	$p < .000$

Note: Respondents were asked to indicate whether or not they were using each of nine online applications: (1) creating or searching a computerized newspaper archive; (2) online receipt of display ads; (3) online receipt or sending of news items; (4) online receipt or sending of display advertising; (5) online receipt or sending of photographs; (6) online receipt of letters to the editor; (7) online search of a database such as Lexis/Nexis; (8) online monitoring of specific news services such as MSNBC or CNN; and (9) online sending or receiving final page proofs. One point was awarded for each use. A one-way analysis-of-variance test showed that with the exception of small and large dailies, all other differences are statistically significant (Scheffé test).

nationally, one might predict significant changes in local communities' applications of online use.

CASE STUDY: SUCCESSFUL ONLINE EXPERIENCES OF A WEEKLY

Editor & Publisher, a leading trade magazine for the newspaper industry, reported on December 28, 1996, that only 10 percent of newspapers with Web sites were making money from them. Robert Mummery, publisher of the *Minnedosa Tribune* of Manitoba, Canada, is one of them. Mummery's newspaper happens to be located adjacent to Riding Mountain National Park, a major tourist attraction across the border in North Dakota. Many people who have grown up in the area have also moved away. For $20 per year, Mummery offers those living away from the area the opportunity to keep up on activities near this attraction. In eight months, 431 subscribers paid the fee. Mummery says he figures that he spends $1,500–$2,000 per year in extra wages to design the Web site, so he is making money. But he doesn't think the site will ever appeal to local residents. "Locals aren't going to go to the Web to get their news," he said. "Why should they when we hand them their papers every week?" (*Editor & Publisher*, December 28, 1996, p. 26).

There are two infrastructure prerequisites for newspaper use of the Internet and World Wide Web. First, the newspaper must have some form of modem. Second, it must have an Internet connection, meaning it must have an account with an Internet service provider (often a local Internet provider, although this is not essential). Figure 6.1 shows that slightly less than half of small weekly newspapers have a modem, while the great majority of the other three forms of newspapers have one. For the Internet connection, only one-fourth of small weeklies have one, while more than three-fourths of the dailies do. Thus, among small weeklies there is a gap between the capability of having an Internet connection and actually having one. Half of the small weeklies that have a modem don't have an Internet connection. For the other three types of newspapers, that gap is much narrower. One reason for the gap is that until recently, local Internet service providers were not available in small communities.

Newspapers can use the Internet or World Wide Web for three purposes—to *gather* news and information, to *deliver* it to a particular subscriber or audience, or to *operate a Web site* where consumers can access the information. Figures 6.2 and 6.3 show that the majority of large dailies already use the Internet/World Wide Web (WWW) to gather information (64%), while only 11 percent of small weeklies do so. The remaining 36 percent of large dailies say it is very likely they will do this in the next five years. Almost half of small weeklies believe they may never use the Internet/WWW to gather news or information. For information and news delivery the numbers are lower, mainly because information providers to newspapers tend to be online while subscribers do not. More than half of small weeklies believe it is unlikely or very unlikely that they will ever use the Internet/WWW to deliver news. Table 6.3 shows results for the intent of newspapers to operate a Web page for subscribers. There are dramatic differences between large and small papers both in the current operation of a Web site and the percentage that believe they will never operate such a site.

Figure 6.1
Percentage Modem and Internet Connection Levels, 1997 Newspaper Survey

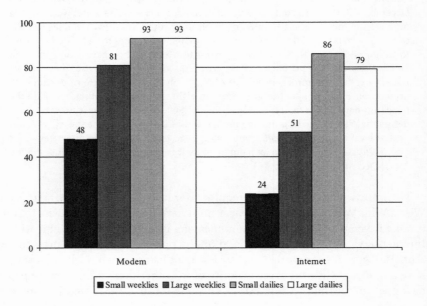

Figure 6.2
Percentage of Newspapers *Gathering* News or Information via the Internet or World Wide Web, 1997 Iowa Survey

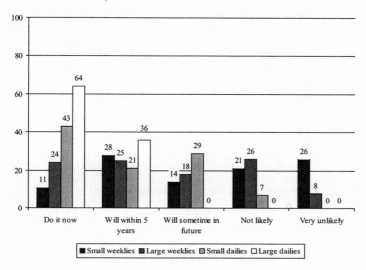

Figure 6.3
Percentage of Newspapers *Delivering* News or Information via the Internet or World Wide Web, 1997 Iowa Survey

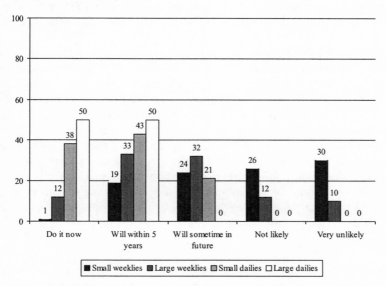

Note: Respondents were asked how likely they thought it would be that they would use the Internet/ WWW to gather or deliver news and information. Choices included (a) we already do this; (b) it is very likely that we will do this in the next five years; (c) it is likely that we will do this sometime in the future; (d) it is not very likely that we will do this; and (e) it is very unlikely that we will ever do this.

Table 6.3
Newspaper Expectations Concerning Operating a Web Page

	Small Weekly (up to 1,807) n = 82	Large Weekly (1,808+) n = 82	Small Daily (up to 6,662) n = 14	Large Daily (6,663+) n = 14
Currently operate a Web page	4%	21%	44%	100%
Are actively investigating whether to offer a Web page or other electronic services	52	65	56	0
Are not interested in offering a Web page or other services	44	14	0	0

Importance of the Internet to the Community

One other indicator of the likelihood that newspapers will play an active role in the online future of rural communities is the extent to which they believe such technologies are crucial to the development of their communities. In Iowa, daily newspapers were more likely to believe that the Internet or World Wide Web will play a central or important role, while more than 20 percent of small weekly newspapers believe it will have very little impact (Table 6.4). Newspapers in all groups are likely to believe that telecommunications technologies will benefit the economies of rural communities. But more than one-third of weekly newspapers believe telecommunications technologies will either have no impact or a negative one (Table 6.5).

These results show that there are important differences between small weeklies and large daily newspapers both in their expectations about the importance of the Internet and World Wide Web in the future of their communities as well as their plans to gather and deliver information and operate a Web site. Because large dailies represent communities of sizable population, it is possible that these differing perceptions may both be accurate—but only if rural communities are not able to participate in this new information technology. Much of the current telecommunications literature concerns how rural communities can take advantage of a technology that does not require a physical location near a metropolitan area. The negative views of some weekly newspaper editors strongly suggest that these individuals will not be out front in the effort to bring these technologies to rural areas.

CONCLUSIONS

What are the parameters shaping how the Internet will affect rural communities and newspapers? First, we must recognize that the future of community newspapers will be determined by the economics of the businesses that serve their trade areas. If the overall effect of the Internet, like RFD, is to lure consumers away from local businesses and to national or even international virtual shopping malls (such

Table 6.4
Perceived Future Role of the Internet/WWW in Communities

	Small Weekly (up to 1,807) n = 82	Large Weekly (1,808+) n = 82	Small Daily (up to 6,662) n = 14	Large Daily (6,663+) n = 14
The Internet/WWW will play a central role and be accessible by almost all citizens	9%	6%	31%	33%
The Internet/WWW will play an important role, but only for a minority of those in the community	32	38	31	17
The Internet/WWW will play a supplementary role, but it won't alter the current basic communication system	38	45	39	50
The Internet/WWW won't have much impact on their community at all	22	11	0	0

Table 6.5
Perception of Benefit of Telecommunications for Rural Communities

	Small Weekly (up to 1,807) n = 82	Large Weekly (1,808+) n = 82	Small Daily (up to 6,662) n = 14	Large Daily (6,663+) n = 14
Strengthen rural communities	60%	60%	77%	79%
Little or no impact	23	16	8	14
Widen economic gaps	16	20	8	7
Other	1	4	8	0

Note: Respondents were asked to choose from three choices: (1) recent advances in telecommunications will serve to strengthen the economies of rural communities by providing them access to the information highway; (2) recent advances in telecommunications will have little or no impact upon the economies of rural communities; and (3) recent advances in telecommunications will serve to widen the economic gap between rural and urban communities because rural areas will be comparatively information poor.

as Amazon.com for books), the critical mass of advertising necessary to support local newspapers will disappear, and they will cease publication. To the extent that certain local businesses might redefine themselves as regional or national by selling their products via the Internet, local newspapers also are likely to be the losers. Such businesses will less likely be concerned about reaching local consumers, and local newspapers are very unlikely to be able to redefine themselves to serve a regional or national audience.

Second, to what extent is it likely that rural citizens will become less oriented to communities of locality and more concerned about communities of interest? Can we imagine a time when the friendships, organizations, and loyalties of rural citizens are not local, but are mediated through the Internet? As some have speculated, it would not matter then where one lives; one could join one's "community" from anywhere. Despite the well-documented trend toward increasing vertical communication flows in rural communities, it would still be a major transformation to move to a stage where horizontal information flows were not important at all. To the extent that communities of locality become less important to rural citizens, rural newspapers would be expected to be less valuable to them. Even if an advertising base remained, lack of interest in content could be fatal for community newspapers, which could then be replaced by advertising shoppers that lack news.

For illustrative purposes, we can present three possible scenarios concerning the Internet and rural newspapers:

- *The Big Dud.* The Internet turns out like other over-hyped information technologies before it; it occupies a specialized niche for universities and corporations, but at the community level it never finds an important use. In this scenario, the community newspaper may use or increase its use of the Internet to get information, but communication to local residents is still via the weekly or small daily newspaper. Local citizens continue to desire to be part of communities of locality and subscribe to newspapers to learn about local schools, sports, deaths, and civic developments. This is the scenario painted by a substantial minority of Iowa weekly newspaper publishers now—they see the Internet as having only a minor or supplementary impact on information in their communities. They may be correct.

- *Information Armageddon.* The Internet turns out to be so important that it transforms rural communities totally. Economically, commerce based on geography declines to the point where local advertising is insufficient to support local newspapers. Even local grocery stores succumb as consumers order from regional warehouses for Federal Express delivery to their homes the next day. Local niche-based businesses demanding specialized information from outside the community develop. Telecommuters locate where they wish, but orient themselves outside the community to their workplace. At the same time, communities of interest grow, reducing interest in local news. Internet virtual friendships and social networks are substituted for local church, civic, and social groups. The rural newspaper loses its reason for existence and disappears.

- *The Advertising Transformation.* The Internet becomes a major medium for commerce. Local retailers, government, and even nonprofit organizations start their own Web sites and begin to reach audiences directly. As clients shift to the Internet, there is less need to pursue them via newspaper advertising and it declines in importance. One might remember that in the 1960s, despite readership in the millions and continuing demand for editorial content by readers, general circulation magazines such as *Look, Life* (its original version), *Colliers,* and *The Saturday Evening Post* all went out of business when national advertisers shifted to television. Could the Internet bring about a similar transformation at the local level? If it did, would businesses communicate directly to consumers, as some now speculate, or would the newspaper simply reinvent itself in electronic form and continue to function as a clearinghouse for both advertising and local news? One might also predict the development of some hybrids—both printed and electronic newspaper forms. Another possibility might be that the local newspaper would continue to look pretty much as it does now, but it also would offer an electronic version of advertising that permits consumers to search for products or order online. Or, the news might shift to an electronic form and the printed version might become only a shopper that continues to reach those who aren't yet connected. Which of these possibilities might happen would be a function of consumer demand and advertiser preference.

There have been some local experiments in providing community news in forms different than the traditional newspaper. Here, we present several illustrative examples of alternative forms that might develop.

- *The Volunteer Newspaper.* As small communities lose the number of businesses necessary to support a weekly newspaper, they continue to have a need to share information about their community and its activities. To serve this information function, several have created alternative forms, such as volunteer newspapers. In one case, community volunteers have collected information, sold some advertising and put out a weekly newspaper delivered free to all local residents and by subscription to others for more than 15 years (for more information, see Abbott, 1988).

- *Local Channel Cable Television.* As a part of contracts with local cable providers, some communities receive equipment and studio space to provide locally produced information programming. Revenues from cable fees pay for this service. If the local newspaper were to disappear, this alternative form could continue to provide information. With a few exceptions, most local cable stations have minimal funding and present an eclectic assortment of programming, little of which concerns community news (some do broadcast public governmental meetings, sports events, etc.).

- *The Paid Subscription Internet Newspaper.* There has been much discussion, and a few experiments, in providing electronic online newspapers. Most experiments have shown that consumers resist paying subscriptions for these services. They might be supported by advertising. It is not yet clear what type of content might be provided, but many of the experimental versions do not provide the same level of news as the traditional printed versions.

- *Specialized Web Sites by Businesses and Community Organizations.* In this approach, each information producer in the community develops its own Web site, and citizens receive information directly from them, rather than using a mediated source such as a newspaper. The problem with this approach is that it would require citizens to log on to many different sites to get information. Also, there would be no one filtering or judging the relative accuracy or value of the information beyond those who initially produce it.

Because the Internet and World Wide Web are still at an early stage in the diffusion process, it is uncertain how they might affect newspapers in rural communities. We close this chapter with six observations about the likely pattern of adoption and the impact of these technologies.

1. *Many Rural Small Weeklies Lack the Technological Base to Provide Internet Services.* Small rural newspapers are less ready technologically to adopt and utilize new telecommunications technologies such as the Internet and World Wide Web. Only half have a modem. Only one-fourth have an Internet connection. Only about half can digitize photos and lay out pages via a computer. Small weeklies also lack a pool of locally available skilled workers who can do these things.

2. *It Is Still too Early to Predict Exactly How Rural Communities Will Utilize the Internet and World Wide Web.* Only 4 percent of small weeklies and 17 percent of large weeklies currently operate a Web site. In diffusion terms, these are innovators that are not necessarily representative of the typical newspaper. How they use the Web may not be a model for the remaining papers. Although most large dailies have a Web site, there is still much debate in the industry about what such a site should contain, and how it might make money.

3. *Small Weekly Newspapers Are Less Likely to Perceive the Internet/World Wide Web as Playing a Key Role in Their Communities.* Because newspaper publishers often represent an important component of the leadership of rural communities, the fact that a substantial proportion of these individuals do not believe the Internet and World Wide Web will have much impact suggests that they are not prepared to help reshape the information environment of their communities. This could be because they feel threatened by the new information technologies and want to defend themselves against them. But it may also be because they genuinely do not believe that new online technologies might one day reshape the information structures of rural communities. Whatever the reasons, the finding suggests that many small rural weeklies will not be providing community leadership in charting a new information future. Daily newspapers, on the other hand, see the Internet/World Wide Web as playing a more important role. A third believe that the Internet/World Wide Web will occupy a central and nearly universal role. These newspapers can be expected to play a more active role in the community in exploring how these technologies might be utilized.

4. *The Internet/World Wide Web May Reshape Who Controls News and What Type of News Is Transmitted.* The arrival of Internet- or World Wide Web-based information systems in rural communities could change local communication gatekeeping in a number of ways. "Gatekeeping" refers to the role of information brokers—those who make decisions about what information to pass along to others. First, and most fundamental, a local Internet information system (or systems) might not be operated by a local newspaper at all. Libraries, universities and schools, businesses and interest groups are already involved in creating their own systems—systems in which newspaper gatekeepers play no role. Second, even if newspapers are the operators of local Internet information services, the real-time demands of such services could change the concept of news. In a traditional small town, word about important events moves rapidly through a series of interpersonal channels—at coffee shops, churches, businesses, and so on. As Tichenor et al.

noted, the function of the weekly newspaper is mainly distributive, providing routine information about community residents, events, and elites. It is not breaking news; it is the community's response to such news. A real-time electronic system such as the Internet is often seen as providing immediate news. How would a weekly newspaper that traditionally has not covered this type of news use this capability? Would breaking news begin to appear, or would weeklies use the Internet in much different ways than our current large newspaper models suggest?

5. *There Might Be Substantial Changes in the Economic Role of Community Newspapers.*
Shifts in local advertising may offer clues about the future of newspapers and their role in communities. A major issue at large newspapers that already have a Web presence concerns how a newspaper might make money using the Web or Internet. Approximately $1.8 billion in advertising was placed on the Web in 1999, less than 1 percent of the estimated $215 billion total advertising placed in all media (Coen, 1999). A number of newspapers, including the Thomson Newspaper Corporation chain (Toronto's *Globe,* and *Mail,* and others), have developed complementary Web sites that carry classified advertising at a 3–4 percent premium. The electronic site is searchable and expands the reach of the classifieds—two factors that appeal to advertisers. Similarly, Albany, New York's *Times Union* was the first newspaper-based site to publish its region's multiple listing real estate service. It charges a 10 percent premium for joint listings (Pavlik, 1998). Some publishers have Web sites that contain nothing but advertising (Dotinga, 1998, pp. 36–37). However, most also provide some news, and the news increasingly is being tailored to the needs of this electronic audience. For large media, such as CNN and the *Wall Street Journal,* Pavlik (1998) reports that their sites "are claiming to have crossed over into profitability—or are near it anyway" (p. 14). For community publishers, the issue is how the Internet or Web can increase its audience or make it more attractive to local advertisers. Because their newspapers virtually saturate the audiences now, electronic forms would not seem to increase the local audience.

Whether the Internet would become a preferred form of communication for local news, and thus attract advertising, is also debated. Mueller and Kamerer (1995) found that in an early trial, Web-based newspaper readers in San Jose, California, did not necessarily like the Web better than traditional newspapers. The authors could find no example of a community newspaper Web site that replaced or even threatened to replace the traditional newspaper. The most likely outcome is that the Web would enable the local newspaper to reach out to new audiences "of interest." These are audiences either within or outside the community that could be served electronically in ways that they are not now served. *Editor & Publisher* magazine has featured a number of examples of sites similar to the one in the case study that focused on unique niches of information for which people would be willing to pay. Former residents who want to keep up on community activities or unique kinds of information about recreational or historical aspects of the community are often cited as examples of information leading to a successful site.

New organizational forms could result from the growth of newspaper Web sites. In southwest Iowa and northwest Missouri, 10 Iowa weekly and 5 Missouri weekly newspapers are presented together on one Web site that represents an enlarged trade area. Riley, Keough, Christiansen, Meilich, and Pierson (1998) suggest that the arrival of the Internet might also bring about a colonization of local media by larger integrated news providers that can use their advertising and economic muscle to take over and manage news. A parallel might be the radio news industry, in which only one or two networks now provide much of the news and virtually all of the programming heard on thousands of local radio stations. Carey (1998) sees the current changes in community media as part of a larger process of globalization. Just as the forces of nationalization in the 1890s led to the rise of national mass media systems supported by mass production and the marketing of goods, today's new technologies will be shaped by a globalization of communication and marketing.

6. *The Web May Result in Changes in the Community Role of Newspapers.* Newspapers have traditionally served a community of locality, those living in a specified geographic area. Bogart (1993) suggests that community newspapers may increasingly focus on serving communities of interest within their geographically defined trade areas. This could impact the traditional role Janowitz (1952) and Park (1929) saw for the press as providing a mechanism for community integration. With a shift to communities of interest, new online newspapers might further differentiate readers living in a geographic area rather than integrating them. This would be similar to specialized magazines, which provide unique information that differentiates the knowledge people have as opposed to local newspapers, which have traditionally provided information that gives people a common base of knowledge. Given the fact that local newspapers already offer saturation coverage of news and advertising to local residents, one might ask what the potential is for increasing this type of information? However, one can easily imagine Web sites providing access to nonlocal information—entertainment, news, travel, and so on. Thus, it seems likely that the arrival of Web sites may lead mainly to more vertical communication coming into the community from outside sources than to a great expansion of horizontal communication. As McLaughlin, Osborne, and Ellison (1997) have noted, such vertical Web sites can create community, in the sense of developing primary interpersonal relationships and group identity, but these communities would be quite different than the local communities now served by newspapers.

NOTE

The authors acknowledge the guidance and review provided by J. Paul Yarbrough of Cornell University in the preparation of this chapter.

7

Rural Libraries: Conflicting Visions and Realities in the Information Age

Eric A. Abbott and Bridget Moser Pellerin

INTRODUCTION

Historically, the public library is a rather recent innovation in community life, and its political and economic evolution has shaped its current structure and functioning in rural communities. The presence of a library alone in a rural community does not lead to an educated population. In fact, the demand for libraries arose from an increase in education and economic status in the last half of the 19th century. As Holt (1995) has noted, although "free" libraries developed that were accessible to everyone and though librarians have taken special pains to reach less advantaged groups, the majority of library patrons and those who support the library and pay the most taxes represent those with higher education.

Despite this, purveyors of new technologies for rural libraries, as well as library trade publications have extolled the transformational possibilities of the Internet, World Wide Web, and other online technologies. One group sees these new technologies as making possible the creation or revitalization of civic communities in which all citizens participate actively in the informational life of the community, with the library as its hub. A second group envisions bridging the gap between information haves and have-nots by linking both rural and urban public libraries to a universal electronic digital virtual library that provides access to a global information system. While access to vast amounts of information as well as increased civic interaction would both be beneficial to rural communities, and may be enhanced somewhat by new technologies, the basic factors governing information seeking and use will direct and limit the extent to which these new technologies might be transformational.

In order to understand the role of rural public libraries in the information age, five important sets of questions are addressed in this chapter:

1. What roles have rural public libraries played in the past and how have they changed over time? Were these roles designed to enhance horizontal community-based communication or vertical outside communication?

2. What visions do librarians and those with the power to shape libraries have with respect to what rural public libraries should do in the information age?

3. What is the reality of how rural public libraries are used now? Who are the primary users and for what purposes are they using the library? Which groups do not see libraries as relevant to their needs?

4. What is the current pattern of connection to the Internet and online services by libraries?

5. What are special constraints and needs of rural public libraries to function effectively with new information technologies?

LESSONS FROM THE PAST

Public libraries began to be formed in large numbers at the end of the 19th century as part of a movement originating in the northeast United States. In the beginning proponents emphasized their value to schools; but later, their value to personal, moral, and civic improvement was put forward (Lawson & Kielbowicz, 1988, p. 30). The original proponents were not specifically concerned about the value libraries might have for isolated rural areas. They were predominately urbanites. However, a series of political setbacks, plus the generosity of Andrew Carnegie, changed their tactics by 1905.

One of the problems early libraries had was how to receive books from publishers and send them to patrons. Although Congress had passed legislation in 1792 providing generous subsidies for mailing newspapers and magazines, the U.S. Post Office prohibited the shipment of books until 1851 due to primitive local transports that could not easily accommodate them. Once they were accepted as mailable, the Post Office assessed them at the relatively expensive rate for irregular pamphlets. From 1863 to 1914, books paid third-class postage, somewhat less than letters but considerably more than periodicals (Lawson & Kielbowicz, 1988, p. 31). During this period, it was too expensive to use the Post Office for books and publishers were forced to use private delivery services to get books to libraries.

Beginning in 1905, the American Library Association began advocating a special library postal rate, and in order to enlist congressional support from states with rural populations, changed focus of the benefits from urban areas to libraries in rural areas. In 1914, the postmaster general shifted library books from third class to fourth class parcel post and in 1928 a special library rate was implemented. The special rate made it practical for libraries to mail books to patrons living in distant communities. The Wisconsin Free Library Commission, which had sent up to 4,000 books a year under the 1914 parcel post rates, was lending about 100,000 volumes by 1937. Borrowers living in villages and along rural routes received their books directly; those in larger towns called at their local libraries for them (Lawson & Kielbowicz, 1988). In 1938, President Franklin D. Roosevelt created the general book rate, which enabled publishers to send books to libraries and individuals cheaply.

Another important factor in the establishment of rural public libraries was the philanthropy of Andrew Carnegie. In the last years of the 19th century and the first 15 years of the 20th century, Carnegie put up the money for 2,800 library buildings for communities that agreed to cover the costs of books and staff (Marcum, 1998, p. 195). As of 1980, there were 6,800 public libraries in the United States in communities of less than 25,000 (Houlahan, 1991). A substantial number of them are still housed in Carnegie buildings. There are a total of 9,050 public libraries in the United States (McClure, Bertot, & Beachboard, 1995). Carnegie believed that access to books and education would provide opportunities for motivated workers to improve their minds and, in the process, their economic conditions. He referred to the public library as the "people's university," a concept that has endured. Carnegie's donations led to the formation of independent municipally owned and operated libraries supported by local tax revenues or philanthropy. Collections and services varied widely.

The special postal rate established for books in 1928 had one other important consequence for public libraries. It gave new impetus to the formation of library systems—an alternative form to the Carnegie independent library. Although the first county library was formed in 1898, this library form grew rapidly once a subsidized mail rate made it easy to move books from one branch to another. California, Louisiana, New Jersey, Ohio, and North Carolina organized county library systems in the first decades of this century using book wagons as well as the mails to reach widely scattered readers. Although launched without the postal subsidy, this movement gained strength because of it (Lawson & Kielbowicz, 1988, p. 31). Whether or not a state's libraries were organized into a system or were mainly independent became a crucial issue in the 1980s and 1990s when decisions were made about how to connect libraries to the Internet. States with strong state and county systems tended to be associated with computer and Internet projects that linked all branches to a central facility and emphasized coordination and economies of scale. Small independent municipal libraries, on the other hand, often lacked the training and resources necessary to connect to the Internet. They were less interested in coordination and more interested in how the Internet and computers could be used to increase patron services and access. In some cases, they focused on the Internet as a community builder, strengthening horizontal community communications, rather than as a way to bring in more outside (vertical) information. These differences in approach are discussed in more detail later.

Although rural libraries have had a tradition of collecting, maintaining, and sharing specific local information about community history, traditions, and celebrations, their primary role in the community has been a vertical one—to bring new knowledge into the community. Those who fought politically for both special library rates and rural free delivery by the Post Office did so based on the idea that rural communities were isolated and needed outside information. Thus, libraries could be thought of as agents of change, bringing in new ideas in fiction books as well as reference materials. In the 1920s, before radio, television, and highways provided alternative routes to information, books provided by libraries played a key role in helping citizens envision life beyond their communities.

CENTRIPETAL AND CENTRIFUGAL INFORMATION ROLES

Interestingly, this outside information played two seemingly contradictory but simultaneous roles in rural communities. The first is what James Carey (1969) has termed a *centripetal* role—one that brings a common culture to rural citizens. In the same way that library novels such as *Gone with the Wind* became knowledge commonly held by many in the community, television later provided *Ozzie and Harriet* and *I Love Lucy*. This common culture was essential to the mass marketing of products and services and led to the creation of advertising messages that millions still can recall decades later. Although the mass media have taken on much of this centripetal role, libraries still support this role by providing a commonly available set of books and resources—most of which bring in knowledge from outside the community.

The second role, which Carey termed *centrifugal*, relates to the provision of unique information that tends to differentiate an audience. In an isolated rural community, there is a common body of knowledge about local customs and culture. When citizens are able to access knowledge from outside the community, they begin to have differences in knowledge and interests that make them increasingly unique. Thus, an important impact of the public library has been to permit people to become increasingly different as they pursue their different information interests. While at first glance Carey's two concepts may seem similar to Warren's horizontal and vertical forces (see Chapter 1), there are important differences. Although both horizontal and centripetal forces result in the creation of a common culture, Warren's horizontal communication was community-based, whereas Carey's centripetal forces are driven mainly by outside mass media and other economic institutions. Both vertical and centrifugal forces differentiate the culture of their audiences and both authors view these forces as coming from outside the community.

In Carey's view, centripetal and centrifugal impacts are going on simultaneously, and both effects must be considered when examining how new information technologies might impact a rural library. On the one hand, a connection to the Internet might strengthen the ability of citizens to increase a common core of community information or to become more like their urban counterparts because of access to a common virtual library. On the other hand, access to a global information system that includes an incredible variety and diversity of material would be expected to result in a substantial increase in differentiation. Sociologist Don Dillman (1991a) believes that the function of the rural public library must now become more centrifugal—to provide specialized information to rural citizens who can then exploit their unique knowledge to telecommute or develop niche-based businesses in their rural communities to replace declining employment in farming and natural resource exploitation. Others hold more centripetal visions of civic communities sharing a common culture.

CONTRASTING VISIONS OF THE FUTURE RURAL PUBLIC LIBRARY

Although there are many different visions of what roles rural public libraries might play with new information technologies, two contrasting ones are highlighted here. The first represents a community-based participatory vision. In this view, the rural library first works as a catalyst to organize the community, identify resources and needs, and serve as the center or information hub for what will become a new and rich community information system. The importance of this organizing role was emphasized by sociologist John Allen from the University of Nebraska's Center for Rural Community Revitalization, who found that, in today's rural communities, "there is no entity with the job of making sure the people in town understand what this new technology is all about and want it. It isn't the phone company, chamber of commerce, colleges, libraries or anyone's particular job" (Olson, 1998, n.p.). Cisler (1995) argues that rural public libraries should serve as a meeting place and facilitator for the creation of new community networks using new telecommunications technologies.

Schuler (1994) describes these community networks as being "intended to advance social goals such as building community awareness, encouraging involvement in local decision making, or developing economic opportunities in disadvantaged communities" (p. 39). They would provide one-stop shopping using community-oriented discussions; question-and-answer forums; electronic access to government employees; and information and access to social services, e-mail, and the Internet. "The most important aspect of community networks, however, is their immense potential for participation" (p. 39). Schuler then describes some of the experiences of 100 communities that, with help of their libraries in 45 percent of the cases, had established "virtual people's parks," local "free nets," and "electronic cafes." While most of these early experiments occurred in cities, Schuler discussed Montana's Big Sky Telegraph as a prototype of the potential of these systems to link rural citizens.

Frank Odasz (1991), creator of the Big Sky Telegraph system, says: "Imagine a community network that is an electronic journal, co-authored daily by all the participants of the community. There are communities that already have this advantage. An idea can emerge, a person in the community can post it on the system and it can become reality in minutes" (p. 87).

A view of such community-based systems also envisions that they would serve as civic networks, funded and controlled by public agencies, that would provide a variety of services, including license renewals, community job information, draft legislation, phone numbers of helping agencies, and so on. While some of this information would come from outside the community, the civic view is that there would be substantial information about the community, its services, and its strengths on these systems; that is, in large part the content would be created and tailored to fit the local community's needs and strengths (Holt, 1995).

The characteristics of such participatory systems, according to Schuler (1994, p. 41), are:

- Community-based: reflecting and serving common community interests of all stakeholders
- Reciprocal: users are both "consumers" and "producers" of information
- Contribution-based: forums and other components build on the contributions of users
- Unrestricted: anyone can use the community network
- Accessible and inexpensive: accessible at a variety of public locations; free of charge or very low fees
- Modifiable: users themselves can modify what is there or add new services.

The second vision, and the one that has dominated most spending to connect libraries, is based on a notion of universal remote access, a virtual library enabling users to tap into the contents and services of many different libraries and other information sources (Holt, 1995). The assumption here is that what the community needs is located *outside* the community. This approach emphasizes the application of the Internet, CD-ROM, and other technologies to make it possible to connect users with vast amounts of information and data, either through terminals at public libraries, "cybermobiles" with wireless transmission capabilities (Drumm & Groom, 1997), or directly from homes and businesses.

What, exactly, is the type of information that rural communities need? To most state library systems, it is not the development of local forums. Instead, it is a connection from rural public libraries to card catalogs, periodicals, newspapers, and databases. How many individual items can be accessed and examined determines progress in these state systems. Wisconsin now boasts a union catalog[1] available on its state system with 24 million holdings from 1,084 contributing libraries (Wilson, 1996). Texas has 75 million total items of all kinds in its 50 automated library catalogs and databases (Martin, 1996). Pennsylvania reports 22 million electronic holdings (Wolfe, 1996).

The main purposes of the massive state systems now being created are: (a) to make it easy to search and locate holdings in any library or database collection; (b) to provide economies of scale, so that libraries can share books, journals, periodical indexes, and other materials. A connection to the Internet on these systems is provided only because this is the evolving mechanism providing the most convenient way to share information. Originally, libraries proposed, and many implemented, massive central computer systems with dedicated lines. Now, most are backing away from these systems in favor of the decentralized World Wide Web systems, in which individual libraries or information providers hold their own information, but share on demand.

This vision has had important short-term effects on small rural libraries. First, state and federal resources have been channeled mainly into digitizing and connecting central systems—state libraries and university research library systems—rather than small rural libraries. Although most state plans contain goals that call for the eventual connection of all libraries in each state, the approach has been to digitize the largest libraries so that their holdings can then be made available to smaller rural libraries. Exceptions usually have been small pilot projects implemented in each state, such as one in the Hudson Valley of New York that hooked up three small libraries (Garofalo, 1995). Second, this approach has emphasized the value of access to information outside the community rather than attempts to marshal and improve the collection and communication of new community information. For state librarians, the solution to the problem of small rural communities is to connect them to huge outside databases.

The two visions have been presented here as ideal types so that their differences can be easily understood. In fact, there are often pieces of these two visions in the plans of many libraries and states. However, it is important to understand that the two visions call for very different designs and investments in rural public libraries. Before deciding which vision might be more appealing, we must recognize that efforts to achieve either vision must build from the current reality of uses and resources in rural libraries. To better understand current uses and how they might affect use of new information technologies, we now examine several national studies of library users.

THE ROLE OF THE RURAL PUBLIC LIBRARY

National surveys over the years have shown that reading books is not a universal activity; neither is visiting a public library. A national study by Knight and Norse (as cited in Sterling & Haight, 1978) found that 50 percent of male adults and 42 percent of female adults had not read a book in the past three months (p. 327). Book reading, like other information seeking, is affected by level of education. Among college-educated respondents, only 21 percent had not read a book in the past 3 months, while the figures for high school graduates and grade school graduates were 41 percent and 76 percent, respectively. A total of 57 percent of college graduates said they had read five or more books in the past three months, compared to 29 percent for high school graduates and 8 percent for grade school graduates. A national W. R. Simmons and Associates survey (as cited in Sterling & Haight, 1978) found that respondents in rural areas were much less likely to report "purchasing a book in the last six months" (p. 330). This finding can be explained in large part by lower levels of formal education among rural respondents, but it also may be related to isolation and lack of access to bookstores. The Simmons study also corroborated a common finding that men are less likely to buy books than women are. In U.S. counties below 35,000 in population, the Simmons study found that 26 percent of women and 17 percent of the men had bought a book in the past six months. This compares with 38 percent of women and 37 percent of men in the 25 largest metropolitan counties.

Two surveys conducted by Bernard Vavrek (1995) provide useful information on the current role served by rural public libraries. The first survey was a 1989 national sample of library users conducted in 300 libraries in nonmetropolitan (populations less than 25,000) communities. Approximately 3,500 usable surveys were collected from adults 17 years of age or older. The second was a 1991 national telephone survey of 2,485 adults 17 years of age or older.

Results of the second survey showed that 45 percent of respondents said they used their public library at least monthly and another 24 percent said they used it annually. Of the 45 percent, 2 percent said they use it daily, 17 percent weekly, and 26 percent monthly. This is similar to other survey results by Westin and Finger (1991), who found that 66 percent of Americans use their library on at least an annual basis. However, Vavrek concluded that those who use a library once a year or less should probably not be counted as users. In fact, he noted that even monthly use does not place a library in a very important position as a community information source.

Of library nonusers, 47 percent said they didn't have enough time to use the library. Another 30 percent said, "I have no need" to use a library. To Vavrek, the latter response could indicate that they (a) are unaware of what a library could provide for them, (b) have alternative sources of information, or (c) use and need less information. Library nonusers were found to read half the number of books read by regular users.

Rural library users in the survey were predominately women (70%) and visited the library mainly to check out books. (In about one-fourth of cases, they said they were checking out books for someone else, such as a child or spouse.) Although a number of other studies have also documented the fact that women are the primary users of rural libraries, it is possible that there are also methodological factors at work in this particular survey. Female library staff administering surveys might tend to request them more from women than from men. In fact, Vavrek's national telephone sample of adults over 17 was two-thirds female, also suggesting a survey bias. Nevertheless, many library studies have noted that the current typical user continues to be female and perceives the public library as a place of books. Bestsellers are more popular among library users than asking reference questions (Estabrook, 1991; Wittig, 1991). DeGruyter (1982) found rural library use about equally divided between reading for entertainment and "trying to obtain useful knowledge." Bundy (1960) found that "[t]o the farmer, a library is an agency for women and children, not geared to the farmer's interests and not planned around his convenience." Michael Marien (1991) summed up perceptions about the role of rural libraries in the information age as follows: "What in fact is the present role of the public library in helping citizens to understand our era of multiple transformations? My tentative answer is that it is virtually nil. Of course, there are some books to explain the multitude of public issues that we face, or the most current ones. But civic education is not a priority; indeed, it does not seem to be recognized as a library function" (pp. 27–29).

A study of the use and evaluation of sources for obtaining information for daily living carried out in New York by Scherer (1987) provides some indication of how libraries compare to other sources of community information. Respondents were asked to indicate if in the past year they had sought out information on any of 21 topics, including chronic disease risk, energy conservation, gardening/landscap-

ing, financial management, better parenting, time management, home management, home-based businesses, and others. Respondents who answered yes to any
items were asked to indicate what sources they used for information. Half did not
seek information on any of the topics. Use and evaluation of sources are shown in
Table 7.1. The "Use" column shows the percentage of respondents (those who had
sought out information on at least 1 of the 21 topics) that reported using each
source. The "Very Helpful" column shows the percentage using the source that
rated it as very helpful. While only 48 percent of those naming a source said they
used their local library, 57 percent of those who did found it very helpful—second
in rank of all 10 sources listed. The final two columns show the relationship between level of education and use and helpfulness of the sources. The correlation
between education and use of the library is the highest for any source, suggesting
that highly educated people are much more likely to use the library when seeking
information. However, a high education does not automatically lead to a rating of
very helpful for libraries or most other sources. Apparently, a library could be
highly rated by any that use it, regardless of education.

These findings reinforce an image of the public library as a place where educated
people go to get information. Given the strong indication that education is highly
related to library use and information seeking in general, one conclusion is that
providing new communication technologies in public libraries is unlikely to have
much effect on those who are not skilled at information seeking or do not have the
habit of visiting their public library. The precursor to more effective library use is
to teach people how to use information to improve their lives. The immediate
likely impact of the virtual library would be to provide an additional source of in-

Table 7.1

1987 Use and Evaluation of Sources for Obtaining Information for Daily Living from a
Random Sample of New York Adults

Source	Use	Very Helpful	Correlations Edu/Use	Significance Test	Correlations Edu/Helpful	Significance Test
Magazines	86%	32%	0.00	n.s.	0.21	$p < .01$
Self	83	33	0.11	$p < .014$	-0.10	$p < .04$
Newspapers	82	23	-0.0	n.s.	-0.03	n.s.
TV	80	20	-0.12	$p < .006$	-0.17	$p < .0001$
Friend	77	26	0.02	n.s.	-0.07	n.s.
Professional	67	58	0.17	$p < .00008$	-0.19	$p < .0003$
Local library	48	57	0.20	$p < .00001$	0.09	n.s.
Other source	48	30	0.12	$p < .005$	-0.02	n.s.
Local business	42	19	0.06	n.s.	-0.02	n.s.
Extension	26	38	-0.05	n.s.	-0.01	n.s.

Note: Data were extracted from Clifford Scherer, (1987), Databook: Getting Information for Daily Living,
Ithaca, NY: Cornell University Department of Communication, pp. 124–131. Used with permission.
Statistical findings in subsequent columns, that is, the correlations of education with use and
helpfulness of sources, are based on a reanalysis of Scherer's raw dataset.

formation for those who already use public libraries. Universal impacts on community building or citizen access seem highly unlikely.

A nationwide survey in 1996 of 1,015 adults 18 years of age or older by Lake Research (funded by the Kellogg Foundation) found that the public overall supports libraries and wants them to continue to provide books, reference materials, and programs for children. Marcum (1998), in analyzing the results, notes that "what becomes dramatically apparent from the opinion poll is that the public sees the public library first and foremost as an institution that benefits children. There is a strong sense that the public library is a safe, educational, and generally "nice" place for children to frequent" (p. 200). Women were much more likely (71%) to support bricks and mortar libraries than men (58%) were. Respondents did not necessarily expect to receive Internet and other information technology services from libraries, nor did they think of the library as the place where they would receive training in how to use these technologies. They also did not express much support for the idea of increasing taxes to provide high-tech services at libraries. Marcum concludes:

Most often, we have discovered that the users of the public library depend upon its physicality. They need to know that a place is there for them and their children. This need presents an interesting dilemma for the librarian who is valiantly attempting to raise money to purchase computers, servers and related equipment so that the library can be the community's on-ramp to the information superhighway. It is in the public library, more than in any other type of library, that we see the delicate balancing act between the social purposes, the information-providing purposes and the cultural purposes of the institution. Each community has defined its priorities in specific ways, often based on the funding that the governed are willing to provide. (pp. 201–202)

BUILDING A TECHNOLOGICAL BASE FOR THE INTERNET AT LIBRARIES

The ability of a rural public library to take advantage of new information technologies is determined by the training and motivation of its staff, its finances, and the extent to which certain functions of the library are automated. It is also determined to some extent by the level of technological advancement and organizational capacity of the community.

Vavrek (1995, 1997) documented the limited training of library staff in rural libraries. Only 21 percent of the librarians in nonmetropolitan libraries (25,000 or less) have completed their first professional library degree. In towns of populations less than 2,500 (rural by definition of the U.S. Bureau of the Census), only 4 percent of the librarians have academic training. States that tend to have many small towns and libraries are often those where training levels are lowest. In Vermont (Klinck, 1996), more than 90 percent of the 204 public libraries are in towns with less than 1,500 population staffed by part-time personnel with no formal library training. In Montana (Terbovich, 1996), a special Internet training program for 110 librarians found that even when given free connect time as a part of the course, most participants did not use it fully, saying they lacked the time. A Florida (Wilkins, 1996) study estimated that 50 hours of training and hands-on experience was necessary to prepare librarians to use the Internet.

Although the Internet can be accessed separately and does not require that the library have an automated collection of its own, there is a synergy in automating services so that patrons can search the library's collection as well as use the Internet. This enables a computer to be used for multiple purposes and means that searches can be done at a single terminal without having to check the Internet for another library's holdings and then march to the card catalog to check one's own library. State systems often begin the changeover to new information technologies by automating their collections and circulation systems and then providing links to other databases. Many rural libraries are not fully automated and a number are not computerized. Mumma's 1991 study of 97 rural libraries randomly selected throughout the United States found that only 10 had an automated catalog, 20 had an automated circulation system, and 33 reported some use of computer programs to assist in cataloging. Twenty-one offered online database searches and eight offered CD-ROM searches. Kentucky (Nelson, 1996) reported that only 18 percent of its public libraries now have automated collections and only half have some sort of publicly accessible technology. In Maryland (Smith, 1996), six of the state's 24 county libraries lacked automated catalogs. In Ohio (Byerly, 1996), 80 of the state's 250 libraries were not automated as of 1995 and thus were the last on the priority list for Internet access. Those with fully automated systems were all to be provided Internet access by January 1997. In Texas (Martin, 1996) only 15 percent of the state's public libraries now have integrated automated library systems, but almost all have microcomputers that could be used to access the Internet with little or no additional investment.

Finances are another serious problem for rural communities. McClure et al. (1995) found that for an initial one-time cost of $1,475 and a recurring annual cost of $12,635, a public library can establish a *minimal* level of single workstation, text-based, Internet connectivity. (The recurring charges include $600 for inkjet cartridges and paper, $6,500 for telephone toll charges, and $5,100 for Internet service provider fees.) Although the recurring costs have declined substantially as more local dial-up access is provided, a number of authors agreed that small rural public libraries have financial difficulties in providing Internet services. A state-by-state assessment in *Library Hi-Tech* in 1996 makes it very clear that unless subsidies are provided for acquisition of equipment, training, and connection costs, many rural libraries are not going to hook up (*Library Hi-Tech*, Special Issue, 1996).

Finally, many studies, especially early ones, reported on problems associated with the lack of a local Internet service provider. In the absence of such a provider, libraries must pay long-distance charges to access the Internet, which causes them either to not connect or drastically reduce use. In one library with such access, the Internet librarian received two messages from her director on the same day. The first said that she had to drastically cut down Internet connect time because the bills were simply too high. The second was a new list of Internet searches the director was passing on from patrons eager to tap this new information source. In the absence of a local service provider, some state library systems have provided an 800 toll-free number that permits libraries to tap into a text-based Internet system. Arizona, Montana, Pennsylvania, Vermont, Wisconsin, and Wyoming are examples of states that have or are providing 800 numbers to reach libraries that would not be able to connect otherwise.

By 1998, local Internet service providers became available in many rural communities so that connections were possible with a local call. In Nebraska, for example, the executive director of the Nebraska Information Network, an association of rural telephone companies, reported in January 1998 that 99 percent of Nebraskans now had access to the Internet without a long-distance call. In many cases, community institutions such as schools, libraries, banks, and local government must pool their communications activities in order to provide sufficient demand to interest an Internet provider (Kenyon, 1997; Warren & Whitlow, 1997). From this minimal beginning, other services and users can be added as more people and businesses find it advantageous to do so.

Internet Connections at Libraries

As of January 1996, 45 percent of public libraries in the United States were connected to the Internet, up 113 percent from 1994. Bertot, McClure, and Zweizig (1996) projected that by the end of 1997 connectivity might exceed 90 percent. However, libraries in communities under 5,000 are significantly (59%) less likely to use the Internet than those serving populations of 100,000 to 1 million (44% of the 1,495 libraries surveyed have legal service areas of less than 5,000). Nearly 40 percent of the libraries Bertot, McClure, and Zweizig surveyed in 1996 as part of a national survey of public libraries and the Internet conducted by the National Commission on Libraries and Information Sciences said they had no plans to connect in the next 12 months. Unless many of these libraries hook up, connectivity would not reach 90 percent. Bertot, McClure, and Zweizig (1996) concluded: "The commission's research prompts concerns that public libraries serving smaller communities of 25,000 or less may not be able to provide public Internet access" (p. 2). Table 7.2, from the Bertot, McClure, and Zweizig final report, shows that a library's service area is closely related to whether or not it has an Internet connection.

Results from the table show that libraries serving 100,000 or more patrons are likely to be universally connected soon. Below 25,000, there is a sharp drop in Internet connections. It should also be noted that a positive answer to the question "Is your library connected to the Internet in any way?" does not mean patron or even regular staff accesses. Studies of large libraries have shown that often one department may be connected while others are not. Small libraries may have a connection via a modem that is used only when absolutely necessary. Often, especially for small libraries, the connection may be to a text-based union catalog and databases at the state library. Many libraries with Internet connections do not have graphic interfaces such as the World Wide Web.

A state-by-state analysis of the online status of libraries in the United States published in *Library Hi-Tech* (1996) shows no case in which a state's rural libraries on

Table 7.2
Public Libraries Connected to the Internet 1994–1996 by Population of Legal Service
Area and Region

Area Served	Internet Connected 1996	1994	Change in Percentage
1 million+ population	82.0%	77.0%	5.0%
500,000 to 999,999	93.1	64.0	29.1
250,000 to 499,999	96.1	76.0	20.1
100,000 to 249,999	88.2	54.4	33.8
50,000 to 99,999	75.0	43.7	31.3
25,000 to 49,999	73.1	27.6	45.5
10,000 to 24,999	53.1	23.2	29.9
5,000 to 9,999	40.6	12.9	27.7
Less than 5,000	31.3	13.3	18.0

Source: Bertot, John Carlo, Charles R. McClure, & Douglas L. Zweizig, (1996, July), The 1996 National Survey of Public Libraries and the Internet. Final Report. Washington, DC: National Commission on Libraries and Information Sciences. Figure 5, p. 14. Used with permission.

their own adopted online technologies in large numbers (pp. 2–3). States in which the largest proportion of libraries is connected have used federal Library Service and Construction Act funds or other special grant funds often supplemented with state resources. Initially, funds have usually been invested primarily in building up a state network and linking it to universities and community colleges. Later, attention has been focused on linking to K–12 schools and rural public libraries. Many of the projects have provided initial incentives to hook up, including training, direct provision of computers and modems, direct provision of connections to Internet service providers, and payments of the first year of connect time. In many cases, libraries must agree to provide services for a period of about five years in order to participate. It is too soon to tell how many rural libraries will continue to provide services after the incentive period ends. The state-by-state analysis shows that the primary motive in almost all state plans is to link public libraries to a network that includes a union catalog, databases, special state information, an e-mail link for librarians, and an Internet gateway. Although community information is mentioned in some projects, it is almost never the driving force behind connection initiatives.

ADOPTION AND USE OF ONLINE TECHNOLOGIES BY IOWA LIBRARIES

Iowa provides a good example of the pattern of adoption and use of online services by rural libraries. Iowa has 485 public libraries. All but 6 percent of them are smaller than the 25,000 legal service population size commonly designated as rural. Three-fourths serve communities of less than 5,000. Most are independent municipal libraries supported by local tax revenues. In July 1995 all 485 libraries were surveyed by mail. A total of 369, or 76 percent, responded. To compare smaller libraries with larger ones, responding libraries were divided almost at their midpoint in terms of legal service size. Those serving fewer than 1,500 patrons (legal service area) were considered "small" and those serving 1,500 or more were considered "large."

Table 7.3 shows the overall adoption of various types of information technologies by the libraries and their accessibility to both library staff and patrons. The first three represent basic technologies available at many libraries. The second three focus on access to television and video-conferencing. The final six relate to applications that would require a computer. Results show that even for small libraries, the great majority have copy machines, VCRs, and facsimile (FAX) capabilities. Very few libraries of either size offer cable television, satellite access, or video-conferencing. More than 90 percent of the larger libraries offer CD-ROM services compared to about two-thirds of small libraries. About half of the larger libraries offer online services, but less than one-fourth of the small ones do.

For the smallest quadrant of libraries—those with service areas of 705 or less—14 percent have no computer. Only one-fourth have a computer with an external communications capability. About half of those with such a capability have an Internet connection. As Table 7.4 shows, as libraries increase in size, Internet connectivity rises dramatically.

Table 7.3
Use of Communications Technologies by Small and Large Libraries, 1995 Iowa Survey

| Technology | Small < 1,500 | | | Large 1,500 + | | | X^2 and p |
	Neither	Staff	Both Patrons and Staff	Neither	Staff	Both Patrons and Staff	
Copy machine	18%	22%	57%	4%	16%	80%	25; $p < .000$
VCR	40	16	45*	23	23	54*	10; $p < .050$
FAX	31	26	42	13	44	44	20; $p < .000$
Cable TV	80	4	16	73	8	18	4.6; $p =$ n.s.
Satellite TV	99	0	< 1	98	<1	< 1	1.8; $p =$ n.s.
Video-conferencing	99	<1	0	96	3	< 1	4; $p =$ n.s.
CD-ROM	37	13	50	9	18	72	35; $p < .000$
Electronic cataloging	86	10	4	54	39	7	40; $p < .000$
Electronic ordering	95	4	< 1	77	23	0	25; $p < .000$
Access to online databases	79	12	9	47	36	17	36; $p < .000$
Online connection to another library	86	11	3	62	26	12	25; $p < .000$
Access to online network	82	15	4	48	43	9	39; $p < .000$

Note: For small libraries, includes six cases with patron-only access; for large libraries, includes eight cases with patron-only access.

Among libraries that do have access to online services, there were no differences between small and large libraries in terms of perceptions of their usefulness in helping patrons find information. Most found them either useful or extremely useful. Small and large libraries also were equally likely to say they were satisfied with online services. When asked about problems faced in using online services, it was the larger libraries that more frequently complained about "lack of simple instructions," "frustration at getting cut off at a location," and problems of "diversity of means of access to different locations." Librarians from the small libraries were slightly more likely (37% to 30%) to report that online training is not available. The results suggest that once libraries are hooked up, the smallest ones find services as useful as large ones and are as satisfied with them.

Five questions were asked of all respondents to learn more about the visions librarians might have about the role of online systems. The first asked how essential they believe telecommunications technologies will be to the future of their library. Response categories were "absolutely essential" (4), "essential" (3), "somewhat essential" (2), "supplementary to other materials" (1), and "not needed at all" (0). The mean response score for the small libraries was 2.51, compared with 3.30 for the larger ones—a statistically significant difference. Large library respondents see these technologies as more essential (although both groups gave answers between somewhat and very essential). Large library respondents also are significantly more likely to believe that the residents of their communities rank access to communication technologies as being very important. On a scale of 0 to 4, with 0 being "not important" and 4 being "very important," the mean score for large libraries was 2.63, compared with 2.29 for small libraries. A third question asked whether or not respondents believe that new telecommunications technologies "will change the role of the public library, making it the 'information hub' of the community, especially in rural areas." Among small libraries, 84 percent agreed with the statement, compared to 74 percent of the larger libraries. When asked why they agreed or disagreed with the statement, the larger libraries were more likely to point out that there are other competing information sources in the community that might

Table 7.4
Computer, External Connection, and Internet Access by Library Size, 1995 Iowa Survey

Legal Service Area	Have a Computer (All Respondents)	Have External Communication (All Respondents)	Of Those with External Communication, Have Internet Connection
1 to 705	86%	24%	52%
706 to 1,499	92	38	62
1,500 to 4,940	95	52	67
4,960 to 200,000	99	89	90

be hubs. The fourth question asked how important online training was in hiring library staff. Larger libraries were slightly more likely to say that such skills were either "somewhat" or "very important" (a combined percentage of 80% to 67%). Sixteen percent of the small libraries said it wasn't important at all, compared to 7 percent of large libraries. The fifth question asked respondents to estimate "the proportion of information accessed by your patrons that will be in electronic form in five years." (Electronic form included videocassettes, audiocassettes, computer CD-ROMs, online information access, etc.) Table 7.5 shows that there are no statistically significant differences between small and large libraries.

Another question in the survey examined sources of information about online information technologies to see whether the most relevant information was coming from horizontal sources within the library or community, or from vertical sources such as the state or regional library systems. Vertical sources included three choices: state government, the state library system, or the regional library system. Horizontal choices were divided into two groups, one focusing on the local library's own internal resources, and the other based in the community. The local library group included the library's own strategic planning process, interest of the local library administrator, or internal staff contributions. The community group included community strategic planning or community colleges. Results in Table 7.6 show that although the relevance scores varied slightly between small and large libraries, the ranking of the relevance of information sources was the same. Both rated the regional library as their top source, the library administrator as the second choice, and the state library as their third choice. Community sources ranked at the bottom. (Values ranged from 1, "not relevant," to 5, "very relevant.")

A linear regression was run to try to identify the overall factors that were most important in predicting whether or not a library would adopt online technologies. Variables included library budget; size of service area; number of computers and telephone lines; attitudes of the library director toward online technologies; and ratings of relevance of community, library, and external sources. Regression uses betas to rank the relative importance of predictors. The dependent variable was a measure of the number of online services provided by the library. Results (Table 7.7) show that the library's overall budget was the most important single factor in determining the number of online services it would provide. Once budget was taken into account, the number of computers was the second most important fac-

Table 7.5
Perceptions of the Proportion of Library Material that Will Be Accessed in Electronic Form within 5 Years, 1995 Iowa Survey

Library	None	1 to 10%	1 to 20%	21 to 30%	More than 30%
Small	6%	19%	30%	26%	19%
Large	2	13	37	31	18

Note: $\chi^2 = 7.4$; $p =$ n.s.

Table 7.6
Relevance of Information Sources for Online Technologies, 1995 Iowa Survey

Source	Small Libraries	Large Libraries	Overall Source Ranking	χ^2 test
State government	3.35	3.45	6	n.s.
State library	3.87	3.93	3	n.s.
Regional library	4.34	4.18	1	n.s.
Library's own strategic plan	3.41	3.49	5	n.s.
Library administrator	3.94	4.16	2	$p < .035$
Internal library staff	3.42	3.80	4	$p < .002$
Community strategic plan	2.92	2.93	7	n.s.
Community college	2.80	2.54	8	$p < .044$

Note: n.s. = not significant

Table 7.7
Linear Regression Predicting Adoption of Online Technologies by Libraries, 1995 Iowa Survey

Variable	Beta	Significance
Budget	.37	.0000
Number of computers	.27	.0005
External sources	.18	.0011
Community sources	-.11	.0489

Note: $R^2 = .40$

tor. Librarians reporting they were highly influenced by external sources concerning online technologies were the third significant factor. Community sources were the fourth and final factor, but in a negative direction, indicating that increased reliance on community sources leads to *less* adoption. Attitudes by librarians were not important. A second regression found that in addition to budget, a perception by library staff that online technologies were very important was associated with a significantly higher range of uses of those technologies within the library (Pellerin, 1997).

The Iowa data suggest several critical-mass points for online technologies. Critical-mass points indicate levels where rapid change in adoption of online technologies can be expected. Libraries with annual budgets of more than $50,000 are associated with a rapid increase in access to online services (see Figure 7.1). Similarly, libraries with legal service areas of more than 1,500 are associated with rapid increases in access to online services. This suggests that there are common patterns

Figure 7.1
Proportion of Library Adoption of Online Technologies by Annual Budget and Size of
Legal Service Area

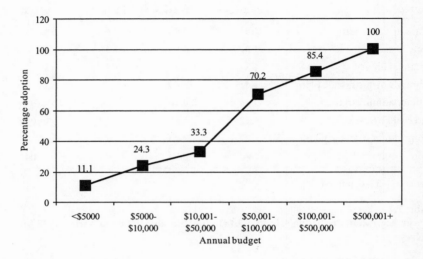

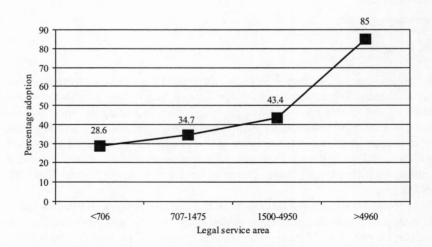

for libraries with small budgets and small legal service areas that are preventing them from adopting online services. Unless subsidies or grants can be provided, it seems unlikely that many of these smaller libraries will hook up. Note that a lack of training and negative attitudes were not key factors in predicting online adoption. This indicates that the primary problems are financial.

THE SPECIAL PROBLEMS OF RURAL LIBRARIES

Rural libraries—those serving fewer than 25,000 patrons—represent almost 80 percent of all public libraries in the United States (Holt, 1995). Dillman (1991a) noted that "rural communities have some very serious information needs quite different from those of the past and no one is meeting them—not the public school systems, not the Cooperative Extension Service, not the public libraries and not anyone else" (p. 31). In Dillman's view, the information needs have changed because of economic and social changes in society. Technologies, then, are seen as tools that can assist rural communities in adapting to this new reality. Mazie and Ghelfi (1995) point out that there are great differences among rural communities that affect their ability to cope with new information needs (p. 7). Four different types of rural communities illustrate this point:

1. *Farming-Dependent Counties.* Mainly in the Great Plains, they have seen employment opportunities drop as people move away in search of jobs elsewhere. The small population levels have resulted in large per capita costs for maintaining services.

2. *Persistent Poverty Counties.* Mainly in the south, 20 percent of people in these counties live below the poverty line. Public services are difficult to provide here.

3. *High-Amenity and Retirement-Destination Counties.* These are located mainly along the coasts, in places with agreeable climates, scenery, and pleasant living. They tend to be growing rapidly (12% during the 1980s) and have a disproportionate share of rural jobs. Retirement areas have grown 34 percent during 1979–89.

4. *Counties Adjacent to Large Metropolitan Areas (1 Million or More).* Here, people move to enjoy lifestyle while they work in the city or suburbs.

While many of the rural communities in the United States do not fall into any of these four groups, the point is that policies for connecting rural libraries need to differ based upon the type of community. If Dillman (1991a) is correct, persons in the first two groups will be most in need of the benefits to be provided by new information technologies. The farming areas are losing their populations as average farm size increases, shopping becomes more regional, and local and natural resource employment opportunities drop. If people are to remain in these areas, they must find new information-based jobs. The persistent poverty areas suffer from few job opportunities and community bases that cannot support needed investments. Yet these two categories are precisely the areas in which Internet services are least likely to be offered and in which public libraries have the fewest resources to adopt. Although 97 percent of even the most remote rural counties (nonmetropolitan counties that are not adjacent to a metropolitan area and do not contain a city with 10,000 or more residents) have at least one public library outlet, those outlets are not hooking up to the Internet on their own (Mazie & Ghelfi, 1995).

Boyce and Boyce (1995) point out that the problems of rural library service "may be characterized in simple terms of cost and distance. Library service costs money. As a public good, its cost is distributed over a large number of people, and its services are available to all of these . . . when the population is dispersed, a far larger area is needed to provide the resources for an effective information service facility" (p. 112). In urban areas, a recent study reported that the percentage of materials in electronic form would increase from 4 percent in 1990 to 18 percent in 2000 (Holt, 1995). While rural libraries would benefit from being able to tap electronic resources, low overall funding levels will make this difficult. Although per capita expenditures in libraries serving fewer than 1,000 patrons are actually higher ($19.03) than the national average ($18.73), funding overall for rural libraries (those serving 25,000 or fewer patrons) is slightly lower ($17.19). Even with these per capita rates, rural libraries lack the resources to invest in new technologies. Low overall revenues are related to the American tradition of local taxation for library support and pressures to keep taxes low on land and farm buildings (Holt, 1995). In 1992, Holt found that in libraries serving fewer than 1,000 patrons, 47 percent of their budget was spent on staff and 23 percent on collections. Nationally, those percentages are 65 percent on staff and 15 percent on collections. The low investment in staff results in a lack of professionally trained librarians that will be needed if libraries are to take on the responsibilities of an information age. Rural libraries have chosen books over trained staff.

Cisler (1995) draws parallels between the Tennessee Valley Authority (TVA) and Rural Electrification Administration (REA) and present-day efforts to serve libraries (pp. 177–179). In 1935, Cisler notes, only 5 percent of rural residents had electricity. The REA was a government-subsidized program necessary to connect rural residents that the private market was not interested in serving. Although Cisler does not believe the current political climate will result in a TVA-like program in the area of telecommunications, without such a program rural areas are likely to lag, he predicts.

In addition to financial limitations, Cisler concludes that "communications technology is much less important to the survival of a rural community than are the people and the policies that they put in place to deal with the challenges they face" (p. 181). Quoting sociologist Kenneth Wilkinson, he notes that rural libraries face organizational and community problems: "Rural places tend to be locations where particular problems and issues appear and not social units where effective collective actions occur" (quoted in Cisler, 1995, p. 181). The inequality among local groups, lack of services and organizational structures, and uneven access to outside resources undermine local capacity for collective action and self-help by libraries and other local organizations.

For Cisler, the rural community of today is not the homogenous social unit of the past. Now, he believes, rural communities suffer from both physical and social isolation. This is a theme that has been advanced most strongly by Robert Putnam, who claims that social capital—the networks and norms of a civil society—is collapsing. Instead of gathering with colleagues and neighbors in civic and social activities that are tied to communities, Americans are giving priority to nonplace-based activities. Influences such as television, the global economy, and two-career families are rendering obsolete the stock of social capital we had built up at the turn of the century (Putnam 1995; summarized by Marcum, 1998, p. 191). A lack of social capital would create difficulties for public libraries in rural areas, which often rely on volunteers and donations to survive. But the views of Cisler and Putnam also lead to a specific vision of what rural libraries should be in an information age—a center for the restoration of the civic community.

A final special factor for rural areas concerns policy factors encouraging alternative sources of information delivery in rural communities. Holt (1995) observes that it is not a given that citizens will turn to rural libraries in the future. Several examples indicate how alternative sources might provide information services. The U.S. Department of Agriculture created a Rural Information Center (RIC) by congressional action in 1987 to help combat the farm crisis by providing information to farmers. Information was to flow through the National Agricultural Library and the Cooperative Extension Service offices to clients. Massive amounts of electronic information were made available through this channel. In 1990, the Office of Rural Health Policy also began using the RIC to pass along information concerning health care delivery services, research findings, personnel policies, financing, and the health status of rural citizens. In addition to working through Extension offices, the RIC also makes its information available directly to users. At present, only about one-third of RIC requests come from Extension. Only 3 percent come from rural libraries (John, 1995). Boyce and Boyce (1995) point out that the provision of agricultural information through Extension rather than libraries is not new: "There is a patron perception that the hard data come from Extension and the fluff from the library" (Boyce & Boyce, 1995). The links between libraries and other local information providers have often been weak. Lynch (1989) believes that libraries will have the most success by interacting with Extension home economists in the human ecology side of information rather than agriculture. The librarian, not Extension staff, has usually initiated existing cooperation.

Holt (1995) also lists other emerging independent third parties that are interested in providing information to rural residents, including CompuServe, America OnLine, Prodigy, and other Internet information services. Holt concludes: "[R]ural libraries already have competitors trying to capture their best customers—i.e., middle and upper income families who pay the bulk of residential property taxes in any political subdivision. Anyone who is certain that for a fee, online library and information services could not replace rural libraries may want to recall how, during the 1970s, 'free' television representatives proclaimed firmly that Americans would never pay for cable television" (p. 200). When cable companies were providing the same programs available via over-the-air free systems, the television representatives were right—the public was *not* willing to pay. However, when cable began providing additional value in the form of independent superstations, such as HBO, Cinemax, and Showtime, sales increased. When libraries are contemplating their information future, they might well consider how they could add value that is not available elsewhere to their product.

CONCLUSIONS

In her book *When Telephones Reach the Village*, Hudson (1984) notes that attempts to place a uniform value on each telephone receiver miss an important point. The one telephone in a distant village has, comparatively, much more economic value and payoff than one telephone in a phone-rich urban setting. Similarly, the value of an Internet connection in a small rural library in a community that lacks other connection points may have much greater economic payoff than an Internet terminal in an urban area that has dozens of Internet access points. In a library context, the value of a book in a library in an urban community with plenty of new and used bookstores, magazine vendors, and other sources of printed materials may not be as great as its value to the small rural library patron in a community with few other information sources. Iowa's small rural librarians may have had this in mind when they were more likely than large libraries to envision themselves as becoming information hubs for the community.

It is this extra value that would justify efforts to help rural libraries provide online access to the Internet and other services. One criterion for funding or subsidizing such a service might well be the lack of alternative access in the community. In those cases, assuming there would also be enough training for the librarian in the effective use of such a system, small rural library online terminals might be able to provide both Carey's (1969) centrifugal information—the bits of information we seek as individuals that differentiate us—and centripetal information, a core of data, culture, and media that provide us with a common culture.

In addition to the fact that small rural libraries might be the only points of access for rural citizens, libraries in general possess two other important characteristics that might provide important value to online users:

1. Librarians have played a vital "consulting" role and this is desperately needed when tapping online information sources. This consulting role involves helping direct patrons to the places that are most likely to provide the answers they are seeking. A strength of librarians is that they know where information can be found and they also have significant insight into the credibility of different sources. In an Internet world, this is extremely valuable. Thus, even when citizens can connect from home, they still need someone who can point them to the most valuable resources and save them endless hours of searches using the World Wide Web's current search engines.

2. State library systems and other networks of libraries have begun negotiating with publishers, journals and mass media to provide electronic access to materials at a relatively low cost—much lower than any citizen would pay. By linking into networks, libraries have increased their bargaining power and have technologies sophisticated enough to meet the conditions set by publishers and media outlets. Universities, which have been pioneers in this area, have begun negotiations with academic journal publishers for the rights to make journal articles available on demand to anyone in their entire network who requests them, at a price significantly

lower than it would cost to provide the physical journal at any one of the libraries. Thus, lower transaction costs made possible by grouping libraries could greatly enhance the future information role of libraries.

These positive attributes of libraries should make them serious players in an online future. However, there also are some factors that would seem to limit the ability of rural libraries to digitize, link to online services, and serve a broad spectrum of citizens. Lack of budget, infrastructure, and training, combined with a tilt toward the highly educated and a frequent focus on entertainment and escape rather than information seeking are all key issues that will have to be addressed if rural libraries are to fulfill a vision as information hubs of the future.

In terms of infrastructure, budget, and training, rural libraries are characterized by the following:

- Lack of budget to buy a computer, modem, and telephone line—the prerequisites of an online system; also, lack of funds for maintenance and replacement. A critical mass of about $50,000 in budget or a legal service area of 1,500 or more patrons leads to more rapid adoption of online services. Below those levels, significant subsidies will likely be needed to set up and maintain online services.
- Lack of local dial-up Internet service providers in rural areas. Although this problem has been remedied in many areas, where it remains it poses a major hurdle for libraries, which need low-cost and flat rate monthly charges for use.
- Lack of training in how to use the Internet by library staff.
- Lack of an automated circulation system to take maximum advantage of new electronic services.

Funding and politics will determine whether schools, Extension offices, government buildings, newspapers, kiosks at shopping malls, or home terminals will become the most-used connection points. When one considers the relative lack of training of rural librarians and the serious lack of financial resources in their budgets, it is hard to imagine that they are poised to become information hubs for rural communities without massive outside intervention and support. At best, they may offer a valuable connection to the virtual pipeline of information.

In terms of types of patrons, the users of public libraries are and will continue to be those who (a) know how to use information sources such as books, magazines, television and newspapers, (b) have enough education so that they know how information can be useful to them, and (c) have sufficient economic resources so that the useful information obtained can have a meaningful payoff. Because the use of libraries is shaped by these three factors, the arrival of new technologies at the rural public library should not be expected to revolutionize library clientele any more than did the arrival of books. In fact, the characteristics of book readers match closely the characteristics of those who are connecting to the Internet. Citizens with low education levels are neither library users nor information seekers from other sources. Those who lack the financial capability of acting on new information seldom seek it out. Because rural areas tend to be populated with more individuals having low education levels, lower levels of information seeking would be predicted regardless of whether that information is provided in books or via the Internet.

Another important factor is current patterns and habits of use. Even among the educated, rural public libraries are typically used by women for books. Populations such as farmers tend to associate public libraries with "fluff," and do their information seeking elsewhere. Changing this public perception of what a rural library can offer will take much more than simply a connection to the Internet. Thus, a transformation of rural public libraries toward either a civic vision or a universally accessed virtual library vision would require a massive public education effort. Such an effort has not been a part of federal, state, or local programs thus far.

Although Schuler's (1994) vision of rural online libraries taking the lead to bring about a resurgence of civic community has played an important role in shaping expectations about what the Internet might do for rural and urban America, most current initiatives at state and federal levels are not directed toward the creation of a civic community with the library as its hub. Despite Schuler's documentation of 100 instances where such civic communities were being created, they constitute only a small proportion of total online libraries and funding efforts. It is the vision of the digitized virtual library system, not the civic community, that is driving library policy and funding today.

However, even if policies and funding were favorably disposed to the idea of creating civic communities with library hubs, fundamental questions would need to be answered. Do citizens feel a need for enhanced local communication? Or are they satisfied with local communication but want to use online resources to connect to distant locations? Much of the discussion in this area takes the position that citizens *ought* to want to create and participate in these networks. But there is little empirical evidence to support the fact that they want to.

Robert Putnam (1995) documents factors that continue to decrease the social capital found in U.S. communities: (a) women moving into the workforce who have less time for civic activities; (b) increased household mobility, which weakens long-term community ties; and (c) viewing of television, which reduces the time available for civic activities. The Internet and online services, despite the possibility that they might be used to reconnect individuals into a common civic culture, could easily also be imagined as another insidious force like television, soaking up massive amounts of time devoted to individual activities that differentiate people.

Although it is possible for the Internet to be used for collective purposes—to build civic discussion and to link to organized activity—Internet studies thus far have found that use of this technology to participate in discussion groups and forums has been minimal compared with individual uses to gain information. While interest groups such as the Sierra Club and the American Association of Retired People can be expected to take full advantage of the Internet to reach and connect with their clients, the majority of users do not tend to be activists seeking to use the system for a collective purpose. Thus, the problem of civic participation would require a remedy that goes well beyond online connectivity. This conclusion is in agreement with both Warren's documentation of the increases in vertical communication and Carey's notion that the major communication forces, both centripetal and centrifugal, are likely to come from outside the community.

It is clear that the future trend is toward digitized information. Great emphasis and significant funding are now creating new information in a variety of electronic forms and converting existing holdings of major libraries to digital form. Soon, an important proportion of the nation's information will be accessible *only* in digital form. Leading libraries such as the Library of Congress and most state libraries are taking the lead in the conversion process. The conversion will make it possible for those who are connected to access documents, databases, reference sources and even books that used to be available only to a selected few.

While rural public libraries and their users can obviously benefit from a connection to the virtual library of the future, to a great extent this is not yet a revolution that has affected them in any direct way. It is a revolution implemented with policies designed to equalize information between the haves and have-nots of society, but its effect at this point has been to greatly enhance access by large automated networked libraries while offering similar access to only a handful of truly rural libraries in most states. While in the future even small rural libraries may be connected and even receive assistance in digitizing and linking their own unique cultural and historical material, the overall trend seems to favor increased *vertical control* over not only the network, but over all decisions about what is digitized and stored and who is connected. State, federal, and even global levels would seem to be placed in the leadership positions for the virtual library of the future.

One lesson from the past was that the origin of policies such as special library rates lay with urban groups who stood to benefit handsomely, but in order to gain support, they sold their idea to policy makers as something that would benefit rural areas. One may be able to understand the current information highway for libraries better by asking how state libraries, research libraries, and other information centers may benefit from creating and building vast interconnected libraries that link to rural areas.

NOTE

1. A union catalog is a catalog of stock in various departments of the library or a number of libraries indicating locations. It may include an author and/or subject listing of all books or selection of books or be limited by subject or type of material.

8

MUNICIPAL GOVERNMENTS' USE OF TELECOMMUNICATIONS: LEADING THE CHARGE OR LAGGING BEHIND?

Erin K. Schreck and Patricia C. Hipple

INTRODUCTION

How do municipal governments seize opportunities in the telecommunications revolution? Do local governments use telecommunications technologies themselves and promote telecommunications development, adoption, and use in the communities they serve? What constitutes success or failure in telecommunications development among municipal governments?

Consider Cedar Rapids, Iowa, the second largest municipality in the state, with a population of about 120,000 residents. This community experienced a dramatic economic and social transformation with the introduction and subsequent growth of its telecommunications industry. Initiated by McLeod Telecommunications and others in the private sector, but spurred by municipal incentives and investment, this telecommunications-induced development has added some 20,000 jobs to the community since 1988. The remarkable economic growth has resulted in a complete restructuring of the local economy, from its former manufacturing base to its current telecommunications and information services base (Pins & Bullard, 1997). Consider, Spencer, Iowa, a community of about 11,000 in the northwest corner of the state. This municipal government recently assumed responsibility for providing telecommunications infrastructure and services to its residents. Spencer Municipal Utilities, a department of the municipal government operated under a separate board of directors, has provided traditional utilities (gas, water, and electricity) for some time. In 1997, however, it expanded to include telecommunications utilities for the community, including cable, Internet, and other data transfer services. Consider Urbandale, Iowa, a community of nearly 28,000 residents with city boundaries contiguous with five other urban municipalities. At this writing, Urbandale city administrators are putting the finishing touches on their new Web site. Dubbed "the 24-hour city hall," the homepage for the City of Urbandale provides its citizens instant access to each municipal department, as well as a calendar of events, an employment locator, a directory of reports and public documents, maps of the community, and links to other government Web sites. Already replete with downloadable permit applications and registration forms, future incarnations of the site will allow electronic commerce with city hall for citizens and staff to conduct government business. And consider Belmond, Iowa, a small rural community of 2,500 in north-central Iowa with its own Web site and homepage. The local chamber of commerce director (an intrepid novice with Web page design) created and maintains the site, and the Belmond Industrial Development Corporation (BIDCO) covers the costs. The site gives Web surfers a description of city government along with links promoting the community as a business location, tourist attraction, and great place to live. Although the city government page is not interactive, there is a direct link to the Iowa Department of Economic Development, and perhaps more importantly, evolving links to more than 80 types of businesses in the Belmond area. Internet subscribers around the globe can locate Belmond's homepage and get a distinct flavor of this community, along with scanned photographs of city attractions, and a reminder not to miss the flight breakfast for the Belmond police reserves the second Sunday in August.

What are the catalysts for these kinds of innovations? From where comes the leadership to guide this development? How do other municipal governments, and the communities they serve, reproduce similar successes? In Chapter 4, Van Wart, Rahm, and Sanders reported the results of a spring 1997 survey on telecommunications use by city administrators and enumerated many of the factors that contribute to success or forebode failure for rural telecommunications development. This chapter compares results from the 1997 baseline survey with a similar follow-up survey of city clerks conducted 18 months later, in the summer of 1998. Our research sought (a) to determine whether telecommunications adoption and use by local government institutions are increasing over time, (b) to examine attitudes and perceptions of local government officials toward telecommunications technologies, and (c) to identify characteristics of local governments and the communities they serve that promote or retard telecommunications adoption. How active are government officials themselves in using telecommunications technologies to enhance government services? What role does the municipal government play in promoting telecommunications development in the wider community? What prerequisites are needed for communities to embark on telecommunications development? Can communities of any size reap the benefits of such development?

MUNICIPAL GOVERNMENTS' CHANGING ROLE

According to the Center for Technology in Government (1998), the role, structure, and processes of government have undergone dramatic changes in recent years. The boundaries once separating government services from business are beginning to blur as traditional roles of the public and private sectors begin to overlap. In Chapter 4, Van Wart, Rahm, and Sanders documented the increasing frequency with which cities and towns in Iowa have assumed the responsibility for providing telecommunications technologies and services to their residents, an example of public entrepreneurship aimed at tackling deficient infrastructure and disappointing service offered by the private sector. Accompanying this trend is an increased awareness of the role information and telecommunications technologies can play to enhance municipal operations and government service provision. The U.S. House Subcommittee on Technology (1993) reports that telecommunications technologies have the potential to transform organizational processes within government to significantly enhance the quality, efficiency, and efficacy of service delivery. Therefore, telecommunications development has immense economic, social, and political implications for the individuals and governments who lead—or follow—these telecommunications adoption trends.

INNOVATION ADOPTION AND DIFFUSION

The adoption of innovations, which Jaakkola (1996) describes as the introduction or application of a new idea or invention that changes the existing order, has been characterized by a number of different models. The most comprehensive is that introduced by Rogers (1962), one of the pioneers in adoption and diffusion of innovations research. Rogers, like others to follow, conceptualized the innovation diffusion process as an interaction among a number of different elements. Innovation decision making can lead either to adoption or rejection according to Rogers (1995). The former he defines as "a decision to make full use of an innovation as the best course of action available," and the latter he defines as the opposite, "the decision not to adopt an innovation" (p. 21). Rogers identifies four crucial elements of the diffusion process: "[D]iffusion is the process by which [1] an innovation is [2] communicated through certain channels [3] over time [4] among the members of a social system" (p. 5). While understanding each of these elements is necessary to understanding the process of innovation adoption and diffusion, Rogers cautions that innovations are continually being adapted to changing circumstances during the adoption and diffusion process. As a result, innovations themselves change over time so that the innovation that was the initial focus of study may take on very different characteristics as it is more widely diffused. And changes in the innovation can result in discontinued use or a renewed appreciation and accelerated use. Rogers advocates a comprehensive approach to the examination of innovation adoption and diffusion that attends to each of these dynamics.

A number of circumstances impinge on successful adoption of technological innovations (Rogers, 1995). Participants in the adoption and diffusion process—especially those for whom the innovation is intended—are as important to adoption success or failure as are the technological aspects of the innovation. Different types of decision makers and stakeholders are involved in the process and each has distinct needs as well as differing views on how to implement and use technologies. Participants' attitudes and perceptions in the adoption and diffusion process can have profound effects on how technological innovations are integrated within an organization. And participants influence the attitudes of one another within the organization (Moore & Benbasat, 1996; Leon, 1996). Especially important are the attitudes of managers and other leaders. Strong support from organizational leaders can provide significant propulsion for the adoption of new technology and promote its continued use by the organization. At the same time, organizational culture, especially environmental factors such as hierarchical structures and rigid communication patterns, can impede technology adoption and diffusion (Kautz, 1996).

ORGANIZATIONAL CULTURE AND LEADERSHIP

Organizational culture frequently makes the difference between organizations that fail and those that succeed in integrating technological innovations (Watad & Ospina, 1996; Leon, 1996). The changes wrought by innovation are often met with resistance, both individual and organizational (Thong & Yap, 1996). Much of this resistance is a natural response to the disruption of organizational processes and operations because new technology frequently requires changing how work is done. This can pose both real and imagined threats to organizational members. Despite the ubiquity of change, many individuals and organizations experience discomfort accepting and adjusting to change. This resistance can be a contributing factor to failures in technology adoption (Hughes, Kristoffersen, O'Brien, & Rouncefield, 1996). Supportive leadership, pro-innovation attitudes, technology advocacy, and the technological knowledge and expertise they engender can increase the likelihood of successful implementation and use of technology. Unfortunately, strong leadership for technology adoption is not present in all organizations and, as a result, many organizations struggle in the absence of an innovation champion. Effective leaders who advocate technology adoption and devise ways to mitigate against the threats felt by participants, however, can overcome even strong resistance (as Abbott demonstrates with the Extension Service in Chapter 11). Championing telecommunications requires the active support and participation of leadership. In local government, this means mayors, city administrators, city clerks, and department directors who are perceived as having power, influence, and authority. Such leadership is necessary for communities of all sizes, but is critical for small communities that may face greater challenges to adoption and use of telecommunications technologies.

THE RURAL-URBAN TECHNOLOGY DIVIDE

While telecommunications proponents argue these technologies' potential for communities of all sizes, a disparity exists between rural and urban communities, with rural communities lagging behind their urban counterparts (U.S. Senate Subcommittee on Science, Technology, and Space, 1995). An underlying condition that reinforces the rural-urban disparity is inadequate infrastructure required for telecommunications development. Despite efforts to address this disparity, many rural areas still lack the infrastructure necessary to realize telecommunications' potential. Consequently, rural communities remain inadequately "technologized" and are therefore at a distinct disadvantage when compared to better endowed urban counterparts. Larry Irving (1997), assistant secretary for Communications and Information of the Department of Commerce, argues that telecommunications services must be made available to rural areas if they are to survive economically; but low population density, distance, and isolation pose challenges to these communities by increasing the costs of installing requisite infrastructure, maintaining technologically complex systems, and providing local services to reap the benefits of telecommunications development.

Because telecommunications development can attract business and amenities to rural areas, helping to revitalize rural economies (U.S. House Subcommittee on Technology, 1993), municipalities that lack the necessary telecommunications infrastructure may be placing their community and citizenry at a severe disadvantage. Moreover, communities that are insufficiently technologized may be at a disadvantage in terms of efficient operation and quality service provision, as well as their appeal to prospective businesses, entrepreneurs, and residents. Irving's conclusions are of particular significance to states such as Iowa, where the majority of communities are small and rural.

THE BENEFITS AND RISKS OF TELECOMMUNICATIONS FOR MUNICIPAL GOVERNMENTS

The potential benefits to be realized through the adoption of telecommunications technologies have been enumerated throughout this book. They include, among others, expanded access to information, improved educational services, enhanced business acumen, stimulation of economic growth, increased speed of response by service providers, and more creative ways to respond to customer needs. Telecommunications has numerous applications for municipal government as well. Because much of the work that government does is related to the dissemination of information, the Center for Technology in Government (1998) concludes that advanced telecommunications technologies are well suited for government work. Telecommunications technologies can play a central role in government administration, finance, tax and licensing, planning and zoning, housing, community services, police and fire protection, sanitation and public works, library services, and more. Telecommunications can promote civic involvement and civic pride—they have been offered as an antidote to lagging participation in the political process by the facilitation of "virtual town meetings" and "cyber-democracy" (Grossman, 1995; Hacker, 1996). Many government applications provide municipalities with opportunities to become more viable and competitive, potentially leveling the playing field between smaller communities and their larger counterparts and making the former more likely competitors for economic development and cultural amenities. The Center for Technology in Government (1997) is one of a number of organizations working to develop "new ways of applying computing and communications technology to the practical problems of information management and service delivery in the public sector" (n.p.).

Telecommunications are not without drawbacks, of course. Several key issues have provoked considerable discussion in recent years (Schmidt, 1998). Privacy concerns arise in light of relatively easy access to a wealth of personal information about individuals, such as addresses, phone numbers, social security numbers, banking and credit transactions, travel patterns, and purchasing habits. Sensitive information contained in medical records, financial documents, and even e-mail messages is at risk as well. For government entities charged with both collecting and protecting vital personal and proprietary information, the risks of exposure via telecommunications technologies pose an ethical dilemma and potential legal liability.

Another problem posed by expanded access to information is the costs incurred for individuals and organizations to become efficient information managers in order to achieve the efficiency and productivity promised by these technologies. One downside of increased access to information is the possibility for information overload, which actually confounds the decision-making process and increases administrative costs. The World Wide Web affords its users immediate access to a tremendous amount of information not easily accessed through traditional information and communication channels—the ease with which information can be obtained is incredible. But without the ability to distinguish good information from bad—accurate information from erroneous—an information glut can be a source of inefficiency. Schmidt (1998) cautions that information overload can complicate planning, policy making, advocacy, and government service provision of all kinds.

Reliance on telecommunications technologies can result in other inefficiencies as well. For example, the convenience of e-mail can accelerate the speed with which written communication is disseminated, but anyone who has been inundated with unsolicited e-mail ("spammed") knows this convenience can become a major annoyance. And the increasing incidence of techno-terrorism perpetrated against individual and organizational Internet users (including the Pentagon and FBI), by means of hacking, espionage, and the introduction of disastrous computer viruses, causes even the most technically sophisticated computer elite to shudder.

The potential for new jobs related to the adoption of telecommunications technologies in some community sectors is mirrored by the risk of job losses resulting from automation and increased government efficiencies. Especially vulnerable to job loss are those lacking the technological skills required of advanced telecommunications for the information and service sector, but small rural communities can ill afford job losses in any sector.

Notwithstanding these drawbacks, and given the growing reliance and dependence on telecommunications technologies in other sectors of the community and economy, it becomes essential to recognize the role high quality advanced telecommunications technologies can play for municipal governments to enhance government services, revitalize rural institutions, and promote economic and community development. Adoption and promotion of these telecommunications technologies promises to make municipalities more responsive to the needs of the community and bring them closer to citizens (Moulder & Hall, 1995). Wise application of telecommunications by city officials can minimize many of the risks while optimizing the benefits of such information technologies to local governments and the communities they serve.

TELECOMMUNICATIONS ADOPTION AND USE BY MUNICIPAL GOVERNMENTS

Are city officials and municipal governments adopting and using telecommunications technologies in their own work, and are they promoting the adoption and use of these technologies in economic development efforts within the communities they serve? To answer such questions we conducted two local government surveys in Iowa, including a telephone survey in 1997 to establish baseline data, a print survey in 1998 as follow-up, and personal interviews with five city officials to elaborate the findings.

In the spring of 1997, a telephone survey of city administrators and clerks from 275 of the largest towns and cities in Iowa was conducted to establish a baseline of telecommunications adoption and use by Iowa municipalities. The sample of possible respondents was selected from the *Iowa League of Cities 1996–98 Directory*. Of the 275 city officials contacted, 265 participated, yielding a 96 percent response rate. The survey examined current and planned use of telecommunications technologies by Iowa's municipal governments including current use of e-mail, the Internet, the World Wide Web, municipal Web sites (homepages), and video-conferencing via the Iowa Communications Network (ICN). The survey also examined whether local administrators view telecommunications as important for their own organizations and for economic development activities in which they are involved, as well as what role these administrators play in local telecommunications development. To assess changes in levels of telecommunications adoption and use by local governments in Iowa and to examine shifts in attitudes regarding the potential of telecommunications to enhance government services, we administered a similar survey via questionnaire to city clerks in the summer of 1998. A convenience sample was drawn from attendees at two centrally located, statewide training institutes for city clerks; 180 city clerks participated in the follow-up survey.

Government Adoption and Use of Four Key Telecommunications Technologies

Table 8.1 presents a comparison of adoption and use of four key telecommunications technologies (e-mail, the Internet, the World Wide Web, and a municipal Web site or homepage) by municipal governments from spring 1997 to summer 1998. Twenty-three percent of Iowa's local governments were using e-mail in early 1997, but that proportion climbed to 39 percent by mid-1998. Access to the Internet increased from 55 percent in 1997 to 58 percent by 1998, access to the World Wide Web increased from less than 17 percent to more than 45 percent, and the percentage of local governments with municipal Web sites (homepages) climbed from less than 17 percent to more than 31 percent. These shifts reflect a 14

Table 8.1
Comparison of Adoption and Use of Telecommunications Technologies by Iowa
Municipal Government Officials, Spring of 1997 and Summer of 1998

	1997	1998
E-mail	23%	39%
Internet	55	58
World Wide Web	17	45
Municipal Web site (homepage)	17	31

Note: 1997 survey questions regarding adoption and use of the ICN (for video-conferencing, meetings, and training) were not asked in the 1998 survey.

Figure 8.1
Municipal Government Officials' Ratings of Local Telecommunications Services in 1997
and 1998, and Projected for the Year 2000

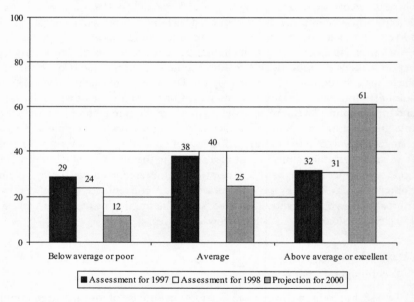

percent increase in e-mail use, a 3 percent increase in Internet access, a 28 percent increase in World Wide Web access, and a 14 percent increase in municipal Web sites—all within an 18-month period.

Quality Improvements in Telecommunications Available to Communities

City officials rated the quality of telecommunications infrastructure and services available to city government and the community generally. Figure 8.1 indicates a general improvement in quality during the 18-month interval between the two surveys. While in 1997, 29 percent reported that services available to their community were poor or below average, that number dropped to 24 percent by 1998. While 38 percent rated services average in 1997, 40 percent rated them average in 1998. And while 32 percent felt services were above average or excellent in 1997, that proportion slipped to 31 percent by 1998. This suggests general improvement in service quality, but the trajectory is lower that that suggested by optimistic projections of service quality for year 2000, when only 12 percent of respondents to the 1997 survey predicted service would be poor or below average and 61 percent predicted service would be above average or excellent. A substantial jump in service quality within the subsequent 18 months to year 2000 would be required to achieve the levels predicted by city officials in 1997.

Increased Government Utilization and Promotion of Telecommunications

As Table 8.2 demonstrates, the majority (59%) of 1997 survey respondents rated local government utilization of telecommunications no higher than 2 on a 5-point scale. Involvement of local governments in promoting the use of telecommunications in the community was also minimal; again the majority (76%) rated their

Table 8.2
Extent of Local Government Utilization and Promotion of Telecommunications

	1997	1998
Extent of Telecommunications Utilization by Local Government		
1 Not utilized at all	25%	23%
2	34	17
3 Somewhat utilized	25	37
4	13	13
5 Greatly utilized	4	5
Unsure		5
Mean score	2.37	2.90
Level of Involvement of Local Government in Promotion of Telecommunications within the Larger Community		
1 Not involved in any promotion	44%	29%
2	33	36
3 Somewhat involved in promotion	12	14
4	8	8
5 Greatly involved in promotion	4	7
Unsure		6
Mean score	1.96	2.66

municipality no higher than 2 on a 5-point scale. By mid-1998, 40 percent still rated their local government's utilization of telecommunications below average (a 19-percentage-point improvement) and 65 percent still rated their promotion within the community below average (an improvement of 11 percentage points). The 1997 mean scores of 2.37 for utilization and 1.96 for promotion climbed in 1998 to 2.90 for utilization and 2.66 for promotion, a marked improvement. But only 4 and 7 percent, respectively, reported their municipalities greatly utilize telecommunications or promote them within their communities.

Telecommunications Benefits for Municipalities

We asked city officials to indicate the relative importance of telecommunications to economic development and quality of life activities in their cities. Figure 8.2 presents a comparison of the relative importance city officials assigned telecommunications for a number of these activities from 1997 to 1998. Thirteen activities were included in 1997, but we truncated these to seven in 1998 while trying to retain the spirit of the original list. Quality of life indicators, such as education, library services, and medical services, consistently rated higher than any economic development indicators. All three quality of life indicators initially rated higher than 80 percent. Although respondents rated telecommunications important for economic development generally, there was more disparity between economic development ratings, with a range from 55 to 79 percent. Business recruitment was rated highly by 79 percent of city officials, while far fewer (58% and 55%) deemed telecommunications important to providing good paying jobs or quality working conditions for community residents.

Consequences for Government Operations and Economic Development

In Chapter 4, Van Wart, Rahm, and Sanders illustrated the proportion of city administrators reporting telecommunications as "important" or "very important" for realizing a number of telecommunications benefits for local governments, including financial savings, time savings, improved customer service, improved external and internal communication, and improved accuracy (see Figure 4.1). Figure 8.3 compares the ratings city officials gave in 1997 and 1998.

Although most ratings were initially high and remained so, with as many as 80 to 90 percent of respondents realizing such benefits, there were notable declines in several areas. Fewer city officials identified time savings, improved customer service, and financial savings as benefits realized through the application of telecommunications technologies. At first blush we might expect appreciation of these benefits to increase as growing numbers of municipalities adopt and use these technologies. That we see a decline instead is indication that municipalities may be experiencing fewer benefits during their time of transition to telecommunications use. The initial outlay of capital to purchase and install equipment, the increased costs, time lost to training, productivity losses, and discomfort as city staff learn and adjust to these new technologies—in short, the expenses asso-

Figure 8.2
Importance of Telecommunications for Economic Development and Quality of Life in
Iowa Municipalities

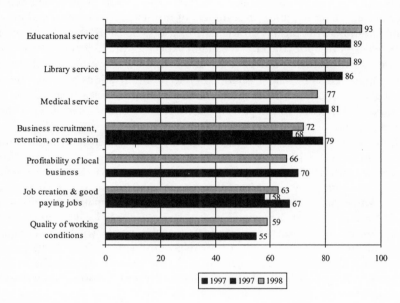

Figure 8.3
Proportion of Government Officials Rating Seven Benefits of Telecommunications
Technologies as "Very Important" or "Important" for Municipalities

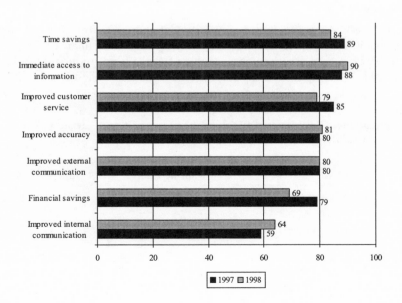

ciated with the learning curve—may squelch some enthusiasm. Likewise, the opportunity costs required for making a commitment to telecommunications and the resultant changes in operations that must be made to accommodate new telecommunications technologies also contribute to a slump in perceived benefit. Considering the quick jump in numbers of municipalities coming online, the slight dip in benefits ratings on the part of users (no more than 3% to 5%) is probably not cause for concern. We might expect to find higher dissatisfaction, frustration, or confusion during times of transition, but overall benefits ratings remain high. A decline in appreciation of telecommunications benefits, coupled with a plateau or decline in adoption and use rates, would clearly signal dissatisfaction and resistance on the part of municipalities, but neither of these situations is apparent here.

Optimism regarding the likely impact of advanced telecommunications for local governments and the communities they serve (reflected in Table 8.3) reinforces this argument. In 1997, 57 percent of local government officials believed recent advances in telecommunications would serve to *strengthen local governments* by providing them access to valuable resources found on the information superhighway, but 31 percent worried they would *widen the gap between local governments* because those municipalities not utilizing telecommunications will have a comparative disadvantage. In contrast, 68 percent of government officials in 1998 thought telecommunications was strengthening local governments vis-à-vis their neighbors and 18 percent believed the gap in the quality and provision of services between rural and urban municipalities was widening.

FACTORS INFLUENCING ADOPTION AND USE

Community Size

Our base-line survey sampled administrators and clerks from the 275 largest municipalities in Iowa and our follow-up survey of city clerks, finance officers, and administrators used a convenience sample of clerks from municipalities larger than 500 in population. As a result of this sampling, the smallest communities in

Table 8.3
The Perceptions of Government Officials on the Impact of Recent Advances in Telecommunications on Communities

	Spring 1997	Summer 1998
Strengthen local economies	57%	67%
Widen the gap	31	18
Little or no impact	11	1
Not sure	2	9

Iowa are underrepresented in these surveys; indeed, communities under 500 in population (which represent 54% of Iowa's municipalities!) may not be represented at all. Nevertheless, because community population size frequently determines the level of physical, financial, human, and social resources available for government services and economic development, we focused particular attention in this analysis on community size as a possible factor influencing telecommunications adoption and use by municipal governments.

Several cautions are in order before proceeding with an analysis of the influence of community size. In isolating community population size as a determinant of telecommunications adoption and use, we are, in effect, using size as a proxy for other community characteristics, such as: infrastructure availability and access to telecommunications appliances (physical capital); tax base, access to grants-in-aid, credit worthiness, and ability to issue bonds (financial capital); technical expertise and computer literacy among citizenry and staff (human capital); and strength of social networks, quality of relationship between elected officials, constituents, municipal employees, and residents, and ability to mobilize community action (social capital). Although capital endowment is usually positively correlated with community size—that is, the larger the community the more capital and the smaller the community the less capital—the relationship between size and capital can also be an inverse one. For example, many metropolitan cities have teetered on the edge of bankruptcy while their smaller suburban neighbors reaped the bounty of proximity to an urban center, population density, high traffic, transportation facilities, and cultural amenities in which they made little capital investment. Even very small, remote communities may be well capitalized if they have exploitable natural resources that provide a handsome return and/or a history of investment in community capacity that can be mobilized for economic development and gain. Recognizing the complexity of community size in these dynamics, we initiated a preliminary exploration of its relationship to telecommunications adoption and use by municipal governments in Iowa. How does size affect the level and rate of adoption and use of telecommunications technologies by municipalities? What other factors influence a community's likelihood to move from being nonusers to light users to heavy users of telecommunications technologies? Table 8.4 presents the breakdown of communities represented in both the 1997 and 1998 surveys by population size.

Figure 8.4 illustrates the disparity between smaller and larger municipalities in their adoption and use of the four key telecommunications technologies and suggests that community population size is indeed a factor influencing government adoption and use. Overall adoption for these technologies increased as community size increased. Larger communities not only adopted more technologies, they adopted them faster than smaller communities. Smaller communities experienced pronounced increases in adoption levels, but they have farther to go to achieve parity with their larger counterparts.

Table 8.4
Number and Proportion of Iowa Communities by Population Size Represented in
Sample of Municipal Government Officials Surveyed in Spring of 1997 and Summer of
1998

	Spring 1997		Summer 1998	
Community Size	Number	Proportion	Number	Proportion
500 to 3,000	166	63%	127	71%
3,000 to 8,000	59	22	32	18
8,001 to 20,000	19	7	11	6
20,000 +	21	8	10	6
Total	265	100	180	100

We asked respondents at what rate telecommunications technologies were implemented and diffused within their local government and whether this pace
seemed reasonable for their municipality to achieve its technology goals. Again,
community size played a role in their responses. Less than 4 percent reported telecommunications technologies were being implemented and diffused very quickly,
39 percent reported a moderate pace, and 48 percent reported a very slow pace.
Over half said their pace was too slow to achieve their community's telecommunications goals. Of those municipalities moving "too slowly," 89 percent were from
the smallest communities (those between 500 and 3,000 in population).

In the pages that follow we present results from a series of cross-tabulations and
correlations to identify relationships between community size and telecommunications adoption and use by municipal governments. We then examine how community size and utilization rates influence, and are influenced by the attitudes of
government officials toward telecommunications technologies and the physical,
financial, human, and social capital endowments of communities.

In measuring the levels of telecommunications use in rural communities, Johnson, Allen, Olsen, and Leistritz (n.d.) developed a typology of communities based
on their adoption and use of technology. *Traditional* communities use telecommunications technologies very little or not at all, *transitional* communities use
technologies for several significant applications, and *innovative* communities use
technologies heavily across multiple applications. We employed this typology to
characterize the various telecommunication technology adoption levels of local
governments represented in our research. For our purposes, traditional communities are those municipalities that use none or only 1 of the 4 key technologies
(e-mail, Internet, World Wide Web, and municipal Web site). Transitional communities use 2 or 3 of these technologies, and innovative communities use all 4.
Johnson et al. explain that traditional communities—those that use little or no advanced telecommunications technologies—are "probably more typical of the majority of small rural communities across the nation's heartland" (p. 2). Figure 8.5

Figure 8.4
Percentage of Municipalities by Population Size with Access to and Use of Advanced
Telecommunications Technologies in 1997 and 1998

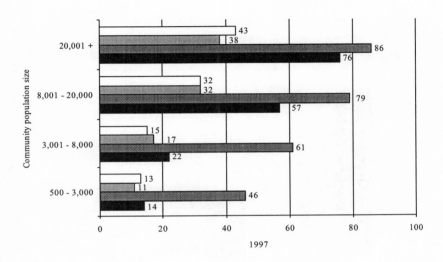

1997

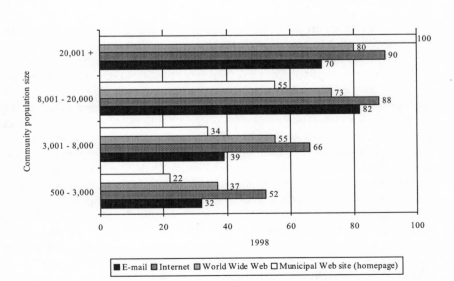

1998

■ E-mail ▨ Internet ▨ World Wide Web □ Municipal Web site (homepage)

Figure 8.5
Change in Levels of Adoption and Use of Telecommunications Technologies by
Municipal Governments between 1997 and 1998 Measured by Community Adoption
Type

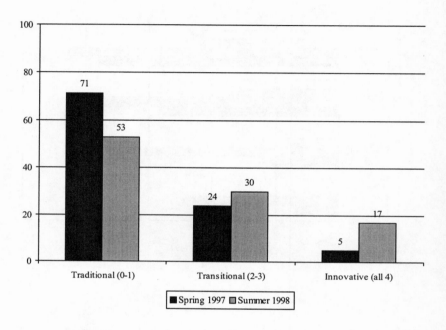

charts the change in proportion of traditional, transitional, and innovative communities within this 18 month interval.

A total of 923 municipalities are recognized by the Iowa League of Cities; 496
(54%) of these have populations less than 500; 385; (42%) have populations between 500 and 8,000; and 41 (4%) have populations exceeding 8,000. Because
small towns comprise the vast majority of Iowa municipalities, we collapsed the
four population categories presented in Table 8.4 into two—small communities
and large communities. As defined, small communities have between 500 and
8,000 residents and large communities have 8,001 residents or more (the largest
municipality in Iowa has under 200,000 residents). Table 8.5 presents the reconfig-
ured population categories for communities whose government officials responded
to the 1998 survey used for this part of the analysis. The convenience sample for the
1998 survey was comprised of 88 percent small communities and 12 percent large
communities, so despite the fact that the number of small communities in Iowa
dwarfs the number of large communities (96% to 4%), large communities are still
disproportionately represented in our data. Table 8.6 presents a simple cross-
tabulation between telecommunications adoption levels and community size and
demonstrates the positive significant correlation between these two variables.

Table 8.5
Number and Proportion of Iowa Communities by Population Size Represented in
Sample of Municipal Government Officials Surveyed in Summer 1998

	Summer 1998	
Community Size	Number	Proportion
500 to 8,000	159	88%
8,001+ to 20,000	21	12
Total	180	100

Table 8.6
Adoption Type of Iowa Municipalities by Population Size

	Small Communities (500–8,000 population)	Large Communities (8,001+)
Traditional (none or only 1 innovation adopted)	58%	14%
Transitional (2 or 3 innovations adopted)	30	33
Innovative (all 4 innovations adopted)	12	53
Total	100	100

Note: Spearman's rho = .34; significant at the .001 level.

Local Government Utilization of Telecommunications

We asked respondents to assess the extent of their governments' use of telecommunications technologies. Few municipal respondents (5%) reported that telecommunications were "greatly utilized" by their city governments; 90 percent of those who reported poor utilization came from the smallest communities (500–3,000 population). While general increases in adoption levels demonstrate that small communities are increasingly adopting, they still have a long way to go to reach parity. Larger communities access a higher number of technologies and achieve higher rates of utilization, and thus, we might assume, reap more benefits from telecommunications. Table 8.7 presents the cross-tabulation and correlations between government adoption levels and utilization rates. The relationship is positive and statistically significant.

Officials' Attitudes toward Telecommunications

As previously noted, we asked a number of questions to measure attitudes of government officials toward telecommunications technologies, including: (a) an

Table 8.7
Adoption Type of Iowa Municipalities by Telecommunications Utilization Rates

	Low	High
Traditional	57%	25%
Transitional	31	34
Innovative	12	41
Total	100	100

Note: Spearman's rho = .30; significant at the .001 level. "Low" utilization comprised of "not utilized" through "somewhat utilized" (1–3 on a five-point scale); "high" utilization comprised "?" though "greatly utilized" (4–5).

assessment of the importance of telecommunications to enhance various government services; (b) their general ratings of the importance given telecommunications for community economic development and quality of life; and (c) their opinion of whether the benefits of telecommunications outweigh the costs for municipal governments.

Importance of Telecommunications for Various Government Services. Respondents rated the importance of telecommunications technologies to enhance 16 different municipal services. School uses ranked highest; 82 percent of municipal respondents maintained computers and telecommunications are "important" or "very important" for education. Similarly, telecommunications were deemed important for library services (77%), police and law enforcement (70%), fire and emergency services (60%), local economic development (60%), and record management (56%) (Figure 8.6). On the other end of the continuum were municipal services such as transportation (23%), garbage collection (22%), fee and fine assessment (20%), parks and recreation (19%), building code enforcement (19%), and snow removal (13%), for which fewer than a quarter of respondents felt telecommunications important. Those municipal functions deemed most amenable to telecommunications, that is, education, library, police and law enforcement, fire and emergency services, and so on, are vertically integrated with federal and state agencies. Those functions deemed less amenable, such as garbage collection, parks and recreation, and snow removal, are locally provided and independent of most external authority.

General Ratings. Fifty-nine percent of respondents indicated telecommunications were "important" or "very important" for providing information to local governments, 64 percent for retrieving information within local governments, and 55 percent for downloading information needed by local governments. In reverse, however, these figures imply that 36 to 45 percent of respondents did not believe telecommunications to be important for such local government functions. These various assessments reflect differences in whether respondents deemed telecommunications important to municipalities generally, or to their own municipality specifically. We asked a simple dichotomous question (yes/no) whether advanced telecommunications were important for government operations within their own

Figure 8.6
Percentage of Government Officials Rating Telecommunications Technologies
"Important" or "Very Important" for Municipal Functions and Government Operations

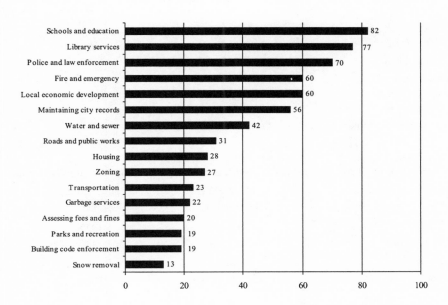

municipality. The majority (80%) answered "yes," but almost all of the 6 percent who answered "no" and 14 percent who answered "unsure" represented the smaller communities. Table 8.8 presents the cross-tabulation and demonstrates the positive correlation between government adoption levels and importance of telecommunications for use in government.

Benefit /Cost Assessment. When asked whether the benefits of telecommunications outweighed the costs, well over half of city officials (59%) said "yes" while just 4 percent said "no." A sizeable percentage (37%) expressed uncertainty, however, nearly all of whom (92%) represented the smallest communities. This level of uncertainty raises the possibility that for the smallest rural communities the investments required for telecommunications development may not pay off in economic or social benefits. Table 8.9 presents the cross-tabulations and positive correlation between levels of municipal adoption and use and belief in their likely payoff. Those adopting more technologies believed the benefits outweighed the costs.

Community Capital

Physical Capital. We asked city clerks what impact their community's infrastructure had on the quantity and quality of telecommunications technologies available locally. As Figure 8.7 illustrates, 17 percent reported their infrastructure

Table 8.8
Government Adoption Levels by Whether Telecommunications Technologies Are
Deemed Important for Local Government Operations

	No	Yes
Traditional	69%	49%
Transitional	28	30
Innovative	3	21
Total	100	100

Note: Spearman's rho = .20; significant at the .001 level.

Table 8.9
Perceptions of Likely Payoff: Whether Benefits of Technology Adoption Outweigh Costs

	No/Unsure	Yes
Traditional	68%	42%
Transitional	21	37
Innovative	12	21
Total	100	100

Note: Spearman's rho = .24, significant at the .001 level.

situation had a "big impact" on quantity and quality, 27 percent reported a "moderate impact," 20 percent said it had "little impact" and 10 percent said it had "no impact" at all. Over one-fourth (26%) of respondents were unsure of the impact infrastructure had on the telecommunications endowment of their municipality.

About 37 percent of those reporting little or no impact were from the smaller municipalities, compared to 63 percent of those reporting moderate or big impact. Table 8.10 presents the cross-tabulation and correlation between municipal adoption and the impact of infrastructure. Despite the apparent influence of community size on infrastructure availability, the results are inconclusive, with a Spearman's rho of only .09, which is not statistically significant.

Financial Capital. Figure 8.8 presents the results of the impact city financial resources had on municipal adoption according to city officials. Sixty-four percent reported that financial resources have a "large impact" and 23 percent believed finances had only a "small impact" on technological innovation and adoption. In contrast, a mere 2 percent indicated that financial resources had "no impact," while 7 percent expressed uncertainty. The smallest communities were disproportionately represented in the latter two categories, "no impact" and "unsure," which may suggest that factors other than financial retard telecommunications development in rural Iowa. Table 8.11 presents the cross-tabulation and correlation between municipal adoption and perceived impact of financial resources. Again,

Figure 8.7
Impact of Existing Infrastructure on Telecommunications Technology Adoption by
Iowa Municipalities

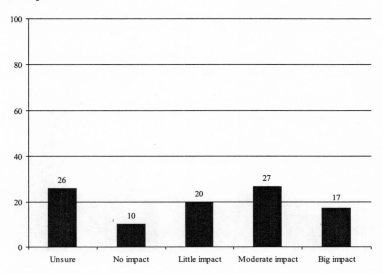

Table 8.10
Perceived Impact of the Community's Infrastructure on Technology Adoption across
Adoption Types

	Little or No Impact	Moderate to Big Impact
Traditional	54%	48%
Transitional	33	29
Innovative	13	23
Total	100	100

Note: Spearman's rho = .09; significant at the .300 level.

findings are inconclusive, with a Spearman's rho of just .05, which is not statistically significant.

Human Capital. Government officials of larger municipalities reported that human resources played a significant role in their ability to adopt and utilize telecommunications technologies, noting that existing city employees lacked the technical
skills required to implement and operate advanced telecommunications technologies. Many communities faced a dilemma—the need for training employees when
they lacked the technical personnel to conduct such training. They perceived a severe shortage of qualified personnel knowledgeable of telecommunications technologies. Complicating this was an applicant pool deficient in technical skills, for
those skilled in telecommunications opted to pursue more lucrative private sector

Figure 8.8
Impact of Financial and Human Resources on Telecommunications Technology
Adoption by Iowa Municipalities

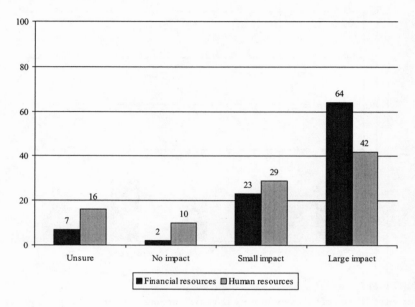

Table 8.11
Perceived Impact of the Availability of Financial Resources on Technology Adoption
across Adoption Types

	Little Impact	Large Impact
Traditional	46%	53%
Transitional	37	30
Innovative	17	17
Total	100	100

Note: Spearman's rho = .05; significant at the .300 level. "Little impact" comprised of "no impact" and "small impact" responses; "large impact" comprised of "large impact" responses.

employment. Figure 8.8 also presents the results of the impact the city's human resources had on municipal adoption. Forty-two percent reported that availability of human resources had a "large impact" and 29 percent believed human resources had only a "small impact" on technological innovation and adoption. In contrast, 10 percent indicated these human resources had "no impact" and nearly 16 percent expressed uncertainty. As with financial resources, the smallest communities were disproportionately represented in the latter two categories, "no impact" and "unsure," but both categories were larger, suggesting that municipal officials attribute even less influence to human resources than financial resources on technol-

Table 8.12
Perceived Impact of the Availability of Human Resources on Technology Adoption across Adoption Types

	Little Impact	Large Impact
Traditional	53%	49%
Transitional	27	33
Innovative	20	18
Total	100	100

Note: Spearman's rho = .03; significant at the .300 level. "Little impact" comprised of "no impact" and "small impact" responses; "large impact" comprised of "large impact" responses.

ogy adoption in their communities. Clearly, these findings intimate that something other than infrastructure, financial, and human resources retards telecommunications development in rural Iowa. Table 8.12 presents results of the correlations between municipal adoption and perceived impact of human resources. Still, findings are inconclusive, with a Spearman's rho of just .03, which is not statistically significant.

Social Capital. To assess the impact of social capital on local government adoption and use of telecommunications, we applied city officials' perceptions of community support for telecommunications. While we surveyed city officials regarding their attitudes toward telecommunications technologies, we did not survey community residents regarding theirs. Government officials reported, however, that community attitudes determined whether resources, especially financial resources, were allocated for telecommunications technologies. Lack of citizen support dissuaded city councils from investing in telecommunications development. Government officials suggested that citizens were more likely to support highly visible projects, while the invisibility of most infrastructure and the opacity of immediate telecommunications benefits earned them little community support; citizens prefer to see where their tax dollars are spent.

Figure 8.9 presents the level of support communities express to their local governments for telecommunications development. Only 10 percent reported strong support from their communities, while 90 percent were uncertain, reported weak, or no support. Table 8.13 gives results of the correlation between municipal adoption levels and community support. The Spearman's rho of .27 is significant at the .001 level. Regardless of the high level of uncertainty, community support is positively correlated with adoption levels.

There may be a reciprocal influence between community support and municipal leadership's attitudes toward telecommunications, in that advocacy for telecommunications by a government official can mobilize community support for development, adoption, and use, while conversely, citizen demand for telecommunications capacity might elevate the importance and value of these technologies in the eyes of an administrator or city clerk. Furthermore, community support can

Figure 8.9
Level of Community Support for Telecommunications Technology Adoption by Iowa Municipalities

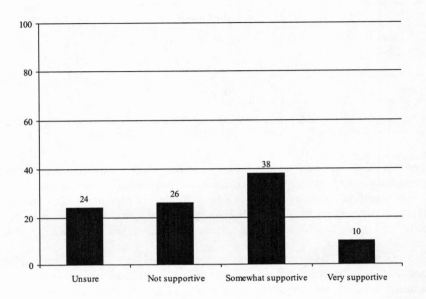

Table 8.13
Iowa Municipality Adoption Type by Strength of Community Support for Telecommunications Development

	Weak	Strong
Traditional	59%	28%
Transitional	29	28
Innovative	12	44
Total	100	100

Note: Spearman's rho = .27; significant at the .001 level. "Weak" comprised of "not supportive at all" and "somewhat supportive" responses; "strong" comprised of "very supportive" responses.

have both subjective and objective measures; support can be expressed through demand for telecommunications infrastructure and service, political support and citizen advocacy, mobilization of community residents and organizations to initiate telecommunications development, and willingness to raise taxes or shift priorities within the city budget to pay for new technologies. Community attitudes toward telecommunications can influence the type, quality, and quantity of technology to be adopted, as well as the speed of adoption and diffusion.

The influence of community support does not negate the importance of leadership, however, nor the importance of championing telecommunications within the community. Local governments report a high level of involvement in promoting telecommunications development within the broader community (Table 8.14). Size did not appear to be much of a deterrent to that level of promotion. One hundred percent of large communities and 86 percent of small communities say they are highly involved in such telecommunications promotion. Only 14 percent report little or no involvement.

In summary, we have identified some key factors influencing telecommunications adoption and use among Iowa municipalities. Table 8.15 provides an overview of these factors and their strength in predicting telecommunications development. These data indicate that community size is the foremost determinant of the level of telecommunications adoption and use by municipalities. Adoption is positively correlated with community size. Increases in technology adoption and use by local governments were visible overall, but more pronounced in the larger population categories. Larger municipalities adopted more telecommunications technologies than did smaller communities. Larger municipalities also use these technologies for economic development more than do smaller communities. Larger municipalities value telecommunications more and deem them more important for conducting city business. And larger municipalities are more likely to accrue and appreciate the benefits of telecommunications for improved government services than are smaller communities.

In addition to community size, the attitudes of municipal officials, as well as the attitudes of community members, play a significant role in determining technology adoption. And there appears to be a reciprocal relationship between leaders attitudes and community attitudes.

Social capital plays a more crucial role in telecommunications adoption and use by municipal governments than do physical, financial, and human resources, perhaps proving that "where there's a will, there's a way." Data provide discrepant evidence regarding the significance of physical, financial, and human capital to telecommunications development. Although the majority of city officials report these capitals are very important to government functions, economic development, and quality of life, they were not statistically significant determinants of mu-

Table 8.14
Local Governments' Involvement in Promoting Telecommunications Development and Use within the Broader Community

	Small	Large
Low level of involvement	14%	0%
High level of involvement	86	100
Total	100	100

Note: "Low level of involvement" comprised of "not involved" through "somewhat involved (1–3 on a 5-point scale); "high level of involvement comprised of "highly involved" (4–5).

Table 8.15
Summary of Variables and the Strength of Their Relationships with Municipal Adoption
of Telecommunications Technologies

Variable	Correlation	Probability
Community size	.34	$p < .001$
Use of technology for government functions	.30	$p < .001$
Community support (social capital)	.27	$p < .001$
Benefit/cost perception	.24	$p < .001$
Importance of use in government	.20	$p < .001$
Infrastructure (physical capital)	.09	*
Financial resources (financial capital)	.05	*
Human resources (human capital)	.03	*

Note: * = Not statistically significant.

nicipal telecommunications adoption. Only social capital—measured by level of
community support—was clearly and positively related to adoption and use.

Does lack of physical, financial, human, and social capital deprive municipalities of the opportunity to benefit from adoption and use of these technologies?
Does the recognition of poor payoff for small communities discourage these municipalities from investing in their potential? The smallest communities represented in our analyses consistently reported less access, less use, lower quality, and
more uncertainty regarding the benefits of telecommunications for their communities. Communities of 500 were our cut-off point for analysis. What then do these
findings regarding the vulnerability of small towns suggest for the 54 percent of
Iowa communities under 500 in population? Can we expect such communities to
benefit from telecommunications commensurate with their investments? There is
a high degree of uncertainty among small municipalities regarding the benefits of
telecommunications for their communities.

Telecommunications are promised to increase overall efficiency and improve
government services. There is, apparently, good reason to question whether or
how these technologies will enhance the efficiency of the smallest rural communities. Officials of such municipalities question whether the benefits will outweigh
the considerable costs of investment in local telecommunications development.
Without substantial collaboration between small communities and assistance
from federal and state programs, and private sector investment in these areas, it is
unlikely that the vast majority of the smallest municipalities will be able to "capitalize" on the telecommunications revolution.

9

TELECOMMUNICATIONS: A COMPLEX PRESCRIPTION FOR RURAL HEALTH CARE PROVIDERS

Patricia C. Hipple and Melody Ramsey

INTRODUCTION

In a 1990 report titled "Health Care in Rural America," the U.S. Congressional Office of Technology Assessment (OTA) identified a number of obstacles that limit access of rural residents to quality health care. These obstacles include: persistent rural poverty, the high costs of health care services, the lack of insurance among many rural residents, the shrinking pool of health care providers in rural locales, the decline in number of health care facilities, especially hospitals, the distance rural patients must travel to providers, and their lack of transportation. In a 1995 follow-up report, the OTA indicated that while "selected improvements" had been made, "the problems rural residents face in accessing health care have not changed substantially"(p. 160). Bashshur (1997) enumerates three general problems confronting the health care sector, problems that differentially impact on the quality of health care available to rural residents. First, health care resources are unequally distributed within and across geographical regions. Second, there is inadequate access to health care, especially by the poor, the isolated, and the confined. Third, the ever escalating health care costs—despite efforts at cost containment—make the price of health care services prohibitive for many. These problems are compounded when, as the Government Accounting Office (GAO) recently reported, rising program expenditures in rural health clinics are not directed toward improving care in isolated areas (1999).

INJECTING RURAL HEALTH CARE WITH TELEMEDICINE

Technological solutions, namely telemedicine, have been promised to alleviate the three most pressing health care problems identified by Bashshur (1997)—unequal distribution, limited access, and high costs. Proponents argue that telemedicine can ameliorate many of the problems that plague rural hospitals and rural health care. Telemedicine networks can achieve the "broader efficiencies and higher quality health care" that linking to an intensive medical infrastructure can afford, according to the OTA (1995, p. 2). Ironically, although telemedicine technologies and practices are touted as a solution to the vulnerability of rural health care systems, rural hospitals are necessary for the adoption and diffusion of telemedicine, for it is unlikely that individual practitioners or clinics have the resources and expertise required to champion the acquisition and use of telecommunications technologies in most rural areas.

This chapter examines the adoption and use of telecommunications technologies for the provision of health care in rural communities and discusses the benefits derived from multihospital collaboration in telemedicine. Using Iowa as a case study, we examine the deployment of telecommunications technologies by health care providers from two vantage points. The first, based on a 1996 survey of rural hospital administrators, focuses attention on computer-mediated communication and interactive video use within hospitals for administrative, clinical, and educational support. The second, based on grant compliance reports and qualitative interviews with physicians and hospital administrators in 1998, focuses on the emergent use of telemedicine by networks of health care professionals in Iowa and the Midwest. Of particular interest is our discussion of the five factors that facilitate telecommunications development for rural health care: visionary leadership, championing of telecommunications, strengthening external linkages, extending vertical and horizontal patterns of exchange, and achieving critical mass. The research reported here is premised on the belief that access to quality health care—a basic human need—is a critical requisite for healthy, viable, and sustainable rural communities, and that telemedicine has an important role to play in individual and community vitality.

RURAL HEALTH CARE NEEDS

Although the health needs of rural residents do not differ substantially from those of their urban and suburban counterparts, there are differences in the scope and scale of rural health problems. According to the OTA, rural residents, in general, are more likely to be engaged in hazardous occupations, to suffer more work-related injuries, to experience chronic illnesses that result from higher levels of exposure to toxic chemicals, and to experience disabilities that interfere with productivity. Rural residents exercise less, tend more toward obesity, and use preventative screening services less regularly. Financial constraints can place the health of rural residents at greater risk. On average, rural residents suffer higher rates of poverty, have significantly lower incomes, are prone to higher rates of unemployment, and are less likely to have health insurance as a benefit of employment. A greater percentage of the rural population under 65 years of age is uninsured and less likely to qualify for Medicaid. At the same time, a disproportionate number of rural residents are elderly and/or poor, leaving a higher percentage dependent on public health insurance such as Medicare and Medicaid. It has been estimated that up to 40 percent of physicians' patient base in rural areas consists of Medicare and Medicaid patients, even though they represent less than 27 percent of the U.S. population (OTA, 1995, p. 177).

The shrinking pool of qualified health care providers in rural communities and the threatened extinction of many rural hospitals exacerbates the problems of distribution, access, and cost. Not only has the number of health care professionals declined in many rural communities, but a large percentage of general practitioners are within 5 years of retirement (OTA, 1995). The National Rural Electric Cooperative (1994) reports that rural residents continue to be more than *twice as likely* than the nation as a whole to face shortages of primary care physicians (in Bashshur, 1997, p. 15 [italics added]). Texas, for example, had 24 rural counties without a single primary care physician in 1995 and 21 counties that had only one; 107 locales were designated health professional shortage areas (HPSAs). Of the state's 254 counties, 55 had no hospital (Zetzman, 1995, p. 3). Iowa's situation is less critical, but still severe. Although no county in Iowa lacks at least one primary care physician, 16 counties have only two or three. HPSAs have been designated in 23 whole and 25 partial counties, and 9 whole and 52 partial counties have been identified as medically underserved areas (MUAs) by the state. Of Iowa's 99 counties, 7 have no hospital (Iowa Department of Public Health, 1999a).

THE IMPORTANCE OF RURAL HOSPITALS TO HEALTH AND COMMUNITY VIABILITY

In addition to fulfilling an obvious and essential medical role, community hospitals play a key economic role in rural locales. Hospitals provide jobs and extend other economic benefits to the community and its residents. The presence of a local hospital is a frequent criterion in decisions to locate new businesses in an area (Bergson, 1994). Vogel (1989) maintains that hospitals are "complex social institutions" that "mobilize some of society's most advanced intellectual and technological resources, and command substantial and growing proportions of the common wealth" (p. 243). As a result, hospitals build human capacity within and create opportunity for the communities in which they locate. Kiel (1993) concludes that hospitals "constitute a major economic, emotional, and symbolic community element" (pp. 625–626).

Belying their importance, rural hospitals are confronted by daunting challenges. Many rural communities have felt the repercussions of diminished economic opportunity for rural residents, the loss of family farms, the loss of jobs dependent on natural resource extraction, outmigration of young families and single youth, and marked shifts in their demographic makeup. All these have had significant consequences for rural hospitals. According to a 1988 report by the National Governor's Association, three events converged in the 1980s to intensify the financial stresses felt by rural hospitals. First, the weakened rural economy led to an increased number of uninsured patients and more difficulty generating revenue through taxation or donations. Second, Medicare's payment schedule failed to reimburse rural hospitals for the full costs incurred in treating eligible patients. Third, a drop in admission and occupancy rates made it difficult for hospitals to cover fixed costs (Mullner, Rydman, Whiteis, & Rich, 1989). Spiraling costs and unanticipated reductions in the patient-day base as a result of cost-containment efforts also contributed to the loss of patient-generated revenue. These reduced the slack financial resources hospitals need to repair deteriorating infrastructures, to purchase new medical technology that would enhance their competitiveness, and to attract and retain high quality health care personnel. Rural hospitals have also been witness to the loss of some of their most financially secure patients. Rural patients who can afford high quality health care frequently seek it in regional or urban health care facilities where a broader array of services is available (OTA, 1995, p. 12). The resulting drop in rural admissions has forced closure of many small town hospitals—nearly one in five (17%) rural hospitals closed between 1977 and 1992—with a concomitant drop in employment opportunities and negative economic repercussions for rural communities. In 1986 the Governmental Research Corporation (in Kusserow, 1989) predicted that more than 40 percent of all hospitals would close or be converted by the year 2000. A disproportionate number of these losses will be rural hospitals. Rural hospitals that have not fallen victim to closure, merger, or acquisition have used a number of strategies to respond to these economic stresses, among them, interorganizational linkages with other health care providers to expand their reach, and investment in telecommunications technologies to enhance their productivity.

Interorganizational linkages may be structural, administrative, institutional, or resource links. Structural links include voluntary partnerships in health care systems, absorption by holding companies, or forming subsidiary relationships. Administrative links usually take the form of management contracts. Institutional links, such as alliances in trade associations and lobbying groups, are less formal and usually noncontractual. Resource links include networking with other organizations and health care entities for mutual exchange of personnel, training, funds, facilities, and information (Goes & Park, 1997). Advocates of interorganizational linkages and multihospital cooperation cite a number of advantages. Among these are service sharing and the elimination of duplicate services; cost reductions and increased efficiencies resulting from staff consolidations; expanded access to physical, financial, and human capital; enhanced purchasing power; improved information systems; expanded market share; and increased political power and ability to influence legislation (Kiel, 1993; Iowa Hospital Association, 1995; Mullner et al., 1989; Gapenski, 1993; Cordes & Straub, 1992). In Iowa, 89 of the state's 117 community hospitals (76%) were part of multihospital systems by 1996.

To stay the erosion of their service base, rural hospitals have also adopted and deployed new telecommunications technologies. Bashshur (1997) explains that these technologies may be as mundane as audio-conferencing via speakerphones and transmission of print material via facsimile machines, or as sophisticated as broad-band interactive video and medical telemetry via land-based fiber optics, T1 links, microwaves, and satellites. While the general benefits of telecommunications technologies have already been articulated (see Chapter 1), proponents of these technologies for the provision of health care maintain that their wise use can ease the performance of difficult job tasks, increase the speed and efficiency of work, and enhance overall productivity. They argue that telecommunications technologies are crucial to rural hospitals because they link them to remote resources that enhance quality and provide their users with a competitive advantage by narrowing the performance gap that is perceived to exist between rural and urban health facilities.

WHAT IS TELEMEDICINE?

Telemedicine has been variously defined.[1] More restrictive definitions are designed to outline allowable services for reimbursement under Medicare and other insurance, identify eligible services to be included within existing grant and loan programs, and clarify services affected by state medical licensure laws. Most of these definitions limit the conceptualization of telemedicine to real-time, interactive video consultations that diagnose or treat patients online. Field (1996) suggests "derivative terms," including teleconsultation, telementoring, telepresence, and telemonitoring, as well as terms related to specific clinical fields such as teleradiology, teledermatology, and telepsychiatry (p. 27). Distance medicine, modem medicine, and "doc-in-a-box" are additional terms that have been used to refer to clinical telemedicine applications. Although the central features of such definitions include electronic transmission of health information over distance, their focus is on clinical applications; they neglect the use of computer-mediated communication and interactive video for health care administration and education. At times our discussion will be limited to clinical applications of telecommunications technologies, but more often we will use a definition of telemedicine similar to that promoted by the American Telemedicine Association (ATA). According to the ATA, telemedicine is "the use of medical information exchanged from one site to another via electronic communications for the health and education of the patient or health care provider and for the purpose of improving patient care" (Linkous, 1999, n.p.). Such a definition incorporates within telemedicine administrative, clinical, and educational applications of telecommunications technologies.

Telemedicine Applications

Health care administrative applications include such things as computerized record management; the electronic input, storage, and transfer of information; electronic billing, ordering, file transfers, and fund transfers; physician credentialing; the transmission of administrative data such as insurance authorizations and regulatory compliance reports; interactive audio and video for teleconferencing; and the use of remote information and resources to support treatment and care of patients. *Clinical diagnostic and patient treatment applications* include things such as two-way interactive video for consultation and diagnostic services; remote emergency-trauma support to assist with field triage, stabilization, and transfer decisions; supervision and consultation for primary care by physician assistants or nurse practitioners when no physician is locally available; and, one-time or continuing provision of specialty care when no specialist is locally available. Clinical applications also include appointment scheduling; medical and surgical follow-up; monitoring of patient status; management of chronic problems and conditions, including medication compliance monitoring; extended diagnostic work-ups; short-term management of self-limited conditions; teleradiology, including diagnosis through remote reading of film, CAT scans, magnetic resonance imaging, and sonographic technology; telepathology; telesurgery; geriatric mental health services and rural telepsychiatry; interactive home-based health care; and, remote health services for prison populations. *Educational applications* for patients and providers include things such as packaged audio and video training modules transmitted over distance, interactive education and training of health care professionals, electronic access to medical libraries and databases, use of computer-mediated communications and interactive video for public health and preventative medicine, and development of "virtual" surgical and diagnostic environments to train medical students.

Benefits of Telemedicine

Field (1996) argues that "telemedicine has the potential to radically reshape health care in both positive and negative ways and to fundamentally alter the personal face-to-face relationships that have been the model for medical care for generations" (p. 3). She cautions, however, that it *has yet to be demonstrated* that "telemedicine could be more practical, affordable, and sustainable than traditional programs, including those intended to sustain or expand rural health care facilities and to attract physicians, nurses, and other personnel to remote areas on a short- or long-term basis" (p. 18, italics added).

The potential benefits of telemedicine for rural communities advanced by its proponents (Bashshur, 1997; Grigsby, 1995; Puskin, 1995; American Telemedicine Association, 1998; Reid, 1999) can be enumerated under several major headings, including: productivity, access, convenience, improved health and safety, efficiency, economy, enhanced appeal of rural medical practice, professional development, and community development.

Productivity. Administrative applications of computer-mediated communication and interactive video can increase speed and efficiency in records management, billing, purchasing, insurance filing, and regulatory compliance, and can reduce expenditures of time and travel through teleconferencing and eased communications.

Access. Telemedicine can extend and improve access to health care services and resources by bringing a wider range of health services to underserved individuals and communities.

Patient Convenience. Telemedicine can reduce the number of medical contacts required of a patient, the need for patient travel and related costs, and the number of work hours lost to seeking health care at a distance. Telemedicine can provide locally the services of a distant specialist and speed speciality service delivery by avoiding long scheduling delays. Telemedicine can allow patients to receive treatment in their home community, or even in their own home, and as a result, minimize the disruption to patients and enhance their ability to live at home longer.

Improved Health and Safety. Rural health care practitioners can communicate with their counterparts at major medical centers through enhanced audio-video links, expanding their access to vital diagnostic and treatment information. By reducing the number of patients' medical contacts while extending specialty services to remote locales, telemedicine can promote greater continuity of care. Quicker response may mean averting the deterioration of a patient's health, while quality improvements in service delivery and patient satisfaction may contribute to faster recovery. In addition, chronic conditions can be better monitored because telemedicine provides a convenient tool for patient education and evaluation.

Health Care Provider Efficiency. Through telemedicine, time and travel costs incurred in administrative meetings can be reduced without sacrificing collegial interactions. Teleconferencing can facilitate networking and collaborative efforts by reducing time and space barriers. According to Zetzman (1995), "rural medical services can be improved by networking rural clinics with each other and with major urban medical services and by the linking of emergency and consult services directly to homes; public health programs will be administratively decentralized closer to their target populations; the quality of research, professional training, and medical practice can be enhanced by improved medical information systems; and medical planning, health policy analysis, and program evaluation can be enhanced by improved methods for gathering, assembling, and interpreting data" (p. 3).

Economy. McCaughan (1995) argues that "telemedicine can be used to reduce the cost of providing health care by ensuring that early intervention, prevention, 'tertiary level' care, and sophisticated case management occurs at the lowest, most cost-effective level of the service enterprise" (p. 13). As a result, local provision of health care to rural residents can be more cost effective. The OTA (1995) estimates the daily cost of a hospital bed in rural areas to be approximately $800, compared with $1,300 at a regional facility (p. 168). Additional savings accrue through reductions in travel expenses, productivity losses, work-time loss, and costs resulting from delayed treatment. According to the OTA, overall costs might be lowered in those cases where patients are diagnosed and treated sooner. (Of course, the OTA admits that earlier diagnosis could encourage expensive treatments and thus increase costs.) The OTA suggests that "costs might also be reduced by staffing [remote] hospitals and clinics with allied health professionals, such as nurse practitioners and physician assistants, who would deliver services where there is no resident physician [because] these providers could be assisted and monitored remotely by physicians using a telecommunications link" (p. 162).

Enhanced Appeal of Rural Practice. Rural medical practice tends to be less lucrative, often fails to provide professional interaction and support, and frequently places high demands on practitioners. Telemedicine, on the other hand, has the potential to generate more income for local providers, increase the scope of practice for primary care providers, reduce professional isolation, and enhance the economic viability of rural practice for current and prospective physicians. Referral channels can be cultivated at the same time rural practitioners keep current with the latest information and developments in their respective fields. According to McCaughan (1995), telemedicine can be used to foster professional interaction with colleagues, solidify professional support structures, and expand referral channels. By making rural practice more appealing, telemedicine can reduce recruitment and retention problems among rural health care providers.

Professional Development. In addition to the professional relationships that are fostered by expanded networks over telecommunications, easy access to reduced-cost training allows rural health care professionals to maintain currency in their fields of practice/speciality. Savings incurred through such cost reductions can be reinvested in additional education, because training resources stretch farther. As a result, Williams and Moore (1994) predict that "health care professionals of all types will be more widely and appropriately trained with advances in distance and on-site instructional media" (p. 3). The quality of training and education can also be enhanced as real-time, interactive preventive and episodic health care training can be provided in local classrooms. And rural practitioners are not the only beneficiaries, of course, because interactive training provides a two-way educational experience that allows urban practitioners to learn from their rural counterparts as well.

Community Development. Because keeping patients in their own communities to receive care can reduce the overall cost of treatment, telemedicine simultaneously increases local hospital revenues while decreasing costs to patients. Telemedicine can ensure local health care jobs; stable hospital payrolls provide important revenues for rural communities. By resisting the drain on health care dollars that may instead go to remote providers, telemedicine allows locally generated health care revenues to be recirculated to enhance the local economy. Sound health care facilities in rural communities elevate the image of the community to locals and outsiders alike, as indicated by Bergson's (1994) findings that the availability and accessibility of hospitals and other health care facilities are important criteria for business site selection by corporate executives. In general, telemedicine-generated prosperity in the community's health care sector provides incentive for further development in the community.

CASE STUDY: BELL ATLANTIC LAUNCHES PHYSICIAN LINKAGE, THE "VIRTUAL MEDICAL PRACTICE"

In 1998, Bell Atlantic launched Physician Linkage, a comprehensive medical informatics program that creates a virtual medical practice for any registered physician. Bell Atlantic has formed strategic alliances with eight organizations—VTel, PictureTel, Pontis EmNet, Image Labs, Spectra Link, Teloquent, Interactive Medicine, and Integration Resources, Inc.—to provide solutions such as videoconferencing, communications systems consulting, picture and image archiving, real time and store-and-forward telemedicine, wireless voice, physician information system access, physician practice management, modality viewing, and electronic patient records consulting. The customizable program allows doctors to admit patients, to update clinical information and medical records, to gain instant access to hospital records and lab results, to access the Internet, and to prescribe medicine. The program also assists physician participation in video-based telemedicine, teleradiology, and distance learning from any PC on any floor of the hospital, their office, or their home. Designed for clinics, labs, pharmacies, insurance providers, and other health care organizations, the network system will allow up to 800 physicians to connect, "knowing the information is encrypted and secure within the Bell Atlantic connection" (Bell Atlantic 1998, pp. 1–4).

CASE STUDY: PHYSICIAN INFORMATION AVAILABLE ONLINE FOR CONSUMERS

While much of the chapter's focus is on the improvement of health care through the electronic transmission of patient medical records, health care improvements can also be realized through the electronic transmission of records on the performance of health care professionals as well. Several states have introduced legislation requiring that information about physicians, including malpractice lawsuits, criminal convictions, and pending disciplinary actions, be available to the public via the Internet. In Arizona, for example, SB 1277 requires both the allopathic and osteopathic boards of medical examiners to post information on a Web site on the health professionals they regulate. Legislation has been introduced in Illinois, New York, and Texas that would provide public access to certain information regarding physicians. Massachusetts is the only state that provides physician data through the Internet, CD-ROM, or in writing (Government Technology, 1999, p. 14).

Risks of Telemedicine

Telemedicine is not without risks, however. The Institute of Medicine (Field, 1996) identified a number of challenges in *evaluating* clinical telemedicine, among them, expense and complexity, rapid change and resultant obsolescence, missing standards that can lead to disappointing applications, lack of ubiquitous and user friendly technologies, distracting technological glitz, and unrealistic requirements for cooperation. We can identify parallel characteristics of telemedicine that discourage rural medical providers and facilities from wholeheartedly embracing computer-mediated communication and interactive video for administrative and educational use, let alone clinical diagnosis, and patient treatment and care. The rapid changes in technology, coupled with their expense and complexity, leave adopters vulnerable to wasted investment in infrastructure, equipment, and training as new generations of equipment and software make existing inventories and related protocols obsolete. The expense involved in adopting unproven, incompatible, and difficult-to-use technologies may yield disappointing administrative and clinical applications until standards are developed and the technology becomes more ubiquitous and user friendly. Exuberant promotion of the "dazzling array" of telecommunications technologies may distract potential adopters from identifying practical, affordable, and sustainable ways to achieve defined quality, access, or cost objectives for rural health care. And the unusual level of cooperation required for interactive telecommunications may discourage individuals and institutions from participation, or escalate the very costs that the technologies have promised to curb.

In addition to the enumerated benefits, telemedicine could have negative consequences for rural hospitals, health care facilities, health care providers, and patients, and these harmful effects could be far-reaching. Here are just a few of the detrimental outcomes that may confront rural health care in the future as a result of telemedicine.

- Many rural health care providers are apprehensive about the dislocation of services, the creation of a subordinated tier of physicians (largely rural), and their displacement by "virtual doctors" and technology. These concerns are currently dismissed as "territoriality" and "turf protection"—ego defects on the part of rural health care providers—despite the very real threats posed by telemedicine technologies and protocols to usurp the autonomy, authority, role, and ultimately, jobs of rural hospitals and health care practitioners.

- Telemedicine could reinforce already existing inequalities between physicians, physician assistants, nurse practitioners, and other allied health care providers. For example, cost containment efforts that see as positive the staffing of rural hospitals and clinics with allied health care professionals rather than physicians, threaten to shift the burden of care to the least remunerated of providers.

- In much the same way, telemedicine has the potential to reinforce a divided health care system, creating a caste of untouchable patients whose only access to medical specialties is electronically mediated.

- Telemedicine has the potential to eclipse the art of health care with blinding fascination for the science and technology aspects of medical gadgetry, leaving rural recipients concerned not about the doctor's bedside manner, but with his or her "bit-side" manner. Such a depersonalized health care system is ultimately deleterious to patient health and well-being.

- The remote transmission and dissemination of patient health care information could make it more difficult for patients to intervene to correct erroneous medical records. Misinformation in one's medical history that has circulated exponentially despite attempts at correction could have deadly consequences for the patients of telemedicine.

- During a period of increased concern regarding patients' rights—the right to know and the right to die—computer-instigated treatment may be seen as an infringement on patient rights. When computer monitoring of a patient's physical condition translates into automated computerized treatment responses, the ability of the patient to remain informed and to provide ongoing consent to treatment is compromised. Remote, rural patients, especially the elderly, will be most vulnerable.

The challenge for rural hospitals and communities will be in their ability to organize effectively to optimize the benefits of telemedicine while minimizing such risks. In Chapter 1 we introduced four kinds of capital—physical, financial, human, and social—on which rural communities are dependent for viability and in which they must invest to reap potential rewards of telecommunications development. Communities that are not "well-capitalized" experience barriers to broader telemedicine adoption and use. They are constrained by inadequate infrastructure, equipment, hardware and software, by logistical problems, by policy and regulatory obstacles, and by performance inhibitors and cultural incompatibilities. The experiences—uncertainties, frustrations, successes, and rewards—of rural health care practitioners in implementing telemedicine in Iowa and the Midwest provide lessons for other rural health care systems and rural communities intent on using telecommunications technologies as means, processes, and ends for health care improvements and community development.

ADOPTION AND USE OF TELECOMMUNICATIONS TECHNOLOGIES BY IOWA HOSPITALS

In 1996, researchers at Iowa State University examined the extent of adoption and use of telecommunications technologies by rural hospitals to explore whether these technologies increased the viability of rural health care facilities. The research focused specifically on computer-mediated communication and interactive video technologies that deliver health care information or services over distance. Administrators of Iowa's rural and rural-referral hospitals participated in a survey designed to identify the types of telecommunications technologies used by hospitals and the extent of each hospital's involvement in the adoption and use of these technologies. The survey was mailed to Iowa's 95 rural and rural-referral hospitals; 69 hospital administrators completed the survey, yielding a 73 percent response rate.[2] Although the questionnaire was sent directly to hospital administrators, they were free to assign the survey to the employee most knowledgeable about the hospital's experience with telecommunications technologies.

Use of Computer-Mediated Communication

All 69 responding hospitals had adopted and were currently using computers, but four did not use these computers to communicate electronically. Of the 65 (95%) hospitals that were using the computers to communicate electronically, 5 percent used the computers exclusively for internal communication within the hospital and 38 percent used the computers exclusively for external communication outside the hospital, usually outside the community. The remainder (57%) of the 65 hospitals used computers for both internal and external electronic communication.

As Figure 9.1 suggests, administrative applications of computer-mediated communication (CMC) are more abundant and appear to be the engine driving adoption in Iowa's rural hospitals. Electronic billing is the most common use, with 54 percent of the hospitals using CMC for electronic billing. Electronic ordering is the second most common activity, with 46 percent of the hospitals placing orders electronically via computer. Forty-one percent use computers for transferring business files, 30 percent for electronic mail, 19 percent for database access, and 18 percent for electronic fund transfers. In comparison, the rates of use for clinical and patient care applications are much lower. Twenty-seven percent of hospitals

Figure 9.1
Percentage Use of Computer-Mediated Communications by Rural and Rural-Referral
Hospitals in Iowa in 1996

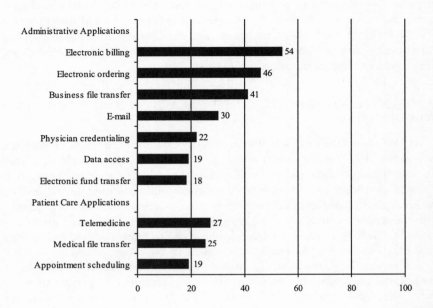

report use of CMC for telemedicine,[3] 25 percent use CMC for transfer of medical files, and 19 percent use CMC for appointment scheduling.

There is a marked difference in CMC use between rural hospitals however, and the larger, regional rural-referral hospitals. As the contrasts in Figure 9.2 vividly demonstrate, use rates for various applications can vary from 3 percent to 76 percent, with higher adoption and use reported by the larger rural-referral hospitals. Only in the case of medical file transfers is that trend reversed; 26 percent of rural hospitals using CMC for this purpose compared with 20 percent of rural-referral hospitals. What remains unclear is to whom the files are being transferred. It may be that greater autonomy of rural-referral hospitals makes them less dependent on external entities to whom smaller rural hospitals may need to transfer files. Alternatively, it is possible that electronic file transfer is more effective for rural hospitals due to their greater distance and isolation relative to rural-referral hospitals, but you would expect to see this benefit reflected in higher use of other electronic applications by rural hospitals.

While these figures indicate the proportion of hospitals using CMC for various applications, they do not tell us how frequently or intensively CMC technologies are used for various types of electronic communication. The fact that 27 percent of rural hospitals report using telemedicine suggests a technological *capability* to do so. However, for most hospitals, *the extent of use* of telemedicine as a proportion of all activities is very low. The American Telemedicine Association (ATA) reports

Figure 9.2
Comparison of Percentage Use of Computer-Mediated Communication between Rural
and Rural-Referral Hospitals in Iowa in 1996

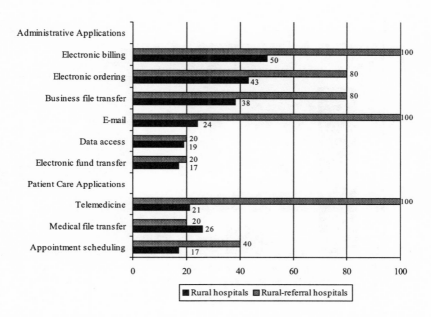

that telemedicine applications represent less than 1 percent of medical encounters
nationally, and pilot projects conducted in Iowa have yielded similarly low numbers of telemedicine consultations via computer-mediated communications or interactive video technologies.

Use of Interactive Video

Interactive video technologies were much less common in 1996 in Iowa's rural
hospitals than CMC technologies. As Figure 9.3 illustrates, only 16 (25%) of the
hospitals reported using any type of interactive video system. Of this number, 9
percent used their own equipment exclusively, 13 percent accessed interactive
video exclusively through another organization such as a community college, and
3 percent (two hospitals) reported use of both their own interactive video system
and that of another organization. The low level of hospital-sponsored usage does
not mean, however, that hospital personnel are not benefiting from interactive
video technologies. In fact, 84 percent of respondents reported that either they or
someone else they knew at the hospital had participated in some kind of class,
course, workshop or teleconference that used Iowa's state-owned fiber optic network, the ICN. The University of Iowa Hospitals, for example, have actively promoted distance education for physicians, dentists, and allied health professionals
via the ICN. Twenty-two percent of interactive video uses were for telemedicine,

Figure 9.3
Proportion of Rural and Rural-Referral Hospitals Able to Access and Use Interactive
Video (IV) Technologies in Iowa in 1996

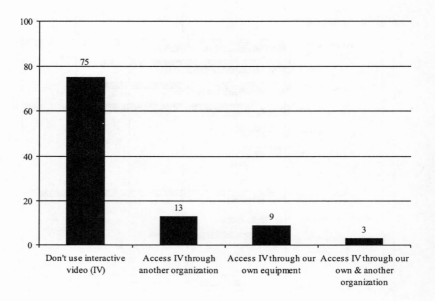

14 percent were for education and training, 10 percent were for informational sessions regarding the ICN itself, 9 percent were about emergencies and traumas, 9 percent covered legislation, and 7 percent were for general meetings or teleconferences.

The most common interactive video uses reported by hospital administrators were for meetings (94%), education (88%), and telemedicine (50%), but it should again be emphasized that these figures represent the proportion of hospitals that *use* telecommunications technologies for such purposes, not the *extent*—frequency or volume—of their use.

Departmental Utilization of Telecommunications Technologies

Respondents in hospitals with telecommunications technologies were asked how many departments use CMC and interactive video technologies. Of the 63 hospitals using CMC, 28 percent reported its use in just a few departments, while only 5 percent reported CMC use in every department of the hospital. Likewise, of the 16 hospitals using interactive video, 31 percent reported interactive video use in just a few departments, but only 6 percent responded that every department within the hospital uses these technologies. Table 9.1 suggests that the departmental spread of CMC technologies has been greater than the spread of interactive video technologies in rural hospitals.

Table 9.1

Proportion of Spread of Computer-Mediated Communication (CMC) and Interactive Video (IV) Technologies across Departments of Rural Hospitals in Iowa in 1996

Departments within Hospitals Using These Technologies	CMC*	IV**
Few	28%	31%
Some	38	56
Many	30	6
All	5	6

Note: *Percentage of the 63 hospitals that reported use of computer-mediated communication technologies; ** Percentage of the 16 hospitals that reported use of interactive video technologies.

Annual Hospital Spending for Telecommunications Technologies

To determine the level of financial commitment to telecommunications technologies, hospital administrators were asked to estimate their annual expenditures for CMC and interactive video technologies, including expenses such as equipment, software, maintenance, and training. Expenditures for CMC by rural hospitals ranged from $5,000 to $300,000, with a median of $27,000. In stark contrast, rural-referral hospitals spend between $150,000 to $1.575 million for CMC, with a median of $575,000. Equally stark is the contrast between expenditures for interactive video. Rural hospitals spent a range of $0 to $70,000, with a median $2,625 expenditure, while the two rural-referral hospitals that invested in interactive video spent $10,000 and $150,000, respectively. Expenditures vary greatly among both rural and rural-referral hospitals, and tend to increase with hospital size. Table 9.2 reports the range of spending for CMC technologies by hospital size. (There were too few cases for meaningful comparisons of interactive video expenditures.)

To identify whether there is a threshold of resources (physical, financial, or social capital) at which hospitals adopt and use telecommunications technologies for administrative and clinical purposes, we ran correlations and analyses of variance on a number of hospital resource indicators[4] and telecommunications use measures. Hospital resources, including number of hospital beds, admissions, and reported revenues, served as the independent variables. The dependent variables consisted of two use scales—the first was constructed from administrative applications including electronic billing, electronic ordering, business file transfer, e-mail, data access, and electronic fund transfer; the second was constructed from patient care applications, including telemedicine consultations, medical file transfer, and appointment scheduling. Hospital admissions has the strongest correlation with the *level of administrative use* of computer-mediated communications; that is, the more patients admitted, the more likely the hospital use of CMC for billing, ordering, transferring files, and so on. The relationship between admissions and administrative use of CMC is statistically significant, with an R^2 of .251. In contrast, number of beds has the strongest correlation with the *level of clinical use* of com-

Table 9.2
Estimated Annual Spending for Computer-Mediated Communications (CMC)

Number of Hospital Beds	Range of Expenditures for CMC
Fewer than 50; (N = 34)	$5,000–$250,000
50–100; (N = 25)	$10,000–$300,000
101–200; (N = 8)	$16,000–$1,000,000
More than 200; (N = 2)	$150,000–$1,575,000

puter-mediated communications. The more beds a hospital has, the more likely the hospital is to be practicing telemedicine, transferring patient files electronically, and scheduling appointments via computer. This relationship, too, is statistically significant, with an R^2 of .243.

A series of cross-tabulations was run to determine the critical point at which adoption and use "take off." The results fail to identify a critical mass of resources or capital needed for adoption of computer-mediated communications or interactive video. Instead of a definitive threshold where we might expect to see a notable jump in the level of adoption and use of CMC or interactive video, a consistent positive relationship is apparent. Table 9.3 presents the results of the cross-tabulations run on two specific variables—electronic billing and telemedicine—illustrating the positive relationship between hospital resources and telecommunications use. As the table indicates, the more resources available to the hospital, the more likely the hospital is to use telecommunications technologies for administrative and clinical applications. Stated another way, the better capitalized a hospital, the more likely it is to adopt and use telecommunications technologies. This suggests that, unlike newspapers and libraries (see Chapters 6 and 7, respectively), critical mass plays no significant role in the adoption and use of telecommunications technologies by rural hospitals. This is perhaps explained by the prevalence of external mandates for telecommunications adoption and use by hospitals. Regardless of their size and resource base, hospitals are adopting telecommunications technologies, in part, to comply with insurance, regulatory, and funding requirements.

Mandated Telecommunications Technologies

Only one (6%) of the 16 hospitals with interactive video technologies reported that use of these technologies was mandated, but almost half (48%) of the hospitals with CMC technologies reported their use was mandated. As Figure 9.4 illustrates, nearly four-fifths of these respondents indicated that insurance providers had mandated this use; 54 percent indicated the mandate came from government agencies, 36 percent from vendors, 11 percent from physicians' credentialing services, and another 11 percent from parent organizations. In many cases, respondents reported that more than one entity mandated use of a technology, so percentages sum to more than 100 percent. Why, the reader might ask, did not *all*

Table 9.3
Proportion of Hospitals Using Computer-Mediated Communications Technologies for
Administrative and Patient Care Applications within the Hospital by Hospital Resources

Hospital Resources	Use for Electronic Billing	Use for Telemedicine
Revenues		
Less than $4.3 million	24%	10%
$4.3 to $8.5 million	50	23
$8.6 million or more	90	50
Beds		
18–36	35	10
38–61	58	25
63–388	68	47
Annual Admissions		
Less than 590	23	9
617–1214	48	19
1,236 or more	95	55

Figure 9.4
Organizations Mandating Computer-Mediated Communications Use by Rural and
Rural-Referral Hospitals (Percent of Hospitals Reporting Mandated Use)

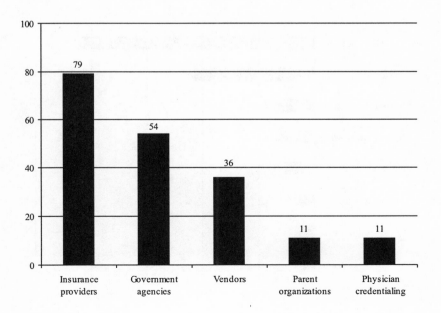

respondents report telecommunications use as mandated when so many respondents reported mandated use? In some cases, hospitals adopted telecommunications technologies in advance of any requirements by outside entities to do so. In other cases, there are differing perceptions of what constitutes a mandate, with strong encouragement or incentives, as well as negative sanctions or disincentives for noncompliance, interpreted equally as a "mandate."

Respondents who reported technologies were mandated said that electronic billing (86%), electronic ordering (43%), and insurance company authorizations (18%) were the most commonly mandated uses (Figure 9.5). Less commonly mandated uses were physicians' credentialing services (14%) and business file transfers (11%). Again, more than one use could be mandated, so percentages sum to more than 100 percent.

Hospital Involvement in the Adoption of Telecommunications Technologies

The extent of mandated use of telecommunications applications would suggest that rural hospitals played little or no role in decisions to adopt and implement these technologies, but our research revealed the opposite. To examine the extent of local hospital involvement in the adoption of telecommunications technologies, the 1996 survey asked hospital administrators who was making decisions about adoption and use of computer-mediated communication and interactive video

Figure 9.5
Purpose of Use of Computer-Mediated Communications by the Thirty Rural and Rural-Referral Hospitals Who Reported Their Use Was Mandated, Iowa, 1996

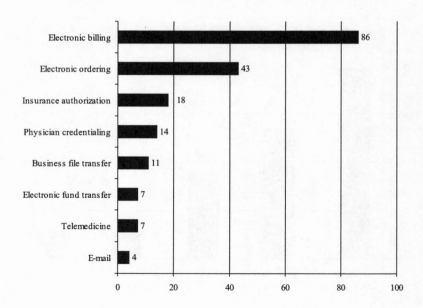

technologies. Vendors and parent organizations were most frequently identified as the external organizations involved in adoption decisions, but schools, government agencies, and the Extension service were also identified. Despite this involvement by external entities, however, Iowa hospital administrators reported *extensive* local hospital involvement in the adoption of CMC technologies and, to a lesser degree, interactive video technologies (see Tables 9.4 and 9.5).

While organizations external to the rural hospital mandated use of telecommunications applications, they did not mandate the types or characteristics of the telecommunications system to be adopted. Those decisions were left to the local hospital, and with them the responsibility and burden of financing, selecting, buying, installing, and maintaining the system and training its users. By mandating use without designating the kind of infrastructure, equipment, and software for that use, outside agencies shifted responsibility for such decisions from themselves to rural hospitals, encumbering rural health care providers with the attendant labor and responsibilities of telecommunications. Vertical organizations (in this case, insurance companies, government, and big hospitals) have

Table 9.4
Level of Hospital Involvement in the Initiation of Computer-Mediated Communications (CMC) and Interactive Video (IV) Technologies in Iowa in 1996

	CMC*	IV**
Local hospital initiated the process	62%	31%
Process was initiated jointly between the local hospital and an outside entity	35	44
Process was initiated outside the local hospital	3	25

Note: * Percentage of the 63 hospitals that reported use of computer-mediated communication technologies; ** Percentage of the 16 hospitals that reported use of interactive video technologies.

Table 9.5
Level of Hospital Involvement in the Implementation of Computer-Mediated Communications and Interactive Video Technologies in Iowa in 1996

	Computer-Mediated Communications		Interactive Video	
	High	Low or No	High	Low or No
Information gathering	86%	15%	38%	62%
Evaluation	88	12	57	44
Decision to implement	91	10	65	36
Installation and use	81	20	44	67
Modification	95	5	66	33

gone electronic and have pushed much of the burden and consequences of tele-communications technologies onto local rural hospitals, institutions least likely to have the resources and expertise needed for successful implementation without costly investment. Chapter 11 describes the experience of the Extension service—one in which a high-quality uniform system was installed that gave every office and almost every staff member immediate and full access to a user friendly and powerful system. In sharp contrast to this, rural hospital administrators describe a "nonsystem" characterized by very uneven technologies and applications. Fragmentation and incompatibility of technologies are common, and mandated changes to the system are not compensated but rather absorbed as the hospitals' cost of doing business.

Perspectives on the Impact of Telecommunications on Rural Communities

Despite the problems experienced by hospitals in the implementation of computer and interactive video systems, respondents see telecommunications technologies as beneficial to their communities and almost 80 percent believe in the potential of these technologies to strengthen local rural economies (Figure 9.6). In contrast, only 8 percent fear that telecommunications technologies will widen the gap between rural communities and their urban counterparts, while another 12 percent perceive little or no impact of these technologies on their communities.

Figure 9.6
Hospital Administrators' Perceptions of Impact of Advances in Telecommunications
Technologies on Their Communities

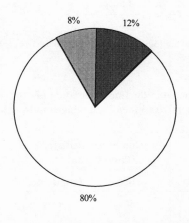

Survey Conclusions

This research confirms the OTA claim that on a general level, nationally, "computers are widely deployed [though] not widely connected" and suggests the OTA is right when it reports that "clinical and administrative health information are rarely commingled." Our survey results also support the OTA (1995) contention that "computers are typically used to organize and administer specific, limited types of health information, [but] they are not linked into an infrastructure that might allow broader efficiencies or higher quality health care" (p. 2).

While most of the literature on telemedicine focuses on efforts to demonstrate the clinical benefits of telecommunications technologies, our concern here is whether or not the rural health care system in the United States can adapt organizationally to these new technologies. But the surveys were directed exclusively to rural and rural-referral hospitals. Our findings on the propulsion telemedicine receives via vertical linkages—agencies outside the rural hospital—suggest that urban hospitals are a significant influence on telecommunications development within many rural hospitals. The case studies that follow provide insight on the telemedicine experiences of three multihospital systems and how they have responded to the organizational challenges of telecommunications technologies.

TELEMEDICINE IN RURAL IOWA

Because Iowa has been at the forefront of telecommunications development through installation of the Iowa Communications Network (see Chapters 1 and 2), health care facilities in the state have provided prime locations to pilot several telemedicine initiatives. Widespread telemedicine links between urban and rural health facilities became possible as a result of 1994 legislation that added hospitals and rural health clinics to the ICN and reiterated public health programs and prison health care delivery systems as appropriate uses. At this writing, the ICN connects all 99 counties in the state and is within 20 miles of every citizen. According to the American Telemedicine Association (Linkous, 1999), there are over 430 video sites connected to the ICN network. Despite state support for telemedicine applications via the ICN, however, fewer than 10 percent of the connected sites are hospitals and only a fraction are rural hospitals.[5] While an additional 480 sites are planned over the next four years, it is unclear how many of these will be situated in rural health care facilities.

Responding to the paucity of evidence to substantiate the purported benefits of telemedicine, the Health Care Financing Administration (HCFA) has launched a number of demonstration projects to evaluate the efficacy, efficiency, and safety of telemedicine and to answer questions regarding the cost of providing health care via telemedicine for Medicare reimbursement. In 1993, HCFA selected Iowa Methodist Medical Center (IMMC) of Des Moines as a national demonstration site for telemedicine reimbursement. Likewise, in 1995, the Midwest Rural Telemedicine Consortium (MRTC) of Des Moines received telemedicine demonstration grants from HCFA and the Office of Rural Health Policy. Although administrative and educational applications are recognized components of telemedicine, the focus of these demonstration projects was on the use of telecommunications technologies—computer-mediated communications and interactive video—for clinical diagnoses, and patient treatment and care. Administrative and educational applications were peripheral to the grants and thus to our discussion.

CASE STUDY: IOWA METHODIST MEDICAL CENTER

Iowa Methodist Medical Center (IMMC) was awarded a $700,000 grant in 1993 to evaluate the efficiency, efficacy, and safety of telemedicine in a broadband environment. The state-owned fiber optic ICN served as the backbone network, connecting IMMC in Des Moines (a tertiary care center), Trinity Regional Hospital in Fort Dodge (a secondary-level regional hospital), and Green County Medical Center (a county primary care hospital). Local connectivity was provided by a commercial telecommunications carrier. The network carried data services, clinical applications, and educational programs between the tertiary, secondary, and primary care facilities.

During the two-year grant period, the three-facility system was used 810 times—for 9 interactive consultations, 9 image transfers (telepathology), 712 store-and-forward transmissions (echo and vascular), and 80 online educational sessions (12 for CME-physicians' continuing medical education, 29 for nurse training, and 39 for outside programs). IMMC reported the following expenses: $40,507 for custom-built telemedicine carts and teleconsult carts; $105,000 per site for ICN electronics and fiber; $15,000 for infrastructure cabling upgrades; and $2,900 per month ($34,800 annually) for local telecommunications line service. The total cost of infrastructure and equipment was in excess of $405,000, which puts the average cost of the 810 uses at $500 per use. Not included in this estimate are personnel expenses incurred during the project, including outlays for technical support.

According to IMMC's final report to HCFA (1995), *connectivity costs* for the state-owned ICN were significantly higher than commercially available services—approximately $105,000 per site as compared to low or no installation costs for leased commercial lines. *Usage fees* for the ICN, on the other hand, were significantly lower than commercially available telecommunications services—$5 per hour per site during the grant period (billed only for actual time used) compared to approximately $400 per hour on a commercial 7.24 leased circuit. As a result, IMMC concluded that long-term payback could be realized through transmission and transport costs. In addition, custom-built teleconsult carts resulted in significant cost savings compared to the purchase of commercial teleconferencing equipment; savings equaled between $10,600 and $18,000 per telemedicine cart. Despite these findings, IMMC concluded that "*the high connectivity costs and high level of technical expertise required will keep most rural facilities from using the ICN for telemedicine purposes*" (1995, pp. 5–6 [emphasis added]).

The pilot project resulted in a number of important findings for telemedicine health care planners and funders. For example, store and forward applications proved more efficient than interactive video consultations because they permitted patient history, clinical images, and audio files to be "packaged" and sent to the consultant in a more comprehensive, integrated manner, and at a slower, and therefore cheaper, transmission rate. These applications eased scheduling of physician and patient time and they rationalized bandwidth allocation for more efficient use.[6] The overall convenience of telemedicine consultation in terms of reduced travel time and costs, resulted in high satisfaction rates among patients and health care providers.

A significant barrier to the diffusion of telemedicine in rural areas is the concern by rural health care professionals that urban providers and facilities will usurp their role. This fear surfaced in the IMMC pilot project as well. IMMC implementers reported that specialists at the secondary-level hospital felt threatened by IMMC's "virtual presence" and expressed concern that they would disrupt referral patterns, with primary care referrals by-passing the secondary care providers and going directly to IMMC. As a result, participation by the secondary-level providers was unenthusiastic, if not obstructionist. IMMC implementers concluded in their final report that "the secondary-level hospital has the unique opportunity to both import and export telemedicine services . . . and . . . the initial network buildout needs to reflect this reality" (1995, p. 13).

IMMC enumerated the following general benefits of telemedicine applications based on their two-year demonstration project. Telemedicine provided

- more rapid intervention in some life-threatening situations,
- a reduction in the lag time between a patient's point of entry into the health care system and the point of actual care,
- a reduction in the time between initial visits and second opinion or specialty consultations and elimination of travel time for rural Iowans,
- an increase in the access to advanced technology and clinical support in decision making for health care professionals in the rural areas,
- a reduction in administrative cost from duplication of records by centrally capturing data and transmitting it through the ICN, and
- increased satisfaction of health care providers by offering increased options in continuing education programs (1995, pp. 15–17).

IMMC predicts that the first telemedicine application to be adopted will generally be educational programs, because existing programs are easily adaptable, and critical questions regarding image quality, control of treatment, licenser or medical staff privileges, reimbursement, security and patient confidentiality generally do not have to be considered. In addition, personnel are receptive to distance learning due to scheduling and travel budget constraints.

CASE STUDY: THE MIDWEST RURAL TELEMEDICINE CONSORTIUM

Established in 1993 with organizational hubs in Des Moines at Mercy Hospital Medical Center and in Mason City at North Iowa Mercy Health Center, the Midwest Rural Telemedicine Consortium (MRTC) was the second health care provider in Iowa to receive an HCFA research and demonstration grant for rural telemedicine. HCFA provided a grant of $1.2 million and the Office of Rural Health Policy (ORHP) provided an additional $2 million for a multiyear study. Beginning with 14 connected sites throughout Iowa, the consortium, at this writing, includes 25 hospitals, four clinics, three long-term care facilities, and two health centers, for a total of 38 sites in 30 facilities. Despite the extent of coverage, however, fewer than 100 telemedicine consults are held each month, an average of 3–4 per site. Based on an estimated 3,600 uses over the 3-year grant period at a cost of $3.2 million in grants, the average cost per use was nearly $900. The vast majority of these consultations are ineligible for reimbursement under HCFA's guidelines because the most frequent and efficient user of the Consortium's telemedicine network is a psychiatrist who provides geriatric telepsychiatry to elderly residents of residential care and skilled nursing facilities. Frequently, these services involve case management and monitoring and maintenance of drug therapies. But HCFA reimbursement requires interactive, face-to-face consultation with patients, and few of the mental health patients served by this provider—many of them suffering from advanced Alzheimer's Disease—are capable of informed interaction and therefore cannot be considered competent to give consent to treatment, a requisite for Medicare reimbursement of telemedicine services.

In their self-assessment of telemedicine services, the MRTC queried patients to assess patient satisfaction with remote telemedicine consultations. Fifty percent of the patients reported that telemedicine was definitely "not the same" as a face-to-face consultation with a medical specialist, but 99 percent reported that, when such issues as time, travel, loss of work, and removal from local community were factored in, they preferred the telemedicine visit to a face-to-face consult with a specialist at a location distant from their homes.

CASE STUDY: HAYS MEDICAL CENTER, HAYS, KANSAS

Championing Telemedicine in Rural Kansas

Because Iowa's telemedicine demonstration projects were initiated by urban health care centers, it is difficult to assess through these case studies the role of visionary leadership and telecommunications championing by rural health care providers. In contrast, the adoption of telemedicine in Hays, Kansas, was initiated by rural health care providers. Situated in west-central Kansas, this community of 20,000 lies 264 miles west of Kansas City and more than 300 miles east of Denver—far from the array of health care specialists available in large metropolitan areas. The citizens of Hays and their rural counterparts are served by the Hays Medical Center. Dr. Robert Cox, the Hays physician who championed telemedicine in west-central Kansas, initially sought to provide relief for the Center's overworked medical staff through interactive video monitoring and call coverage. However, he soon realized the opportunity to tap expertise not available locally and to identify appropriate technology that might make the jobs of Hays's health providers easier.

Relinquishing his private practice to become a full-time advocate and coordinator for telemedicine technologies and services, Cox enlisted participation from the University of Kansas, the area health education center, Hays Medical Center, Essential Access Community Hospitals/Rural Primary Care Hospital (EACH/RPCH), and individual physicians. Resources for telemedicine came from Kansas University Medical Center, the State of Kansas, Mead-Johnson Nutritionals, Kansas Health Foundation, Rural Utilities Services, the U.S. Department of Commerce, Hays Medical Center, and EACH/RPCH.

According to Cox (1997), three telemedicine applications met with limited success in Hays: emergency department support, teleradiology, and interactive home health care. *Emergency department support* via telemedicine relieved pressure on overworked physicians, reduced errors due to fatigue from prolonged work hours, and improved the overall work environment. As a result, it improved Hays's ability to recruit and retain physicians and other professionals. *Teleradiology services* gave Hays physicians the ability to obtain immediate consultation on radiology interpretation services, but interstate licensing restrictions and liability concerns hamper broader utilization of these services. *Interactive home health care*, the interactive monitoring of fragile, home-bound, and remote patients via cable television connections, has resulted in reduced emergency room visits and hospitalizations, but its deployment has been limited by the lack of two-way cable access into many homes. Cox reports a marked improvement in patient self-esteem and sense of independence as a result of home-based telemedicine, arguing that when it is available, patients are able to live at home longer and delay moves to extended care facilities. Cox foresees the efficacy of interactive patient monitoring for nursing homes and other extended care facilities, and envisions expanding telemedicine services to rehabilitation, speech pathology, and oncology patients as well.

Despite the success of telemedicine applications in Hays, Kansas, serious impediments to its wider utilization remain. Cox identifies three types of barriers: economic, regulatory, and knowledge-based. Obtaining affordable, adequate bandwidth at remote sites continues to be a challenge. Restrictions on interstate licensing and credentialing, coupled with concerns about medical liability also continue to limit providers' enthusiasm for telemedicine. The failure of private and public insurers to reimburse for telemedicine services makes promised financial benefits elusive, especially in view of the equipment costs, line charges, technical support, and maintenance costs. Many physicians fear patient loss due to a disruption in traditional referral patterns and so eschew telemedicine to protect their turf. And, paradoxically, while new technologies promise to make the jobs of health care providers easier and more efficient, rapid advances in technology make it difficult for health professionals to stay current with telemedicine technologies. Cox reports that after nearly 10 years of telemedicine use in Hays, community awareness and acceptance continue to lag.

The strength of telemedicine thus far in Hays, according to Cox, lies in teleconferencing and distance learning, because unlike clinical applications, interactive meetings and training engender less resistance. The benefits of telemedicine in reduced travel costs and reduced dislocation of local services are realized by more staff, due in part to the unrestricted access to these telemedicine applications. For example, 40 percent of the interactive video use is for health administration meetings and 40 percent is for continuing education and coursework. Only 20 percent of the use is for clinical applications such as cardiology and psychiatry.

Notwithstanding the investment in infrastructure and equipment, Cox argues that the cost of a telemedicine visit is only half the cost of an on-site visit, and additional patient savings are realized in time and travel reduction. Ultimately, Cox argues, the costs of rural telemedicine to Hays, Kansas, are easily offset by savings in patient transfer to urban hospitals.

CONCLUSIONS

Proponents in Iowa and Kansas report success in the adoption and implementation of rural telemedicine, citing life-saving speed, expanded access to advanced technologies and clinical supports, and the long-term potential of reduced costs through the elimination of redundancies in infrastructure, equipment, and services. They also report high levels of satisfaction on the part of patients and providers who appreciate the savings in time and travel-related expenses that telemedicine allows in its provision of diagnostics, treatment, and educational opportunities. But customer satisfaction claims must be tempered by the knowledge that actual use rates of telemedicine in pilot projects have been disappointingly low, averaging just 3 per day in Iowa despite the more than $4 million investment thus far. While per use costs (between $500 and $900) are likely to decline as telemedicine technologies and protocols diffuse more widely, current expenses are likely to frustrate rapid adoption and use.

In Iowa, the existence of a state-supported telecommunications system (ICN) provides a financial advantage, limiting infrastructure usage fees to $5 per hour rather than the $400 charged by commercial enterprises. This means that although the initial investment is high, ongoing expenses can be controlled because of cost efficiencies made possible by state subsidization of use through the ICN. At the same time, connectivity costs in excess of $105,000 will continue to discourage use by many rural health care providers.

Hays Medical Center has experienced sufficient success implementing telemedicine to warrant expansion. But like facilities in Iowa that have piloted telemedicine, most of the financing for infrastructure, equipment, and technical expertise comes from outside, further strengthening the vertical hold that external entities have on the rural health care system. The case studies from Iowa suggest that large urban medical facilities have been the catalysts behind telemedicine development, especially infrastructure investment and system design. In addition, large urban medical facilities have been the primary beneficiaries. Rural access to, let alone control of, telecommunications technologies, is still severely limited. A survey of approximately 2,400 rural hospitals conducted for the federal Office of Rural Health Policy found that only 20 percent of rural hospitals had access to telemedicine services and 60 percent reported no plans to provide or access telemedicine services in the future (Jones, 1996, in Field, 1996, p. 20). The costs associated with obtaining telecommunications technologies continue to be prohibitive for most rural areas. Based on Iowa Methodist Medical Center's conclusion that the expense of telecommunications technologies and technical support may remain beyond the reach of rural hospitals and health care providers, it is questionable how the benefits of telemedicine experienced by large urban medical facilities can ever accrue to their smaller rural counterparts.

Telemedicine pilot projects have confirmed the barriers that continue to impede adoption and use of telecommunications technologies by rural health providers. These include infrastructure weaknesses, user *un*friendly technologies coupled with insufficient local technical expertise, interstate licensing prohibitions, liability fears, privacy concerns, reimbursement restrictions, threats to physician and service dislocation, and overall lack of affordability. Indeed, education via computer-mediated communication and interactive video seems to be the only application for which there is agreement that current telemedicine technologies will meet with continued success.

Administrative applications drive rural adoption and use of telecommunications technologies. For rural health care facilities, the result of most computer-mediated communication and interactive video transactions is to access vertical linkages with large medical centers, insurance companies, and state agencies. Most clinical applications are initiated by urban facilities where resources and expertise are more likely available to support the demands of start-up, implementation, and maintenance of telemedicine systems. While rural facilities are the purported heirs to telemedicine's bounty, most of the benefits accrue to urban medical centers.

Economies of scale provide the simple answer: the more resources available to a hospital, the more likely it is to adopt and use telecommunications technologies for administrative, clinical, and educational applications. In general, rural hospitals have fewer resources than their urban counterparts, making them less likely to invest in telemedicine and reap the purported benefits. Despite enthusiasm for rural telecommunications expressed by hospital administrators, there are disappointingly low rates of adoption and use. Notwithstanding reported satisfaction by patients and providers for telemedicine, community awareness and acceptance continue to lag. The expense and effort required to overcome the barriers currently frustrating diffusion of telecommunications technologies threaten not to buoy, but to sink rural health care facilities.

Whereas virtually all rural hospitals are computerizing, our evidence suggests it is unlikely that hospitals could assume the lead in rural communities, championing the use of telecommunications technologies and making resources (infrastructure, equipment, training, technical assistance, strategic planning, etc) available to foster general telecommunications-generated development in rural areas. Although their own continued health and viability should economically benefit the communities in which they reside, it is doubtful that rural hospitals will be able to share with the broader community telecommunications resources developed for telemedicine. The data suggest that most individual rural hospitals are dependent on external entities for much of the technological know-how required for adoption and use. Consequently, they lack the resources needed to make these technologies more widely available. At the same time, multihospital consortia are unlikely partners in telecommunications-generated community development for the individual locales in which their members operate.

While the design of computer-mediated medical technologies and telemedicine applications must take the culture of the medical profession into close consideration, the culture of rural communities and their residents should be a critical focus as well. The goal in introducing telemedicine to benefit rural areas should be to do so in a way that respects rural values and promotes the stability of rural institutions. Because rural areas are already more vulnerable, telemedicine interventions must be designed in a way that builds capital and capacity within rural communities rather than siphoning these resources to urban centers through modem connections.

Rural providers must be wary that over-reliance on technology can become a dependency that increases vulnerability. Telemedicine may alleviate many of the problems confronting rural health care. Conversely, telemedicine investments may temporarily mask structural problems that a technological fix cannot cure. Telemedicine purchases require a budget shift from payrolls to infrastructure and from personnel to technologies. Immediate investments in telemedicine technologies do not benefit rural economies, but rather urban communities where the technologies are manufactured and marketed. Planned obsolescence and compatibility problems could result in costly, even deadly delays in service and bankrupt vulnerable facilities with ever changing technology requirements. Even in urban settings, the cost of telecommunications technology and the availability and cost of requisite expertise continue to be major impediments. Unless telemedicine is designed with affordability the goal, rural areas will be excluded by expense, resulting in a truly "limited access" information superhighway.

NOTES

1. See Weisfeld, 1993; Grigsby, Adams, & Sanders, 1995; Perednia & Allen, 1995; Bashshur, 1997; U.S. Office of Technology Assessment, 1995; Puskin, 1995; Field, 1996; and Linkous, 1999.

2. Of the 69 hospitals represented in the research, 64 (93%) were rural hospitals and 5 (7%) were rural-referral hospitals. Rural and rural-referral hospitals are defined by the Association of Iowa Hospitals and Health System as hospitals located outside a Standard Metropolitan Statistical Area. Rural-referral hospitals, of which there are only 7 in the state, have operating characteristics similar to urban hospitals, despite their rural location. The 64 rural hospitals represented by the study averaged 54 beds, including convalescent units. In contrast, the 5 rural-referral hospitals averaged 228 beds.

3. Here we use telemedicine in its more restrictive sense, as real-time interactive video consultations to diagnose or treat patients online.

4. Due to the high intercorrelation between admissions, bed, and hospital revenues, regression analyses were deemed inappropriate.

5. Education was the state's top priority in establishing the ICN. Their first goal was to connect schools, creating at least one point-of-presence (POP) in each county. Telemedicine applications via the ICN were not affirmed until much later. It is, therefore, not surprising that hospitals represent only 10 percent of connected sites.

6. Bandwidth is a measure of the information carrying capacity of a communications channel such as fiber optic cable (Zetzman, 1995, p. 2).

10

FARMERS, COMPUTERS, AND THE INTERNET: HOW STRUCTURES AND ROLES SHAPE THE INFORMATION SOCIETY

Eric A. Abbott, J. Paul Yarbrough, and Allan G. Schmidt

INTRODUCTION

For more than a century, a never-ending stream of new agricultural technologies coupled with a strong public policy oriented toward helping farmers increase output has resulted in a steady increase in farm productivity. Early technologies such as the steel moldboard plow made it possible for farmers to tame prairie soils and establish themselves on new lands. Introduction of the steamboat and the construction of canals, railroads, and highways allowed them to market their produce and delivered products that farm families increasingly wanted and could afford. The majority of U.S. rural communities established themselves as trade centers to service both the personal and technical needs of a predominately agricultural clientele, and to a large extent their geographic locations and service areas were determined by their ability to attract and serve enough farmers to support themselves economically. Initially, a strong agriculture led to significant community development. The huge Catholic Basilica in rural Iowa's Dyersville, one of only a few globally located in rural areas, plus hundreds of other beautiful churches across the Midwest were constructed with manpower and donations from local farm families that had prospered from farming around the beginning of the 20th century. Their pews, and seats in newly constructed schools, were filled by these same families. In 1930, 56 percent of all rural residents lived on farms (Economic Research Service, 1997, p. 7).

During the first four decades of the 20th century, new agricultural technologies flowed at an increasing rate from land grant colleges and universities, agricultural businesses, and creative farmers themselves. With the conspicuous exception of the automobile (which saw rapid rural adoption in the 1910–30 period) farmers were relatively slow to accept these new products and ideas. Consequently, the numbers of farms and persons employed in farming and the population of rural communities were relatively stable between 1910 and 1935 (Census 1996, His. Stat. Series A54, p. 11; K1, 4, p. 457; K174, p. 467), despite the disruptions of World War I and the Great Depression.

With an improving economy in the late 1930s and demand created by World War II and its aftermath, farmers began to adopt the backlog of farm innovations with increased enthusiasm. Between 1940 and 1960, use of chemical fertilizers and the numbers of farm tractors tripled, farms using milking machines quadrupled, and the number of grain combines increased fivefold (Census 1996, His. Stat. Series K184–187, K189, K193, p. 469). Rapid adoption of new technologies led to soaring increases in farm productivity, but it did *not* lead to rural community prosperity. Instead, improved means of farm production coupled with significant off-farm employment opportunities began a trend toward a capital-intensive agriculture that would drastically reduce the number of farms from 6.5 million in 1935 to 2 million in 1997 (Economic Research Service, 1997, p. 4). This, in turn, led to long-term changes in the economies of rural communities. To prosper, or even to survive, rural communities had to develop industries and businesses that were not directly linked to agriculture. Most of those that did not or could not adjust have faced economic decline. In 1960 two-thirds of the 3,045 counties in the United States were economically dependent on farming; in 1997 only one in six (556) met this criterion (Economic Research Service, 1997, p. 10). Among counties that remain dependent on farming are many that appear stuck in a state of "persistent poverty."

Farming should not be seen as being independent of rural nonfarm business and industrial development. It is a two-way street. Nonfarm income has become critical to the survival of farm households, especially smaller noncommercial farms. In the cornbelt states, 85 percent of total farm income is now from off-farm sources (Sommer, Hoppe, Green, & Korb, 1995). And farm-linked businesses (those that provide farm inputs or market and process farm outputs) currently account for 17 percent of all employment in nonmetropolitan counties and have been the most rapidly growing sector of rural economies during the past decade.

The major premise of this chapter is that, to a great extent, the impacts of two new technologies—computers and the Internet—are being shaped by ongoing transformations in the social and economic structures of rural areas. While not denying the possibilities for individual transformations, we believe that existing economic, education, and employment patterns have powerfully shaped how computers are being adopted and used by farm households, and that the same patterns are beginning to be seen as the Internet is adopted. Rather than seeing the computer and the Internet as devices that will revolutionize agriculture and rural communities, we believe their impacts will accelerate current trends.

By utilizing a unique Iowa dataset that followed the adoption and use of computers and computer-based information systems (CIS) by farmers, spouses, and children over a 15-year period, we have been able to document important changes in the functions of computers. Our results show that while computer use initially had and continues to have significant benefits for large-scale farmers, there has been a great increase in use for nonfarm applications during the 1990s. Increased use in schools, off-farm and home-based employment, entertainment, and communication has opened up new productive possibilities. The effect has been a gradual change of what computers mean to farm households over time.

To understand these changes and to link their consequences to rural community development requires further examination of changes in the structure of agriculture, changes in the economic structure and interests of farm households, and changing demands for and means of satisfying the information needs of farm operators.

THE CHANGING FARM ENTERPRISE

The number of farmers has declined to just 2.8 million—less than 2.5 percent of the nation's workers—yet they are still of vital importance for the economic development and prosperity of many rural areas (Economic Research Service, 1996; Census Bureau, 1997, p. 403). The entire food and agriculture system constitutes 16 percent of total gross domestic product. Along with the numerical decline in the number of farmers has come fundamental change in the *structure* of farming, with growth in both large commercial/industrial farms and very small noncommercial enterprises, and decline in medium-sized family farms. These structural changes have important implications for patterns of technology adoption and use.

A second factor concerns major shifts in the *information behavior* of farmers. As information has become more important to agriculture, more information has been provided, and farmers have spent more time searching for information. A third factor concerns the *geographic isolation* of farmers, and the extent to which highways and information technologies may have reduced this isolation. Because one key promise of the Internet is the linking of such isolated groups to the vast storehouse of knowledge on the network, how well it reaches this group may serve as an indicator of the Internet's success. A fourth factor concerns the extent of *public and private investment* in reaching farmers. Historically, government and private businesses have made linkages to them a priority. Through Extension, magazines, radio programs, and more recently satellite-based electronic teletext systems, farmers have been the focus of much communication attention and past experimentation with new technologies. We will now discuss each of these factors in more detail.

Changes in the Structure of Farming

Farming as an industry has changed substantially in ways that have had important impacts on the use of information technologies. The nation's 2 million farms have been classified by Tweeten (1984, p. 6) into three groups: (1) Large commercial or industrial farms that utilize sophisticated management and technical assistance and access diversified sources of debt and equity capital. While 2.4 million farms produced half the nation's agricultural production in 1900, 83,000 farms produced half in 1993 (Economic Research Service, 1996, p. 15). (2) Traditional medium-sized family farms, which have been declining steadily in number for decades. These farms are often too large and demanding of time to allow the operator to find stability in off-farm employment, yet too small to use sophisticated management, marketing, and diversified sources of capital available to large industrial-type farms. (3) Noncommercial farms with less than $40,000 in annual sales (more recently, this has been increased to $50,000). These 1.5 million farms comprise 73 percent of all farms, yet produce only 10 percent of agricultural output. These farmers are a diverse group. Wheeler (1985) termed this group "double tillers" because they often work off the farm in addition to farming. One large portion of these is the semiretired—farmers whose farming activities have been reduced in scope over the years but remain active in some aspects. A second group consists of those whose income comes mainly from off the farm, but they prefer to live on a farm for lifestyle reasons (Yarbrough, 1987, p. 3). Each of the three types of farms has different levels of access to information, and the payoff in terms of scale from information seeking is much larger for the industrial farms. Because off-farm employment has become essential to the existence of many farmers, information technology use for both on-farm and off-farm uses must be considered.

Changes in Farmers' Information Behavior

Along with changes in farm structure have come changes in farmer information behaviors. Yarbrough (1990) has identified three main causes of changes in information behavior:

During the past 50 years, farmers of all types have greatly increased their information processing activity. . . . Three factors underlie this change. First, the production technologies, economic systems and regulatory systems confronting managers have grown more complex. Dealing with this information demands more (and better) information. Second, the knowledge base itself has grown. This growing knowledge base contributes to increased complexity of business operations in the first instance. It also provides the answers needed to cope with this growing complexity. And, finally, the information "out there" is more widely available because of improved communication systems. (p. 80)

Because more information is "out there," to keep ahead of competition, one must access and synthesize more and more information.

Geographic Isolation of Farmers

A common policy objective concerning information technologies has been to use them to end the isolation of farmers caused by the U.S. settlement pattern that scattered farms across the rural countryside. Rural free delivery of mail (Yarbrough, 1990), a special library book rate (Lawson & Kielbowicz, 1988), the telephone (Shipley, 1985; Kielbowicz, 1987), and radio (Kielbowicz, 1987) were all presented as technologies that would help bridge geographic barriers and end the isolation of farmers and other rural residents. Similarly, a major tenet of policy concerning the Internet has been that by fostering universal access, it makes geographic location irrelevant. Yarbrough (1990) sums up this aspect: "There is spec-ulation among information and development scholars that new information technologies—especially advanced telecommunications technologies—have the potential to facilitate other transformations in rural economies and to offset urban business advantages. The premise of this speculation is that capabilities of these technologies make the location of business operations relatively unimportant" (p. 104). For example, Dillman and Beck (1988) in summarizing Cleveland (1985) de-scribe "the dissolution of the hierarchy of geographical location by the informa-tion era and the passing of remoteness as one of the major trends of our times. Information technologies have the ability to overcome remoteness which in the past has excluded rural people from important societal decisions and activities" (cited in Dillman & Beck, 1988, p. 33) Communication scholars gathering at a spe-cial Aspen Institute Conference in 1988 focused on similar benefits of information technologies for farmers rural community businesses (Bollier, 1988, pp. 17–19).

Public and Private Efforts to Use Technologies to Reach Farmers

Concern about delivering information to farmers in rural areas was present even when lands were first settled in the United States. As noted in the chapter on the Extension service's use of information technologies (Chapter 11), one of the first acts of the Iowa territorial assembly in 1838 was to specify methods by which farm-ers could collect and share information about how to make agriculture more pro-ductive (Bliss, 1960). Edwards (1994) provides detailed information about efforts by private agricultural newsletters and magazines to reach farmers in the 1840–1856 period.

Efforts to use computers to assist farmers began in 1958 when the Wisconsin Dairy Herd Improvement Association began using a mainframe computer to provide farm record keeping and management services. In the 1960s, Virginia Polytechnic Institute and Michigan State University began using mainframe com-puters to provide agricultural information and databases to Extension staff (Whit-ing, 1981). By the late 1970s and early 1980s, there was a rapid increase in agricultural information delivery via "videotex" or "teletext." Teletext systems broadcast a fixed menu of information or information choices that scroll across or are indexable on a television screen. Videotex systems provide more possibilities for users to request what they want via a search engine, communicate with others, or even submit data for analysis. Both types of systems utilize mainframe comput-ers to provide information and some sort of decoder, receiver, keypad or computer at the receiver's location. As these devices became more sophisticated, most of the simpler scrolling and keypad versions were abandoned, and receiving devices be-came computers, although often they could be used for only one purpose. In the area of electronic videotex/teletext systems, Durand (1983) noted that "the most favoured group in public service applications of both teletext and interactive videotex" has been farmers. A brief summary of some of the electronic applica-tions directed toward farmers is provided in Table 10.1. Green Thumb, a videotex

Table 10.1

Videotex/Teletext Applications Directed toward Farmers

Application Name and Research Source	Initiated	Target Audience
AGNET (Whiting, 1981)	1977	Farmers, Extension, and agribusiness in Nebraska, North Dakota, South Dakota, Wyoming, Montana, Washington and Wisconsin
Green Thumb (Case et al., 1981; Clearfield & Warner, 1984; Rice & Paisley, 1982; Warner & Clearfield, 1982)	1980	Farmers in Kentucky
Instant Update (Whiting, 1981)	1981	Farmers in Iowa and nationally
GRASSROOTS (CanWest Agricultural Research, 1984; Universitel Field Trial Survey Results, 1984)	1982	Farmers in Canada
AGRI-VIEW (Pfannkuch, 1988)	1982	Farmers in Iowa
Infotext (Gonzalez, 1988; Vedro, 1983)	1982	Farmers in Wisconsin
FirstHand (Ettema, 1983, 1984a, 1984b)	1982	Farmers in North Dakota
British Prestel Farmlink (Tatchell, 1987)	1984	Farmers in United Kingdom
Dataline (Abbott, 1989; Abbott & Yarbrough, 1992). Became DTN-Data Transmission Network; now nationwide	Mid-1980s	Farmers in Midwest

Note: For a summary of early research on the use of electronic applications to reach farmers and Extension, see Whiting (1981).

trial in Kentucky, and AGNET, based in Nebraska, were two of the early efforts to get farmers to adopt a computerized information service.

As in many other areas, media hoopla (Abbott & Eichmeier, 1998) associated with farmers, computers, and other innovations has often painted an overly optimistic view of both adoption rates and the transformational powers of these electronic technologies. In 1981, *Successful Farming's* computer editor predicted that the great majority of his magazine's subscribers would have a computer by 1990 (Vincent, 1987). An often quoted article published in *Successful Farming* in 1983 (Schmidt, Rockwell, Bitney, & Sarno, 1994) predicted that 80 percent of farmers would be using a computer by 1990. The actual figure was slightly over 15 percent (Abbott & Yarbrough, 1992). Similarly, the transformational impacts of computers on farmers were exaggerated. Iddings and Apps (1990) stated the dream: "We can imagine beautifully organized databases filled with data on cattle and crops. We can envision spreadsheets distilling mountains of figures into easily usable information. Long-range plans, market reports, and electronic mail march across the farm video displays in our minds" (n.p.). They then acknowledge that at least as of 1990 when they wrote their article, the dream had failed to match reality. These findings indicate that the adoption of computers and the

Internet by farmers, and the transformational impacts they might have, do not represent an "automatic" process affecting every farmer. Rather, the process will be neither rapid nor even, and will be shaped by patterns of information seeking, management expertise, age, education, capital, access to technologies, and a variety of other factors.

IOWA LONGITUDINAL SURVEYS

Data in this chapter come from a longitudinal study of information technology diffusion and use among Iowa farm households. General information about computer and computer-based information system adoption and use comes from three independent random samples of 1,000 Iowa farmers, each surveyed by mail in 1982, 1989, and 1997. Detailed information on computer use among adopters comes from a special 1990 survey of 136 Iowa adopters obtained from previous random sample surveys.[1]

Overall Computer Adoption

The overall pattern of adoption of computers by Iowa farm households is shown in Figure 10.1. Over a 15-year period, from 1982 to 1997, adoption has increased from 2 percent of households to 46 percent. The figure shows that most farmers have moved through the adoption process, with only 20 percent still giving com-

Figure 10.1
Farm Computer Adoption Progress in Iowa, 1982 through 1997

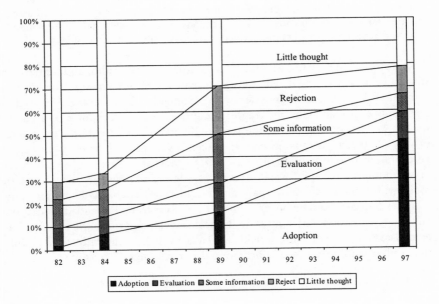

puters "little thought." About 20 percent are either seeking information or actively deciding, while 10 percent have rejected computers. Computers are being used for both farm and nonfarm purposes. Their uses and impacts for farming purposes will be considered first.

Patterns of Information Seeking by Farmers

Although some may hold a perception of farmers as being information poor because of their geographic isolation, many farmers complain that they are overwhelmed by information coming from print media, electronic media, and salesmen. As the case study suggests, it is not unusual for a farmer to receive 10–20 magazines per month, and those with farms larger than 500 acres often receive many magazines free—paid for by advertisers or companies with products or ideas to sell.

CASE STUDY: ONE FARMER'S MAIL

Rogers and Case (1984) illustrate the widely diverse use of farming information with an anecdote from an interview with a Green Thumb videotex system user. The interview was interrupted by the mail carrier; there was too much mail to fit in the mailbox. Among the 35 items delivered were three newspapers (local, state, and a national business paper); 13 magazines; and a requested research report. The farmer commented that it required about 3 hours each evening to read his mail, but he felt this information was the most profitable part of his farming job. He said his grandfather had believed hard work was the key to farming success. His father believed close attention to marketing was the key. This farmer argued that today gathering and processing information are the most important behaviors for farming success. He said a useful change in Green Thumb would be to provide weather maps of the Ukraine (the interview was before the breakup of the Soviet Union). He felt the information would benefit himself and other farmers involved in futures contract trading.

As Table 10.2 indicates, general farm magazines, farm radio, and newspapers still lead the way as mediated sources of agricultural information for farmers. Computer-based information systems, on the other hand, are very seldom used. In 1997, only 21 percent of farmers reported any use of online computer-based information sources for agricultural information,[2] and only 8 percent used them often or very often. This compares with 95 percent using general farm magazines, 49 percent of whom do so often or very often. Farmers also are heavy users of meetings and interpersonal information sources. Almost all reported discussing farm topics with other farmers in the past month, and 93 percent had talked to an equipment dealer, salesman, or buyer. Almost three-fourths had attended at least one farm supply or cooperative meeting in the past year, and more than half had attended an Extension meeting or field day. Thus, in terms of the entire farm population, online computer-based information is not yet a significant source.

Table 10.2
Agricultural Information Source Use by Farmers Based on the 1997 Iowa Farmer Survey

Information Source	Use at All	Use Often or Very Often
General farm magazines	95%	49%
Newspapers	88	43
Agricultural radio programs	84	41
University Extension publications	83	29
Specialized farm magazines	78	30
Agricultural television programs	78	20
Farm organization publications	77	27
Dealer's magazines	74	18
Private information and management services	41	16
Computer-based information systems	21	8

Note: N = 302.

Factors Related to Computer Adoption and Impacts

The patterns of adoption and use of computers by farm households have been shaped by four important factors: (1) changes in computers; (2) structural changes in agriculture and rural life; (3) individual and farm business characteristics; and (4) management orientation and information seeking behavior.

Changes in the Innovation Itself. In 1982, computers were relatively expensive, required extensive training, and were limited by a lack of quality software. Users often had to create their own programs. In that year, an entry level system including computer, monitor, printer, modem, and software cost about $5,000. By 1997, an entry level system cost between $1,000 and $1,500 (1982 dollars). In addition, the 1997 unit is hundreds of times more powerful in raw computing power, comes with a multiple-gigabyte hard drive (in 1982, most computers had no hard drive), and includes a graphic user interface software system that is much more powerful and easier to use.

Changes in the Structure of Agriculture and Rural Life. Several structural trends have encouraged computer adoption. Businesses and agencies linked to farmers have computerized, providing a synergistic environment encouraging adoption and use. Banks require periodic financial reports to monitor farm loans, and these are beginning to be submitted on disk or online. Government farm agencies, feed dealers, and other specialists are increasingly using computers and farm computer databases. Other community businesses and institutions have also been computerized. Schools, which perhaps had one or two token computers in 1982, now are beginning to integrate them into the curriculum. Even in 1982, farm households began adopting Apple II computers to match those at school so their children could use them for school work. By 1997, the impact of school adoption of computers was even more pronounced. The structure of games and entertainment has also changed. While sales of Leggos® and other noncomputerized games have fallen over the past five years, computerized entertainment games have become commonplace. Meanwhile, farm spouses, who increasingly work off the farm to supplement farm income, are finding that the use of computers either at home or at a business is important. As rural main streets, hospitals, and government offices computerize, spouses must computerize.

Individual and Farm Business Attributes. Our results (see Table 10.3) show that computer adopters share a number of characteristics. There is a curvilinear relationship for age, caused by both lifecycle and generational factors. In terms of lifecycle, households with children are much more likely to have a computer. Thus, the youngest farmers who are not yet married or have very young children, and older farmers whose children have grown up and moved away, are less likely to have a computer. There are generational factors at work as well, with a tendency for older farmers to be less likely to adopt. This has been frequently noted in other studies (Putler & Zilberman, 1988; Batte, Jones, & Schnitkey, 1990; Lazarus & Smith, 1988; Iddings & Apps, 1990). But over the 15 years of our study, the generational factor has diminished as computer adopters who were in their 30s and 40s are now age 45 to 65 and continue to use computers. The strong tendency to continue computer use as one grows older means that age will eventually cease to be a factor in computer adoption. Education and farm scale are two additional powerful predictors of computer adoption. Those with some college have been at least twice as likely to adopt as other groups, and this has been found in many other studies (Putler & Zilberman, 1988; Batte, Jones, & Schnitkey, 1990; Lazarus & Smith, 1988). Scale of the farming operation, reflected in terms of gross farm income in our table, is a third important variable. Other studies using income (Jarvis, 1990), acres farmed (Batte, Jones, & Schnitkey, 1990), and herd size (Lazarus & Smith, 1988) have reported similar findings. Farmers themselves recognize the importance of scale of farming to productive computer adoption. When the 1997 farmers were asked if their farms were too small to justify the purchase of a computer, more than half with gross farm incomes of less than $20,000 agreed, and 40 percent of those with incomes from $20,000–$39,000 also agreed. One-fourth of those with incomes of $40,000–$99,999 agreed. But only eight (7%) of the 114 farmers with incomes of $100,000 or more agreed.

Management Orientation and Information-Seeking Behavior. Management orientation—an indication of the farmer's management knowledge and skill—is a summed score of the use of various management techniques. Results in Table 10.3 show that a high management score is closely related to computer adoption. Seventy percent of the "high" group now has a computer—slightly less than double the rate for the other two groups. High agricultural information seekers also are more likely to have a computer. These two scores indicate that high levels of agricultural knowledge and management skills lead to much greater computer adoption. One conclusion that can be drawn from Table 10.3 is that computers are still diffusing very unequally in the farming community. To the extent that adoption of these devices creates greater profits, they may be expected to increase inequalities

Table 10.3

Percent Adoption of Computers by Iowa Farmers by Farm Income, Education, Management Orientation, and Information Seeking for 1982, 1989, and 1997

	Iowa 1982	Iowa 1989	Iowa 1997
Total Sample	2%	15%	46%
Farm Sales			
< $40,000	1	9	38
$40,000 to 99,999	1	10	34
$100,000+	4	23	56
Education			
< High school	0	0	20
High school	1	10	33
College	4	29	62
Age			
34 or younger	2	13	39
35–44	5	33	66
45–54	1	25	62
55–64	2	7	40
65+	2	4	18
Management Orientation			
Low	1	6	39
Medium	2	20	42
High	5	29	70
Information Seeking			
Low	1	10	34
Medium	2	15	51
High	4	19	54

Note: Farm income is the farmer's gross farm income and is an indicator of farm size; Management Orientation: an indicator of the management skill of the farmer and includes the use of management techniques such as record keeping ability, cash flow analysis, forward contracting and hedging, and enterprise accounting; Agricultural Information Seeking: the summation of the extent of use of agricultural media and interpersonal sources for information, including use of TV, radio, newsletters and other specialized publications, magazines, newspapers, Extension materials, and interpersonal contacts with professionals and neighbors.

in rural areas. Although one cannot predict whether universal adoption will one day reduce these inequalities, the data at present suggest that poorly educated and lower income farmers may never adopt these devices in large numbers. Thus, far from the universal benefits projected by technology enthusiasts, our results suggest that the ultimate impact of these two technologies may be to drive some farmers out of business.

Farm Versus Nonfarm Computer Uses

Our 1990 results in Table 10.4 show farm and nonfarm uses split about evenly in terms of hours spent on the computer. The 1997 data, divided into those adopting before 1990 and those adopting in 1990 or later, shows that the earlier group continues to devote substantial time to farm applications, with a slight increase in hours since 1990. The later group, however, is spending much less time on farming applications—almost half as much. For nonfarm hours, both groups show rates of use higher than the 1990 group. The data indicate that nonfarm users have become relatively more important for *both* groups, and that newer adopters are spending much less time on farm applications.

Diversity of computer use for a farming purpose is explained by the scale of the operation (larger scale operators use almost twice as many applications as small-scale operators) and the degree of management skill and agricultural information seeking. Farmers who use their computer at least half of the time for farming have management orientation scores that are 50 percent higher than computer adopters who use their computer less than half the time for farming. Diversity of computer use for nonfarm applications is explained by a cluster of family attributes (children present, family's off-farm computer experience, spouse employed off-farm).[3]

Perceptions of benefits of computers also show the difference between those who use computers heavily for farming and those who use them primarily for nonfarm purposes. Perceptions of benefits from *farm* use are strongly skewed toward the highest farm income group. More than 70 percent of those with farm incomes of $200,000 or greater reported they either derived a modest or substantial

Table 10.4
Changes in Farm and Nonfarm Computer Use: Number of Hours of Computer Use, 1990–1997

Means Hours Computer Use Per Week	1990 Survey	1997 Survey	
		Adoption before 1990	Adoption 1990 or later
For farm applications	3.1	3.8	2.0
For nonfarm applications	3.0	4.6	3.9
Total hours	6.1	8.4	5.9

gain from use of the computer. For the $100,000–$199,000 farm income group, only half perceived they had gained benefits. Only a small percentage of lower income farmers using computers thought there had been a net benefit for their farming operations (Abbott & Yarbrough, 1993; Abbott & Yarbrough, 1992). Lower farm income households saw their benefits mainly in nonfarm uses.

An analysis of who is using computers in the farm household reinforces the conclusion that there have been important shifts over time. Because there are often many users, earlier studies have pointed out the importance of examining the computer behavior of multiple members of the farm household. Iddings and Apps (1992) found that only 15 percent of Midwest farm computer adopters were the sole users of their computers. Our data on multiple computer use in farm households shows two interesting trends. First, use by multiple members of the household is high. Second, across time, spouses and children have increased their computer use substantially. By 1997, the spouse was as likely as the farmer to be named the primary operator of the household's computer, and children were becoming the primary users more frequently (see Table 10.5).

The increase in children as primary users is due to new adopters, and not to any long-term increase in use by children. Farms with computers purchased before 1990 have only 9 percent of children as primary operators, while households with computers purchased after 1990 have 29 percent. Similarly, in households adopting before 1990, farmers themselves don't use the computer in 12 percent of cases, but this increases to 30 percent among households adopting after 1990. This finding is influenced to some extent by lifecycle factors, because younger farmers adopting earlier are less likely to have children at home.

Use of the computer by family members other than the farmer can have an important effect on the pattern of overall computer use. In many cases, spouses or children use the computer for nonfarm purposes. In our 1993 regression analysis of factors shaping computer use, we found that the presence of children was the single most important factor shaping the diversity of nonfarm uses (Abbott & Yarbrough, 1993, p. 8). The 1990 data in Table 10.6 show how different farm household members use the computer. The farmer predominates in farm use, but also uses the computer for many nonfarm purposes. Children use it mainly for games and education. Spouse use is spread over a variety of applications. Our 1997 Iowa data show that having a computer at the farm is associated with a 10 percent

Table 10.5
Use of Farm Computers by Iowa Farmers, Spouses, and Children: 1982, 1989, and 1997

	Farmer			Spouse			Child		
	1982	1989	1997	1982	1989	1997	1982	1989	1997
Primary user	56%	39%	38%	10%	30%	39%	10%	26%	22%
Also use	33	26	38	40	33	36	50	46	42
Total	89	65	76	50	63	75	60	72	64

Table 10.6

Percentage Computer Use by Iowa Farmers, Spouses, and Children: 1990 Random Sample of Computer Adopters

Uses	Farmer	Spouse	Child
Use for farming	78%	32%	11%
Nonfarm employment	36	31	5
Education	27	22	63
Household records	37	32	3
Games and entertainment	39	25	61
Word processing	68	61	43

increase in the likelihood of off-farm work for both the farmer and spouse, and with a 10 percent increase in the likelihood of the farmer operating a small business in addition to farming.

Despite the importance of nonfarm computer use, a 1992 study (Masuo, Walker, & Furry, 1992)) found that the countryside is not yet dotted with "electronic cottages" filled with white-collar telecommuters. She concluded:

One major contribution of this study is to dispel the predictions by futurists, high technology companies, and researchers of white-collar workers that a massive migration of white-collar workers from office to home is occurring. Quite the contrary, this study finds that home-based workers living in the nine states studied (excluding the largest metropolitan areas) are primarily older male business owners who are more likely to work as home maintenance contractors, truck drivers, or house and office cleaners than as home-based clerical workers. (p. 260)

A study by Rursch (1988) included in-depth interviews with multiple members of Iowa farm families that had adopted computers. Results showed many different uses and purposes. Spouses often used computers for off-farm employment, although some were responsible for farm records. Children frequently used computers for entertainment and educational purposes, although there were several cases in which the child figured out how the computer could be used in agriculture and set up programs for this purpose. Although the farmer most often used the computer for agriculture, there were cases in which it was being used to learn a new skill that could lead to off-farm employment (i.e., getting out of agriculture altogether or supplementing farm income). Interestingly, Rursch's results showed that within the farm household, there was little sharing of information about computer use. The farmer was often not aware of uses by spouse or child. There was little cross-training within the family.

Rursch also found there was little communication between the farmer and neighbors, possibly because they did not own a computer, but also because farmers have always been very reluctant to share financial information about their farms with others, and this is often how computers are used. Rursch concluded that there was surprisingly little sharing of information that might cause others to adopt computers or improve their computer skills.

Our longitudinal data show that farmers' communication about computers has steadily increased across time, indicating that Rursch's 1988 findings now have less validity. Table 10.7 shows that communication with educators, farmers, and nonfarmers has increased substantially since 1982. By 1997, almost all farmers know another farmer who is a computer user.

Farmer Adoption and Use of Computer-Based Information Systems

Earlier, we noted that computer-based information technologies such as video-tex/teletext systems have been available to farmers since the late 1970s. Thus, the arrival of the Internet at the farmgate, rather than being viewed as revolutionary, is more accurately seen as an enhanced computer-based channel available to farm households. In the beginning of the chapter, we noted that there has been an emphasis on reaching farmers with high quality information in both traditional and electronic forms at reasonable prices.

Data Transmission Network's DTN system (one of the systems listed in Table 10.1) is the most commonly used agricultural teletext system, with about one-quarter of Iowa farmers subscribing to it. The great majority of these farmers receive weather and market information via a small satellite dish that continually

Table 10.7
Spreading the Word about Computers: Farmers Discussing Computers with Educators, Other Farmers, and Nonfarmers in the Past Year

	1982	1984	1989	1997
Percent of farmers discussing computers with educators	32%	43%	40%	41%
Mean contact score: educators	0.7	1.0	0.9	0.8
Percent of farmers discussing computers with other farmers	34%	53%	61%	64%
Mean contact score: other farmers	0.6	1.0	1.3	1.4
Percent of farmers discussing computers with nonfarmers	46%	56%	56%	61%
Mean contact score: nonfarmers	1.0	1.2	1.4	1.5
Percent of farmers who know other farmers with computers	39%	67%	78%	93%
Mean score: number of farmers known with a computer	0.9	1.6	2.2	3.5

Note: Mean score for educators, farmers and nonfarmers is based on a 0-4 scale: 0 = never; 1 = once; 2 = twice; 3 = three times; 4 = four or more times. For the number of farmers known who have a computer, the score was: 0 = none; 1 = 1, 2 = 2, 3 = 3, 4 = 4, and 5 = 5 or more.

updates information on a dedicated computer terminal in the farmer's house or office. Three percent also receive DTN by subscribing to a special World Wide Web site on the Internet. By the late 1980s, DTN and similar services were beginning to be named as an important source of *perishable* market information by adopting farmers (Abbott, 1989).

By 1997, enough farmers had begun to adopt the Internet to warrant its inclusion in our survey as a possible information source. Because we were interested in being able to separate its use for farming from nonfarm use, we asked Internet adopters to indicate whether or not they made any use of the system for a farming purpose. Figure 10.2 shows the overall trend in adoption of both types of computer-based information systems. Results for 1982 and 1984 include DTN and several alternative videotex/teletext systems used exclusively for agricultural purposes. The 1989 data separates DTN from other online alternatives. The 1997 data includes DTN plus both Internet groups (farming and nonfarming). To avoid duplication, the 2.3 percent of farmers who have both DTN and use the Internet for a farming purpose are shown in the DTN group. Therefore, the "farm purpose" Internet group (5.1%) shows only the *increase* in overall computer-based information system users (7.4% of all farmers use the Internet for a farming purpose). The figure shows a steady increase in the use of computer-based information systems over time. By 1997, slightly less than one-third of farmers are using either the Internet, DTN, or both for farm information.

Figure 10.2
Percent Use of *Any* Teletext, Videotex, or Internet Source by Iowa Farm Households

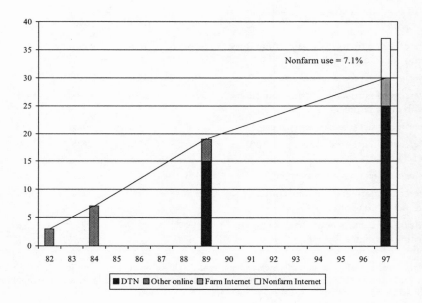

Figure 10.3 shows Internet adoption progress for 1997 Iowa farm households. Of the 302 farmers responding to the 1997 Iowa survey, 18 percent said that one or more household members were using the Internet or a general computer online service. About 40 percent of farmers said they hadn't given much thought to hooking up to the Internet. Another 15 percent said they had rejected the idea of hooking up to the Internet. That leaves about 25 percent who are gathering information or actively considering whether or not to adopt. These figures are similar to farmers' reports of their progress toward adoption of computers in the mid-1980s. Ten years later, only 20 percent of farmers now say they haven't thought much about getting a computer, and only 10 percent have rejected the idea.

Nonadopters of the Internet were asked to indicate which of four reasons for not adopting might apply to them. About half the nonadopters indicated that they have "little need" for the Internet on the farm, and about half indicated that the Internet was "too costly for my budget." About 30 percent cited as a reason for nonadoption that the Internet is changing rapidly and it is best to wait a while be-

Figure 10.3
Percentage Adoption of the Internet: 1997 Iowa Sample

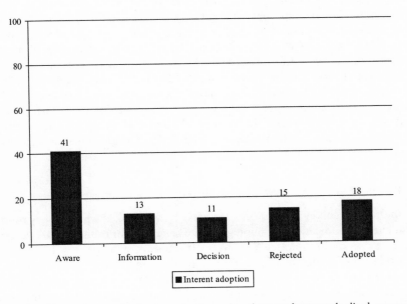

Note: Methodologies for assessing Internet adoption and use have not been standardized, so current estimates often vary widely. Some with higher estimates ask about any online activity, while others specify the Internet. Some samples include only respondents 18 and older, while others include 16-year-olds. This is an important distinction for an innovation that has been most heavily adopted by younger persons, especially students. In this study, the specific question used to assess adoption was "Does anyone in your household use the Internet or a general computer service such as America OnLine, CompuServe, or Prodigy?"

fore getting access. Twenty percent agreed with the fourth item, which stated "I don't want Internet access on the farm."

About half of those adopting the Internet (47%) say they use it for farming. Twenty percent of adopters use e-mail daily, and a total of 60 percent use it at least weekly. Two-thirds say they use the Internet for "family entertainment or education." Half say they use the Internet "to look up information about commodity prices or the weather," and slightly over half use it "to look up information about businesses or to purchase products." The only category of use showing very low levels was reading or contributing to electronic news groups. Only 14 percent said they ever did this. In short, farm households use the Internet for a variety of farm and nonfarm purposes, but the patterns of use vary among households. About one in five of these users access the Internet through a site away from home.[4] This means that about a third (43 of 134) of computer adopters were connected to the Internet by July 1997.

Because farm computer users showed very different use characteristics depending upon whether or not the computer was used mainly for a farm or nonfarm purpose, Internet users were also segmented according to whether or not the Internet was used for a farm purpose. Results in Table 10.8 compare farmers who have not adopted either computers or the Internet to Internet adopters. For the Internet group, results are broken down by whether or not they use the Internet for a farming purpose. Results show that in those households that use the Internet for a farm purpose, the farm operators are highly educated, having 1–3 years of college on average, and they have a 50 percent higher management orientation score than nonadopting farmers. In contrast, those households using the Internet mainly for a nonfarm purpose tend to look much more like nonadopting households in the survey, with lower management orientation and agricultural information-seeking scores. Internet adopters tend to be younger than nonadopting farmers and are much more likely to have children under the age of 18 living at home. Nonfarm Internet users have children present in 65 percent of households, compared to 21 percent for nonadopters.

Perceived Economic Development Benefits from the Internet

In the 1997 survey (Table 10.9), farmers also were asked whether they thought that telecommunications technologies were beneficial for economic development of Iowa's rural communities. They were offered three statements and asked which one best reflected their own view. They could also select "other" and write in an additional view. Table 10.9 shows that being a computer adopter or Internet adopter was associated with being more optimistic about the potential of telecommunications technologies for rural economic development. Computer- and Internet-adopting farmers are significantly more positive. Nonadopters were more likely to say either that the technologies would have no effect, or that the effect would be negative.

Table 10.8
Farmers Without a Computer or Internet Use Compared with Internet Users for Farm
and Nonfarm Purposes by Age, Income, Farm Size, Education, Presence of Children,
Management Orientation, and Information Seeking: 1997 Iowa Survey

	Farmers with No Computer or Internet	Internet Adopters No Farm Use	Internet Adopters Farm Use
Age	57	46	49
Gross farm income	3.4	3.8	3.8
Median acres owned or rented	320	500	449
Education	3.3	3.8	4.0
Percent children present	21%	65%	44%
Management orientation	2.9	2.8	4.7
Agricultural information seeking	20.2	18.6	24.1

Note: Gross Farm Income: 1 = under $20,000; 2 = $20,000–39,999; 3 = $40,000–99,999; 4 = $100,000–199,999; 5 = $200,000+; Education: 1 = 1–8 years; 2 = 9–11 years; 3 = 12 years; 4 = 13–15 years; 5 = 16 or more years; Percent children present: the percentage of households with children 18 or younger still in the home; Management orientation: an indicator of the management skill of the farmer and includes the use of management techniques such as record keeping ability, cash flow analysis, forward contracting and hedging, and enterprise accounting; Agricultural information seeking: the summation of the extent of use of agricultural media and interpersonal sources for information, including use of TV, radio, newsletters and other specialized publications, magazines, newspapers, Extension materials, and interpersonal contacts with professionals and neighbors.

PROJECTING FUTURE ADOPTION PATTERNS FOR FARM COMPUTERS AND THE INTERNET

Rates of farm Internet adoption are similar to general U.S. trends (see Figure 10.4). A national survey taken by IntelliQuest in June 1997 found that 19 percent of adults in the United States and Canada (defined as those 16 years of age or older) were online users (for members of Iowa farm households at this same time it was 18 percent). A second survey taken in April 1997 by FIND/SVP reported 16 percent hooked up (again for both the United States and Canada, for those 16 or older) (Nua, 1999). By February 1998, an Iowa Farm and Rural Life poll found 22 percent of farm households had adopted the Internet or World Wide Web (Lasley, 1998). In October 1998, a survey by IntelliQuest found 28 percent of the U.S. and Canadian samples were online adopters. A second survey by CommerceNet/Nielsen found 30 percent adopting in August 1998 (Nua, 1999). If we can extrapolate from these national surveys (see Figure 10.4), approximately 30 percent of farm households might have adopted the Internet by the end of 1998.

Table 10.9

Perceptions of Benefits of Telecommunications Technologies for Rural Economic
Development by Adoption of Computers and the Internet: Iowa 1997 Survey

Which One of the Following Best Reflects Your Own View?	Computer Nonadopters	Computer Adopters	Internet Adopters
Recent advances in telecommunications will serve to strengthen the economies of rural communities by providing them with access to the information highway.	58%	75%	76%
Recent advances in telecommunications will have little or no impact upon the economies of rural communities.	20%	8%	4%
Recent advances in telecommunications will serve to widen the economic gap between rural and urban communities because rural areas will be comparatively information poor.	17%	9%	13%
Other	5%	8%	7%

Note: Chi-square test comparing computer adopters and nonadopters shows the two groups are significantly different (p < .01).

To project future patterns of adoption and use of computers and the Internet, one must understand both general conditions that determine computer adoption and use, as well as why certain groups are adopting and not adopting. General conditions would include (a) the price of computers and Internet access and (b) the availability of hardware and software. Because adoption of computers by farm households has been similar to that of U.S. households in general, one might project that the similarities will continue. As the functions of computers and the Internet in homes and businesses evolve, they will attract different kinds of buyers and have different uses.

For farmers, Tweeten's division of farmers into three groups is relevant. For the commercial/industrial farmers, adoption of computers and the Internet have been wise investments that most perceive have paid off handsomely. Their farming practices are too large for them to manage without good record-keeping and decision-making tools. The proportion of these farmers (gross farm income over $100,000 per year) adopting these technologies is approaching 60 percent. Adoption in this group continues to be more rapid than in the group of smaller-scale farmers. In terms of the Internet, general adoption trends will follow computer adoption (more adoption by households with children, higher education, higher income, and management orientation). However, there are some special factors involved in determining adoption of the Internet. In our 1989 analysis of farmer use of computer-based information systems (Abbott, 1989), we found that in ad-

Figure 10.4
Farm and National Surveys of Internet Adoption: April 1997 through October 1998

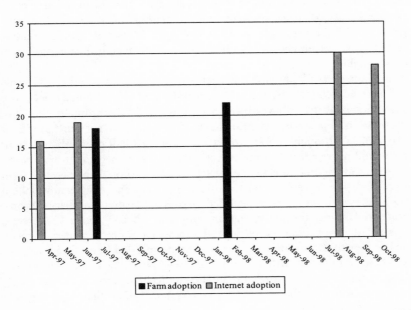

dition to the advantage of scale and the other factors mentioned above, adoption depended on whether the information obtained was advantageous. This, in turn, depended on:

- the extent to which the farmer's crops or livestock were subject to rapid fluctuations in market prices (the more volatility, the greater the adoption);
- the extent to which farmers can control when they market (dairy farmers must sell their milk as it is produced, while corn farmers can await higher prices before selling). The more control, the better the chances for adoption.

In short, information for decision making is valuable *only* if one can act and has a significant opportunity to gain from that action.

There is also a tradeoff in types of technology that will be a factor in levels of Internet adoption versus continued use of videotex/teletext systems such as DTN. DTN provides rapid downloading of market and weather information 24 hours a day for about $33 per month without tying up the household telephone. Farmers could elect to subscribe to DTN via the World Wide Web for about $25/month, but they would have to provide their own computer and modem plus pay a monthly Internet Service Provider fee (about $20/month). The advantage of the Web would be extra value in the form of two-way communication (e-mail capability comes with the Web version) and greater potential interactivity with data. The Internet also could be used for many nonfarm purposes. Two important disadvantages of using the Web version would be first, tying up the phone line when connected, and second, probably slowing down delivery speed because many rural household phone lines cannot handle high speed data transmission. Large commercial farmers already have computers and modems and could use the Internet to coordinate with shippers and negotiate better prices. They might find it useful to switch, or they simply might get both Internet access with continued reception of DTN via a satellite.

For medium-sized family farms, investments in computers and the Internet may be worthwhile, but the payoffs will come more slowly. Continued price drops for computers and software have made it more possible for this group to adopt, but fewer of these farmers say that adoption has led to a substantial net benefit for their farm. In 1987, Yarbrough (1987) projected that at least half of the then current farmers would never adopt a computer. They lack the management orientation and information seeking skills to use the technology and/or they operate a farm whose scale of operation is too small to justify use of this technology. Many of them will retire or leave farming and be replaced by younger farmers with larger farms. Therefore, over time, the percentage of adopters will rise above 50 percent. Long-term patterns of adoption will depend on whether or not farmers can experience a clear benefit. Concerning the Internet, medium-sized farms find they have less need to use e-mail to coordinate farm activities. Many may not have the proper computer equipment. Therefore, one might predict that farmers in this group would continue to use the existing teletext system rather than adopting the Internet.

The third group, noncommercial farmers, has very little justification for buying a computer for the farm. However, as we have seen, their nonfarm uses for computers are much more important, as many rely on computers in nonfarm jobs or for education. This group has accelerated its adoption of computers since 1989, but not for farm purposes. Computer adoption by this group will hinge on advances made in nonfarm uses, probably for a combination of education, entertainment, and nonfarm employment. This group's behavior probably will most closely resemble that of the general public.

CONCLUSIONS

This chapter has examined two innovations—the computer and the Internet in farm households. The personal computer has been widely available for 20 years. The Internet, on the other hand, has been widely available for less than 10 years—5 years in its popular World Wide Web format. While there have been many grand pronouncements about its future, to date the Internet has had little impact on farm households. We have argued that because the Internet is computer-based, it will, to some extent, follow the patterns established by computers.

We divide our conclusions into three categories: (a) impacts of computers and the Internet on the farming enterprise; (b) impacts of computers and the Internet on nonfarming activities; and (c) possible community development benefits or drawbacks of these technologies.

Impacts of Computers and the Internet on the Farming Enterprise

Computers and the Internet will likely exacerbate the trend toward concentrating agricultural wealth and power. The large-scale farms comprising Tweeten's commercial farm sector will benefit substantially from computer and Internet use for farming purposes for three main reasons:

1. *Economies of scale.* In general, the larger the farm, the greater the economic payoff from adopting and using computers. Computers permit farmers to manage more land or livestock; having more land and livestock, in turn, makes it worthwhile to develop sophisticated decision models and accounting systems. The great majority of Iowa large-scale farmers have no doubt about the benefits they can derive from using a computer. Thus, the computer has become one key to making it profitable for already large farmers to grow even larger.

2. *Intellectual tools—education and knowledge.* Knowledge about how to make money from agriculture—by careful management of resources, by integrating record keeping and decision making, and by exploring multiple alternatives—is not spread evenly among farmers. Large-scale farmers have management orientation scores that are twice as high as others. This means they have the knowledge to use computers effectively to make money. In a volatile and competitive farming environment, this knowledge is critical to staying ahead and will lead to greater profitability and larger size. Their knowledge and skill drive their computer and Internet adoption and use. Simply delivering computers and the Internet to farmers without these skills would have little effect.

3. *Taking advantage of opportunities to exercise knowledge to increase profit.* Commercial farmers are actively seeking ways to use their computers and the Internet to link beyond their communities to access favorable credit, lower-cost inputs, better market prices, and contacts with agricultural experts. As agricultural enterprises increase electronic links for marketing, inputs, and regulatory purposes, farmers who are large enough to negotiate favorable terms and sophisticated enough to know where to look for the best deals will benefit most. These are often large commercial farms. They are using their computers effectively to negotiate with bankers, suppliers, and marketers.

The impact of computers and the Internet is likely to bring about an acceleration of the trends Tweeten presented—decline of the middle-size farm and growth of large commercial ones.

Impact of Computers and the Internet on Nonfarming Activity

Our data have shown a great increase in the adoption and use of computers for nonfarm purposes by members of Iowa farm households, including a substantial increase in the use of the farm household's computer by the spouse and children. In Iowa, the spouse is now as likely as the farmer to be named the "primary operator." New computer adopters (post 1990) are using their computer for nonfarm purposes, while earlier adopters continue to use their computers heavily for farming. Two general factors have brought about this increase: (1) computers have become more powerful, cheaper, and able to handle many different kinds of tasks, and (2) computers have become more integrated into the life of the community in ways that stimulate adoption and use.

School adoption of computers and the Internet has been a very important factor leading to their adoption at home (often the same computer brand to permit children to work on the same machines). Off-farm employment opportunities now include computer work. A computer on the farm can mean valuable training that can lead to off-farm work, or the off-farm work can provide the training and insights needed to stimulate home adoption. Computers can also lead to "take-home" work from an off-farm job, or even the establishment of a nonfarm business operating out of the home.

Nonfarm uses of computers and the Internet have a much wider variety of potential effects on the farm household—from business to distance education to entertainment. At this point what data we have in this area are meager. Rursch's 1988 in-depth household studies found that in some cases, computers were being used to exit farming altogether, while in others, they permitted the farmer or spouse to operate a nonfarm business that allowed the farmer to continue to farm. Nonfarm uses fluctuate in the household as children age. Young children tend to use computers and the Internet for entertainment or education. Older children may use the computer for a job or e-mail with friends. Rursch found that in some cases, when the child left for college, so did the computer—the ultimate nonfarm impact. Although our data indicate that most spouses and farmers use their computer for word processing, it is difficult to place a value on this. Are they able to derive any unique advantages by doing this? Systematic measurement of some of these nonfarm impacts is needed, because this is now driving adoption among small farm households.

Nonfarm uses of computers by farm households closely resemble patterns of their counterparts in urban households. Our results parallel those of Dutton, Rogers, and Jun (1987) and Watkins and Brimm (1985), finding that general adoption patterns are strongly skewed toward high education and high-income households with children. Specifically:

- There is a strong tendency toward multiple users of computers in farm households. Although farmers use the computer for both farm and nonfarm purposes, spouses and children tend to use the computer more for nonfarm purposes.
- Spouses now are as likely as farmers to be named the primary operator of the household's computer. Uses, in addition to word processing, tend to be income generation or education.
- Children tend to use the computer primarily for education and entertainment.

Community Development Benefits or Drawbacks

The tremendous changes in agriculture, of which computerization is becoming a part, have had and will continue to have important implications for rural communities. The commercial farm sector, which utilizes sophisticated management techniques combined with access to capital, was said by Tweeten (1984) to represent "an extension of the urban-industrial process to the farm" (p. 19). Noncommercial farms, with their extensive reliance on off-farm employment, represent "the extension of the farm into urban-industrial society" (p. 19). The medium-sized farms are squeezed between these two sectors. Impacts of computers on rural communities will be shaped by what happens in these sectors. Overall, we know that those who have adopted the computer and Internet believe that they are important for the economic development of communities, and as a result they might be expected to advocate policies favoring the growth in use of these technologies.

Korsching, Hipple, and Abbott pointed out in Chapter 1 of this book that there is a difference between *economic development* and *community development*. The continued growth of commercial farms in rural areas has brought about economic development—more dollars flowing back into the hands of these farmers that can be invested in improvements to these farms. But the economic development that benefits these large farms, and the computers that make it possible for them to manage ever larger operations, may not lead to community development.

In the long term, larger farm operations and more mechanized agriculture mean fewer farmers in rural communities—farmers who once constituted the majority of workers in these communities. Now, they are a distinct minority. Their loss means fewer church members, fewer volunteers in community organizations, and fewer shoppers at the grocery store and pharmacy. Larger farm operations are also likely to continue a trend of orientation *beyond* the community that began when good roads were first built in rural areas. Good roads provided choice—the ability to shop beyond the local community, and the ability to buy seed corn, chemicals, and other inputs at a regional center. The computer, and especially the Internet, will make it possible for farmers to buy and sell nationally or even globally. While electronic connections enable farmers to e-mail each other and connect to the local community, the forces that shape the agricultural industry today are increasingly *vertical*. To understand them and negotiate with them, farmers with sufficient size to leverage their buying and selling will increasingly orient themselves to these vertical institutions. The large-scale farmer of the future may come to care increasingly less whether or not there is a local community. The farmer will have become the ultimate "cosmopolite" citizen—one who is oriented outside the community.

On the other hand, medium- and small-scale farmers may be able to use the "nonfarm" capabilities of the computer to find ways to continue to live in rural areas and prosper. With 66 percent of spouses and 56 percent of farmers in noncommercial farms working in nonfarm employment and operating a farm as a lifestyle choice, finding good jobs is a major challenge. To the extent that computers can enable such farm operators and spouses to work from their homes or train for better jobs in the community, they may strengthen the ability of these households to continue to live in rural areas and enjoy the lifestyle advantages of rural living. Improved distance education opportunities or access to doctors and clinics via the Internet and telemedicine would be other capabilities that might attract and hold families on farms. These families, unlike their counterparts with large commercial farms, are more oriented to communities because they depend on them for jobs.

Their computer skills and employment capabilities may contribute to community development. Because these families may have selected a rural lifestyle, they may also have selected their residence with a nearby community in mind.

NOTES

1. Each initial sample included approximately 1,000 farmers. All were surveyed by mail. When those sampled who were not farming or deceased were eliminated from the list, return rates averaged 65–75 percent except for the 1997 sample, which was 44 percent (Ns: 1982 = 544; 1989 = 649; 1997 = 302). One additional random sample was surveyed in 1984, but is not included in most of the analysis in this chapter to focus on three roughly equal time periods. Each of the random samples then became part of panels that were restudied in 1983, 1984, 1986, 1987, 1988, 1998. In 1999, another survey of all computer adopters is being conducted.

2. Online computer-based information sources were defined in the survey as "computer-based information systems where you use a computer to obtain information, such as CompuServe, America OnLine, @gOnline, and AGCAST." Data for use of farm videotex/teletext systems, another form of computer-based information, will be discussed later. These are more widely used.

3. For a multiple regression analysis of factors predicting both farm and nonfarm computer use, see Abbott & Yarbrough (1993).

4. They have access at schools, businesses, or other sites. National data show a similar trend.

PART III

MAKING THE CONNECTIONS

11

What Happens When Information Technologies Are Forced on Rural Community Organizations? The Case of Iowa State University Extension

Eric A. Abbott and Jennifer L. Gregg

INTRODUCTION

The process of computerization and adoption of new information technologies in rural communities is not one that happens quickly or uniformly. Some sectors of the community adopt more quickly than others and some are more successful in using these new technologies than others. Organizations are one place where adoption and use of these technologies can be studied. In this chapter, we examine one specific organization—Iowa State University Extension—that serves rural communities. In particular, we study the process by which this organization has been linked to the Internet, and the impacts of that connection for the organization and the community. The adoption of a wide-area network by Iowa State University Extension is used as a case study. In the case study, we pay close attention to three key questions:

1. What were the forces that brought about the adoption of a sophisticated wide-area network linking employees to the Internet and the World Wide Web?

2. What have been the major impacts of the new system on the functioning of this organization 2 years after adoption, both for internal and external communication tasks?

3. What have been the major impacts of the adoption of the system for the community?

FORCES SHAPING ADOPTION OF INTERACTIVE TECHNOLOGIES BY ORGANIZATIONS

Michael Dertouzos, author of *What Will Be* (1997), argues that the adoption of computers and other telecommunications technologies is part of the emergence of

"The Information Marketplace." This marketplace will arise as businesses and organizations discover more efficient and effective ways to find customers, do business, keep records, and communicate internally and externally. Changes will occur as organizations adopt "electronic bulldozers"—word processing systems, spreadsheets, knowbots, and other programs that enable them to do their work better (p. 253). Although Dertouzos acknowledges that the progress of these "bulldozers" will be slow because it takes time for people to change their habits and procedures and learn new ways of doing things, the concept of a bulldozer suggests a rather uniform and inevitable result.

When one corporate executive minimized the impact of new information technologies, pointing out that they were affecting only 5 percent of the economy, Dertouzos responded:

Some 95 percent of the economy out there are changing like wildfire under the impact of the Information Age, and you good people are oblivious to it all!! . . . A company that ignores the full reach and impact of the Information Marketplace is planning its future on willful ignorance. It will view as irrelevant new technologies that might improve its business, like the automization (sic) of office procedures; group work for task force deployment; data sockets and e-forms for sharing information with employees, customers, and suppliers; recruitment of employees over the nets—in other words, most of the tools (and associated benefits) discussed in this book. (p. 247)

The information marketplace will touch essentially all human activity. However, Dertouzos argues that the impact will not be uniformly significant—it could be very deep in sectors and activities where automation, group-work and the other areas can benefit from its adoption (p. 246). The information marketplace, he writes, "will transform our society over the next century as significantly as the two industrial revolutions, establishing itself solidly and rightfully as the Third Revolution in modern human history. It is big, exciting and awesome. We need not fear it any more or any less than people feared the other revolutions, because it carries similar promises and pitfalls. What we need to do, instead, is understand it, feel it, and embrace it so that we may use it to steer our future human course" (p. 306).

Dertouzos's vision for organizations in rural communities is clear: their managers must recognize how to reinvent their organizations to take advantage of the electronic bulldozers. Those that lack sufficient managerial skill or capital to do so are likely to be eliminated. Because large well-capitalized organizations and businesses can afford to have both good managers and the ability to invest in new technologies, they are likely to improve their competitive position. This leads to an "unfortunate instability," he says. "The painful conclusion is that left to its own devices, the information marketplace will increase the gap between rich and poor countries and between rich and poor people" (p. 241).

In *Computerization and Controversy: Value Conflicts and Social Choices*, Rob Kling (1996b) takes a different approach. Kling argues that projections made about the social consequences of computerization are based on overly rationalized views of organizations and markets and relatively romantic views of technology and human behavior. In fact, organizational patterns of computer adoption should be seen as the result of the promotion, lobbying, and financial support of groups whose interests are served by computer adoption rather than as the result of any inevitable revolution bulldozing its way through society. These supporting groups engage in what Iacono and Kling (1996) term "technological utopianism." In the same way that the Moral Majority raises funds and lobbies to support its causes, computer advocates such as the National Science Foundation, National Telecommunications and Information Administration, and industry groups provide funding and lobbying activities that have shaped not only the actual adoption of types of technologies, but also the cultural expectations that such technologies are needed. This approach has been used successfully, for example, to mobilize parents' groups to demand millions of dollars worth of computers for schools without any substantial evidence that they are resulting in increases in learning.

Mobilized by the belief that more computerization is better, coalitions of administrators, teachers, and parents are banding together to push for extensive computerization in classroom settings and a basic shift in the educational programs in public schools, universities, and libraries. Advocates of computer-based education promote utopian images of information-age schools where students learn in cooperative, discovery-oriented settings and where all teachers can be supportive, enthusiastic mentors. In fact, however, the deployment of technology in schools has generally not affected the day-to-day values and practices of teachers and students due to its threat to the existing order. But utopian visions resonate with parents, teachers, and school administrators who are concerned about how education will meet the challenges of the future (Iacono & Kling, 1996, p. 96).

The societal expectations set in motion by technological utopianism advocacy groups provide enthusiasm for the general computerization movement and organizing rationales (e.g., transition to a new information society, participation in virtual communities, and societal renewal) for unbounded computerization. In addition to advocacy groups, mass media add to cultural expectations about these new technologies. Kling (1996b) notes: "I am struck by the way in which the news media casually promote images of a technologically rich future while ignoring the way in which technologies can add cost, complexity and new dependencies to daily life. The glossy images also ignore key social choices about how to computerize and the ways in which different forms of computerization advance different values" (p. 23).

Studies have shown that within large organizations with hundreds of thousands of employees, workers were frequently segmented into coalitions that held conflicting views about which goals the organization should emphasize and which strategies would best achieve them. "While the computer projects favored by specific coalitions often had important elements of economic rationality, they often helped to strengthen the power of their champions as well. In fact, systems champions often exaggerated what was known about the economic value or necessity of specific projects" (pp. 119–120).

Both Dertouzos and Kling discuss studies dealing with what has come to be known as the "productivity paradox." Why, after spending so much money on computerization, aren't organizations becoming more productive? A major study by Stephen Roach, chief economist of Morgan Stanley, found little evidence of gains as a result of investments in computers in the United States during the 1980s (Roach, 1989). While most scholars agree that the paradox is a complex question for which there are as yet no clear answers, researchers have looked in different places for evidence. Dertouzos—citing a Xerox study of 630 companies—focused on poor management skills to explain it (pp. 270–271). Rogers (1986), in a study of adoption of computers in a California school system, noted that the "champions" who advocated computer adoption often were not actually the same persons who used them (p. 141). This led to situations where users were not well trained or did not even have a use for computer equipment that suddenly arrived in their classrooms. Kling includes the possibility that computerization was wrong in the first place as well as the possibility that the studies didn't measure productivity adequately.

Another researcher, John King (1996), focuses on the time required to increase productivity. He compares the current investments in computers to a period between 1890 and 1920 when electric power replaced steam power in factories. During the period, studies showed no evidence of gains in productivity per worker, he found. This was because electric power caused the entire way organizations were structured and run to change. When new organizational structures and systems were gradually developed, productivity per worker again shot up. King argues that we are now in a similar period for computerization. In King's view, there must be a confluence of three factors for productivity gains to take place: (1) the new technologies themselves, (2) the know-how needed to apply them successfully, and (3) the wholesale substitution of the new technologies and new methods for older ways of doing things (p. 247).

This kind of change takes a long time. Therefore, it may be another decade before the clear benefits of computer adoption will be demonstrated. In a study of adoption of several information technologies by offices and factories, Yates (1989) found a "near universality" of decades-long diffusion times for new information technologies among complex organizations (p. 250). Futurist Paul Saffo (as cited in Fidler, 1997) applies a 30-year rule to adoption of new technologies: "First decade—lots of excitement, lots of puzzlement, not a lot of penetration; second decade—lots of flux, penetration of the product into society begins; third decade—'Oh, so what?' Just a standard technology and everybody has it" (p. 9). Because the microcomputer only became widely available beginning in about 1980, at this point we are nearing the end of the second decade of adoption. In the third decade, according to Saffo, we should see routinization of use of computers.

HOW COMPUTERIZATION CHANGES ORGANIZATIONS

To this point, the discussion has focused on general issues relating to the underlying processes driving computerization of organizations. There also has been study of how organizations change as a result of computerization. One area of change of particular relevance to this research concerns the extent to which organizations become "flatter" or less hierarchical as a result of computerization and Internet communication. The reasoning is that the ability of lower-level employees to communicate to upper-level employees is enhanced by e-mail, thus permitting middle managers to be bypassed. In addition, the tendency of higher status individuals to dominate communication can be reduced when everyone shares the same electronic message system (Becker, Tennessen, & Young, 1995). Organizations would be expected to become more interactive, as many employees can be consulted for input before decisions are made, and new communication networks might form.

Ahuja and Carley (1998) define a *virtual organization* as "a geographically distributed organization whose members are bound by a long-term common interest or goal and who communicate and coordinate their work through information technology" (n.p.). In such organizations, they argue, communication structures should emerge via an amorphous web of connections, changing constantly in response to communication needs. However, a study of network communication among 66 members of a virtual organization showed that hierarchies of communication continued to be very important. Those identified as authority figures received more messages, as did those who demonstrated expertise in discussions. Thus, although messages did not necessarily follow a strict authority pattern, definite hierarchies did develop. They concluded that some of the early assumptions made by researchers about how organizations change in response to becoming virtual may not be as clear as originally thought. Other researchers (Dubrovsky, Kiesler, & Sethna, 1991; Kiesler & Sproull, 1992; Saunders, Robey, & Vaverek, 1994) have found that although messages tend to be equalized among members of a computer-mediated communication group at the beginning, when e-mailers do not know each other, a hierarchy tends to develop when the status of message recipients becomes known.

A second area of particular interest concerns adoption of new technologies that are imposed on employees from above. Participatory research, the human relations school of sociology, and diffusion research all suggest that an innovation that is neither desired nor understood by employees has a poor chance of being adopted and used (Rogers, 1995). Much organizational computerization is of this type—new machines and software simply arrive one day at an office. On the other hand, critical mass theory (Markus, 1987; Rogers, 1989) would predict that an innovation that is placed simultaneously in every office, giving all staff the opportunity to communicate electronically, would be quickly adopted because its potential utility would be maximized. With many information technologies such as facsimile (FAX) machines and e-mail, the utility of the device is determined by the number of people one wishes to communicate with that have a compatible technology.

A third question involves the relationship between organizational innovation in a community and spread effects (direct or indirect impacts on other community organizations as a result of adoption of a technology) for other community organizations or individuals. In a sense, the county Extension office and other local organizations function as participants in a community's information network—part of its social capital. Individual organization members often serve as bridges between themselves and community groups or citizens. When an organization such as Extension adopts a new technology, are there "trait-making" effects, demonstrating the benefits of the technology to other organizations? By trait making, Hirschman (1967) is referring to the ability of others in the community to understand and build upon the innovations introduced by an organization or change agent. Do citizens link to the new technology and benefit directly or indirectly from it? Tennessen, PonTell, Romine, & Motheral (1997) see a clear role for Extension in this area: "There are tremendous opportunities for Cooperative Extension on the Internet. These opportunities are for improved functionality of the Cooperative Extension system, and new opportunities for communities that sustain the Cooperative Extension system. . . . Many local Cooperative Extension offices may find a new role in facilitating community Internet development as part of its greater role in community outreach and life-long education" (n.p.).

The authors discussed above provide a base for the research questions that began the chapter. In our case study, we focus on the forces and interests that brought about the adoption by Iowa State University Extension of a wide-area communications network. We also examine not only adoption, but also the extent to which employees have actually changed communication patterns and perceptions as a result of computerization. Finally, we examine possible spread effects to the community from the network adoption.

Before presenting the case study, it is important to understand some of the history of Iowa State University Extension as well as current issues confronting the organization. Among them are the horizontal and vertical aspects of the Extension organization and how they have influenced organizational structure and innovation.

STRUCTURE AND INNOVATION IN THE IOWA STATE UNIVERSITY EXTENSION

Extension has a long history of providing specialized agricultural, youth, family, and community development information services to communities using a variety of information technologies. The need for localized agricultural information in Iowa and other states became apparent as soon as settlers began plowing farmland. Iowa's first territorial legislature, which met in 1838, passed an act providing for the incorporation of agricultural societies designed to help farmers solve their problems through discussions, competitive exhibits, and wide dissemination of known agricultural information (Bliss, 1960, p. 3). This legislation, and much that followed later, emphasized community organization and cost sharing as two key principles underlying what would later become the Iowa State University Extension. A third important principle—the linkage of county agricultural societies to agricultural colleges and research farms—came with the passage in 1862 of the Morrill Land Grant Act, which provided federal funds to each state for the purpose of establishing colleges of agricultural and mechanical arts to train farmers, engineers, and others for technical employment. In Iowa, even before passage of the Land Grant Act, funds were appropriated in 1858 for an agricultural college. Because the college at that point had no research results to report on, the legislature provided that college officials would collect agricultural knowledge from the county societies and pass them on.

In Iowa, technologies for communicating information to and from rural communities were varied. Local fairs and demonstrations led to the establishment of regional and eventually the first state fair in 1854. Although entertainment was a part of these fairs, educational programs, including demonstrations and lectures, were commonly offered. When Iowa State College of Agricultural and Mechanical Arts began accepting students, President A. S. Welch suggested that groups of 50 farmers or more invite faculty to travel to their communities and provide a 3-day educational institute. The first one was offered in 1870. An alternative mode of delivery of agricultural information was agricultural newspapers and magazines, which first appeared in the state in the 1840s. By the Civil War, 50 to 60 agricultural journals were being published, several of them in Iowa (Edwards, 1994).

Techniques for delivering agricultural information effectively were sharpened considerably through the efforts of Perry G. Holden, an Iowa State College faculty member who specialized in helping educate farmers about corn. In short courses, Holden used "a cartload of boxes, charts and other illustrative and demonstrational materials" (Bliss, 1960, p. 32). Later, in 1904, Holden used "corn trains"—special trains that stopped at local communities. Farmers would fill the special coach, and Holden used diagrams and charts mounted at one end of the car to deliver his message. In the 1906-07 year, Holden's special trains stopped at 670 towns, passing through 96 of the state's 99 counties, traveling nearly 10,000 miles, and lecturing to 127,000 farmers (p. 44). Other efforts were made to organize farmers' clubs, youth clubs, and a number of other local organizations.

On July 1, 1914, the Smith-Lever Cooperative Extension Act ushered in a new cooperative relationship between state Extension services and the U.S. Department of Agriculture. Organizationally, this represented a partnership between federal, state, and local groups that continues today. Although it links to vertical organizations outside the local communities, there is considerable local influence and control over the hiring and supervision of county employees. Thus, Extension is something of a hybrid—it is a government agency whose overall mission and practices are designed and controlled at the state and federal levels, but with substantial local input. In Iowa, originally, a local county had to raise $1,000 each year and register at least 200 farmer members to receive state and federal funds (Bliss, 1960, p. 226). At the local level today, Extension fosters 4-H youth clubs, advisory groups, and a number of services for families. Thus, it has had both local and outside information functions.

Three Types of Extension Information Flows

Groves (1978) found that three types of information patterns had developed within Extension. *Administrative information* flows were essentially hierarchical, flowing out from the campus of Iowa State University to each county office. In the early years, counties hired and evaluated their own Extension agents, which gave them more power as information players. But in the past few decades, all Extension staff have become Iowa State University employees and are hired and evaluated from Ames. Thus, the information flow has been strongly top-down.

A second information pattern concerned *content*—specific recommendations and research findings. Groves found that this type of information also followed a hierarchical pattern, with county directors communicating mainly with specialists in Ames. This pattern has become more pronounced as agriculture has become more specialized. In Extension's early years, county agents were expected to answer most of the questions of local residents. They were generalists—trained to assist in a wide variety of areas. By the 1960s, it was clear that the specialized nature of agriculture was making it impossible for county agents/directors to keep up. Area offices were then set up with specialists who could handle the highly technical questions about swine, pesticides, integrated pest management, and family and youth issues. County directors also communicated directly with state specialists at the University. By the late 1990s, Iowa State University Extension had begun to place specialists directly in local county offices, but they served multiple counties. County directors have come to serve more as facilitators, fielding local questions and then getting answers either from local or state specialists. Content information patterns thus in general have moved through county directors to multicounty or state specialists.

The third information pattern concerned *educational techniques*—what works or does not work when communicating with Extension clients. Groves found that this form of communication, unlike the other two, was often horizontal, flowing between county Extension offices.

How might a new communication technology change these flows? If the nature of the communication system had been to constrain the flow of information vertically within the organization, a wide-area network that permitted unlimited flows both horizontally and vertically would cause an increase in horizontal flows. Anyanwu (1982) found that county directors often tried to use either their own knowledge or locally available materials to answer questions. He found a predictable pattern of response to requests for information that was due to a combination of experience, organizational structure, and ease of use of communication channels.

This approach is followed in order: by solving problems by use of readily available prepared materials; talking to another Extension person; personally researching the problem and later providing the information to the client; referring the client to another Extension person; talking to someone outside of Extension; referring the client to someone outside of Extension or to the library; and finally, telling the client "I don't know" (p. 68).

Would a wide-area network increase contact with specialists or other county directors to answer questions? And to the extent that gatekeepers (county directors, secretaries, area directors, etc.) bottled up or controlled traditional information channels, the new system might be expected to "flatten" the hierarchy by permitting local staff to communicate directly with those at Iowa State University. However, if, as Ahuja and Carley (1998) note, communication flows are determined by functions they serve and status labels within the organization, then a wide-area network would only serve to maintain the previous communication structures, although it might increase the efficiencies of these flows by eliminating phone tag, permitting transfer of formatted documents, and making it much easier to send one message to many people.

Technological Innovation within Extension

Following a pattern established by its early history, Extension has continued to use new information technologies to provide information to staff and clients. In Iowa, for example:

- In the late 1970s and early 1980s Extension integrated pest management and other specialists used programmable hand-held calculators to help farmers improve management and record keeping (Whiting, 1981).
- In October 1982, Extension launched AGRI-VIEW, a teletext market information service that used the vertical blanking interval (unused lines on a regular television set) along with a decoder. Farmers could buy the decoder and then receive specialized market information to view on their television screens for no charge (Pfannkuch, 1988).
- In October 1983, Extension began sending home and family news releases to Iowa newspapers electronically. Case studies of adopting newspapers showed that there were both positive and negative aspects to delivering information in this new way (Abbott, 1986). Speed was an advantage, and releases did not need to be rekeyboarded to be used, but there was often confusion at the media concerning how to download the releases and where to route them, and a 24-hour automatic "kill" policy designed to discard old news often eliminated them from the system before an editor even saw them.

The organization has also been impacted by decisions at the federal level. The U.S. Department of Agriculture's decision in 1987 to establish a Rural Information Center (RIC) system to make publications and other information available in electronic form spurred state Extension leaders to take similar action. Although envisioned to operate primarily through local Extension offices, local officials and citizens began using the RIC system directly as well. From 1987 to 1991, the RIC operated via an 800 phone number. In 1991, an electronic bulletin board was added at the National Agricultural Library. In 1994, an Internet gopher began providing materials electronically. The national Rural Information Center Health Service was added in 1990 (John, 1995).

Extension information reaches clients in a variety of ways. In 1992, for example, 11 percent of Iowa's farmers said they had either visited the Extension office or called an agent to get a copy of the annual Iowa Corn Yield Test Report. However, 90 percent said they could recall seeing the report reprint included as an insert in a farm newspaper. Only 8 percent said they had not seen it from either source (Narigon, 1992, p. 56). Seventy-three percent of states make at least some Extension publications available through public libraries, and 69 percent provide them free (Fett, 1993).

Given the organization's history of innovation with information technologies, it is not surprising that it was a leader in adoption of high-speed Internet and World Wide Web communications at the county level. Building on our understanding of how Extension is organized and functioning, and our earlier consideration of the ways in which new information technologies might be transformative, we now turn our attention to the case study. The case study begins with a brief organizational history of communication technologies and flows leading up to installation of a wide-area network, and then utilizes data from three studies at four time periods to examine how installation of the wide-area network was received by Extension staff, and how it changed communication patterns within and outside the organization.

THE ADOPTION AND IMPACTS OF A HIGH-SPEED, WIDE-AREA NETWORK ON IOWA STATE UNIVERSITY EXTENSION

Iowa State University Extension launched its first computer-mediated communication system in 1984, when EXNET was inaugurated. Much of the initiative for establishing EXNET came from integrated pest management specialists at Iowa State who began creating a computerized pest monitoring network linking area Extension offices in 1980. Agricultural specialists who were using programmable calculators to help farmers improve management and decision making also saw the potential for computers and e-mail. In 1982, a committee was formed, comprised primarily of university faculty and staff but with county office representation as well, to consider how computers and computer-mediated communication might be useful. At the time, only 7 percent of county Extension offices had a computer (Orr, 1985). The committee's recommendations were endorsed by Extension administration, which was looking for a way to speed up and improve communication with Iowa's 100 county Extension offices. A formal proposal was developed in June 1983, and the hardware to operate EXNET arrived in March 1984. The system began operating in May 1984 (Lee, 1987).

To encourage county offices to adopt, the central office offered to help pay for the cost of an Apple III computer (the model adopted by central administration) along with a modem and a full range of software (word processing, spreadsheet, database). For most county offices, it would be the first computer in the office. Adoption and use of EXNET was rapid but not universal in its early years; 70 percent of counties had adopted by the end of 1984, but those with fewer economic resources or unsupportive Extension councils did not invest. To serve both adopters and nonadopters, communication messages were sent out in the traditional mailed form as well as the new electronic form. By early 1986, all but 11 of the county offices had adopted EXNET (although some elected to buy IBM computers instead of the Apple III). In that year, central Extension administration decided there could be significant savings if the entire system could switch to EXNET for internal messages. Thus, counties not yet on the system were told that they either had to hook up (using a combination of county resources and subsidies) or they would stop receiving news releases, market reports, and routine Extension directives altogether. During a transitional period, some counties did without these messages, but eventually all were connected to the system (Lee, 1987).

Overall use and reliance on the EXNET system depended in part on support from office or area champions, individuals who pushed for increased use of the system. Counties with such support showed higher use patterns than the others (Lee, 1987). Despite differences in level of use, patterns of use in most county offices were relatively uniform. A secretary would be selected to log on to the EXNET system, usually once a day, and download or upload messages. The messages included general messages for all Extension staff, messages for specialized groups, and even messages directed at particular individuals. County staff agreed that many messages were not relevant to their offices at all. In almost every case, *all* messages were downloaded, printed, and then bound in a three-ring notebook for inspection by Extension staff when they had time. Because a secretary downloaded the messages, it was felt that a person in this position did not have the authority to decide if a message was relevant or not, and this is why almost all of the 100 county offices immediately printed out everything that came to them electronically. In 1994, when in-depth interviews were conducted in 15 county offices (Gregg, 1997), bookcases full of these three-ring notebooks filled with all the EXNET messages ever received were found. No one had ever specified how long they should be kept. Often, secretaries would notice what seemed to be important messages, and they would place these on the desk of the person or leave a note that they should check their messages. Although a minority of directors and field specialists embraced the system and acquired their own EXNET accounts so that they could communicate directly, the general pattern was to receive and send messages only indirectly. Because of this pattern, EXNET was used mainly for routine messages. The phone or FAX machine was used when something urgent or important needed to be communicated.

In early 1994, county offices were still in a text-transmission mode. Text could be rapidly sent via EXNET or a FAX machine, and the central office had been sending electronic text news releases to media for about 10 years. However, planning was well under way for a new system, EXNET-IP (IP for Internet Protocol), that would provide a major change in Extension communication. Recommendations at least two years earlier had been made to increase the capabilities of Extension "to use modern information and communication technologies" (Recommendations . . . field operation, 1992). Also emphasized was the goal of having "an appropriate computer for every member of the field staff, to provide a modern computer network to link Extension staff together, and to link them to a variety of appropriate databases" (Recommendations . . . field operation, 1992).

By 1994, several more incentives had been provided to encourage EXNET use. In the early 1990s, every field specialist was given a Zenith laptop computer with modem and software. Although these specialists reported that they often did not receive much training in how to use their laptops, this did encourage them to get their own EXNET accounts and some of them began checking their e-mail as they traveled around the state. Some directors also began acquiring computers. By early 1995, two-thirds of Extension employees had their own EXNET accounts, but they were also continuing to get e-mail through the three-ring notebooks in their county offices.

A New Wide-Area Network

The process of getting a new system that came to be known as EXNET-IP began in 1993, when Iowa State University Extension received grants from the National Science Foundation and the U.S. Department of Agriculture to connect all 107 county and area Extension offices through a high-speed wide-area network. The network would link all offices with each other and Iowa State University, and through the University provide a link to the Internet and the World Wide Web. Installation of the new system was begun in mid-1994 and continued through 1995. In addition to external linkages, each office would have its own local area network, making it possible to share printers, messages, and files on a local server. To maximize the benefit, county offices would need to provide a computer for every employee. At that time, only about two-thirds of employees had their own computers; the others shared or did without. The grant provided a server at each office, connected each computer to the server and provided software including the first World Wide Web browser, MOSAIC, and Eudora for e-mail messages.

The power of the new system and the high hopes for it were reflected in official statements made at the time of its installation. Because the new system was capable of high speed transmissions of all types of information and came with graphics software (MOSAIC), it now became possible to transmit publications, newsletters, reports, visuals, and even audio and video clips over the system. While we currently take much of this for granted, it was a revolutionary change for county offices and the rural communities they served in 1994. In many Iowa rural communities, installation of EXNET-IP made Extension the state-of-the-art center for Internet telecommunications. At the same time, a companion development in Iowa, the state-owned fiber optic network, provided two-way real-time video to one school in each county.

Robert Anderson, Jr. (1993), Extension director when EXNET-IP was implemented, emphasized access to information, sharing with other agencies, and quick response time as important values of the system:

By connecting all of our staff to this global computer network (called Internet), ISU Extension will improve service to clients, develop new patterns of work and expand cooperation with other agencies and organizations. Improved services will occur by more sharply targeting responses to client needs and by providing more up-to-date information faster. Staff will spend more time doing creative problem solving with clients and less time searching for information and writing reports. Expanded cooperation with other agencies will happen through broader access to knowledge and by connecting our staff to the resources of other organizations. . . . Increasingly, staff will search a centralized database for information, download relevant sections, and either print them out, or send them via electronic mail or FAX to their clients. (p. 1)

EXNET-IP's coordinator, Tom Quinn, envisioned that the system would make it possible for:

- Iowa residents with access to a computer and modem to dial into the databases;
- clients to leave electronic mail messages requesting information or assistance and print out publications on topics of interest;
- county Extension council members to review the minutes of the last council meeting or indicate their date preference for the next council meeting;
- 4-H members to enroll their projects from home or school computers; and
- Extension to conduct electronic straw polls of Iowa residents to determine their program preferences. (Anderson, 1993, p. 1)

The new system had the advantages of being comprehensive and providing a uniform high-speed system for every county office. It was envisioned mainly by Extension administrators and others who were beginning to understand the major changes to digital communication planned by the U.S. Department of Agriculture through its Rural Information Center project. It was also seen as a means of continuing to provide services in an era of tighter budgets and fewer employees. Even in 1999, many other state systems that had not adopted a comprehensive high-speed communication system continued to struggle with dial-up limited systems in some offices, while other offices had moved independently to much more modern systems.

Study Shows Initial Opposition to EXNET-IP by County Staff

Despite the high-level visions, the announcement of the new system was not universally applauded, especially in county offices. A series of in-depth interviews conducted with all staff at 15 county offices from March to April 1994 before the system was installed at their offices showed that many believed the money spent on this system should have been used for "more important" things such as local staffing and programs. It is clear that had a vote been taken of county directors, it is unlikely that the system would have ever been installed. In addition to criticism of the system, local staff also were unhappy that the new system was being forced on them, and that they would have to find the funds to buy the computers necessary to see that every employee had one. In addition, there were other criticisms due to delays in installation.

This opposition to the system provides an interesting opportunity to test several ideas concerning information innovations in an organization:

1. Would opposition to the system lead to its failure, or lack of use, as predicted by participatory theorists?
2. Would useful characteristics of a superior information technology, combined with a "critical mass"—the ability to link to all Extension employees—lead to successful adoption and use?

As the system was being installed, training of local staff was provided at each office, and additional training was provided at Iowa State University. Because installation was not on schedule, some of the training at Iowa State occurred before installation of the system. Training covered how to send and receive e-mail, how to use the World Wide Web, and how to send and receive documents.

In March and April 1995, staff in the same 15 counties were reinterviewed to assess preliminary use of the new system because their offices were the first to be connected. In late fall 1995, just as most of the other offices were hooked up, a benchmark survey of all county and area Extension staff except those in the 15 counties was undertaken to assess EXNET use before the new system. Responses were received from 353 staff, or 77.5 percent. The final survey in late 1997, also of all 535 county and area staff, assessed the impacts of the new system. (A chronology of major events is shown in Figure 11.1).

The second survey repeated a number of questions from the 1995 survey and included special new measures of system use and evaluation. A total of 405, or 82 percent, of county and area staff responded. Data from both surveys were designed to assess several areas of system impact:

- What was the level of adoption and use of the system after 2 years?
- How extensive was the use of system features by employees?
- What value do Extension employees place on a system they criticized so harshly in 1994 and 1995?
- How has the system changed communication channels used to reach various groups?
- Is there any evidence of a change in the amount of horizontal or vertical communication as a result of the new system?

Figure 11.1
A Chronology of Major Events and Studies Concerning the Iowa EXNET and EXNET-IP Systems

1984	Initiation of original EXNET system; all but 11 counties were hooked up by 1996.
1994	March and April: Pre-change study done in 15 counties; 80 in-depth interviews conducted. Mid-year: Actual installation of EXNET-IP wide area network begins.
1995	EXNET-IP installed in most counties by the end of the year. March and April: Post-test of study in same 15 counties; in-depth interviews conducted. November and December: Mail survey of all Extension county and area staff except those in 15 counties; 353 surveys returned.
1997	November: survey of all Extension county and area staff; 405 surveys returned.

Levels of Adoption and Use of EXNET-IP

In 1995, 81 percent of Extension county staff had a computer; in 1997, that figure was 92 percent. Staff with their own e-mail accounts had increased from 66 percent to 95 percent. In terms of critical mass theory, the 1997 figures are close enough to universal to suggest that the organization could convert much of its communication to electronic forms. Critical mass theory suggests that the higher the level of adoption of a new information technology, the greater will be the use of the technology and the benefit to the organization from its adoption. Beyond access, frequency of use of the system increased dramatically when the original secretary-operated dial-up system was replaced by one in which an employee, at the touch of a mouse, could check mail. Table 11.1 shows the increase in accessing the system and number of messages sent and received in an average week.

These results show a substantial increase in use of the e-mail system, with 60 percent of staff now checking their mail four or more times per day and another 24 percent at least 2 to 3 times per day. With frequency of use this high, the system can now be used efficiently for perishable information. In the 1995 interviews, respondents said they would use e-mail only if they personally knew that the other person was a frequent user and would check for messages. Otherwise, they used FAX or phone. Now, much of the routine and even important messages can be carried via

Table 11.1
Frequency Accessing E-mail Account, World Wide Web, and Number of Messages Sent and Received, 1995 and 1997 Surveys of Iowa Extension County and Area Staff

Frequency of accessing e-mail account	1995 Survey	1997 Survey
More than 4 times per day	1%	60%
2 to 3 times per day	19	24
Once per day	44	8
Once or twice per week	19	6
Less than once per week	10	1
Never access my e-mail	8	0
Average number of messages *received* per week	27	71
Average number of messages *sent* per week	4	24
Frequency of accessing World Wide Web browser	*	
More than 4 times per day	*	8
2 to 3 times per day	*	21
Once per day	*	221
Once or twice per week	*	26
Less than once per week	*	15
Never used it	*	9

Note: * This question was not asked in 1995; WWW was too new.

e-mail with the knowledge that employees are checking for messages several times each day. The savings from this change are substantial for local offices that used to pay thousands of dollars per year in FAX bills. Now, e-mail, whose costs are absorbed by the state system, can be substituted. Results also show a threefold increase in messages received and almost a sixfold increase in messages sent.

Results concerning use of the World Wide Web (first through MOSAIC and later through Netscape or Microsoft Internet Explorer) show that use of the browser is at a much lower level than e-mail. With all the public discussion and attention to the Web, it is interesting that e-mail is much more important for many users. These findings are similar to other user studies (Buehler, 1997). In analyzing the reasons for the failure of Knight-Ridder's multi-million-dollar Viewtron experiment (a videotex system linking homes interactively to news, entertainment, and e-mail via a computer in the home), Fidler (1997) noted that while *providers* of the system envisioned delivery of news and entertainment, *users* of the system were more interested in e-mail with one another (p. 155).

Another set of questions involved the *depth* of use of the features of the system. Fifteen items were listed, and respondents were asked to indicate whether or not they had used each one. The items were designed to measure adoption of more advanced capabilities of the system. These questions were not asked in 1995, because few of these features were available then. Table 11.2 shows that some features are now in common use, including the ability to use Acrobat's PDF coding (a software program permitting sophisticated publications to be made accessible in their original format to a variety of clients). This is especially important because county offices must use Acrobat to download many current Extension publications. The high numbers associated with receipt of news releases and attached files also indicates an ability by most staff to create a document in Microsoft Word or a similar format and then transmit that document along with its text, tables, and charts to another user. The 80 percent of respondents who have set a bookmark have learned one of the basic building blocks for effective use of the Web. However, only a few have created their own Web page. More than two-thirds have created an e-mail list, which is one of the ways in which e-mail can save time and provide efficient communication to groups.

A final question concerning the depth of integration of electronic forms into county offices concerned the extent to which five types of materials—reports, publications, meeting notes, correspondence, and newsletters—are archived in electronic or printed form. Results in Table 11.3 show that correspondence was the category most likely to be archived in electronic form, and publications were least likely. However, the significant thing to note is the general percentage of electronic archiving that is occurring across topics. More than one-third of reports are stored either totally or mostly in electronic form. Almost half of all correspondence is stored electronically. This represents a transformational change over the last five years.

One additional reason why the extent of electronic archiving and availability of publications is important concerns the overall flow of publications and other materials from the top of an organization to the bottom. In Extension, at the federal

Table 11.2
Adoption of Specific Features Available on EXNET-IP, 1997 Survey of Iowa Extension County and Area Staff

Feature	1997 Percentage Adoption
Set a bookmark for a Web address	81%
Successfully sent or received an attached file	77
Downloaded a news release	73
Used e-mail list	69
Created an e-mail list	67
Used Acrobat reader to access a publication	62
Downloaded Acrobat reader (necessary to view Extension publications prepared in PDF mode)	59
Used Telnet to access your account from a remote site	46
Trained anyone to use a software package	42
Set e-mail filters	40
Communicated electronically with someone in the media	40
Used a LISTERV	39
Participated in a discussion group on the Internet	23
Successfully downloaded and played an audio or video file	22
Published a newsletter through e-mail or a Web page	12
Created a Web page	11

Table 11.3
Electronic Archiving of Extension Documents, 1997 Survey of Iowa Extension County and Area Staff

Material	Almost All Electronic	Mostly Electronic	Half Electronic Half Paper	MostlyPaper	Almost All Paper
Reports	14%	20%	36%	19%	11%
Publications	5	9	26	40	19
Meeting notes	12	22	28	24	15
Correspondence	17	28	33	13	8
Newsletters	11	15	29	27	18

and university levels, extensive conversion to electronic formats has been occurring, and administrators envision the majority of Extension materials being in electronic form in the next few years. One might think of state and federal offices as being at the top of an information technology-rich organizational pyramid. As material moves down the pyramid, one often finds a less information technology-rich environment. This may mean that persons at lower levels are unable to access certain high technology materials at all (such as what is on the World Wide Web), or that they must convert it to some other form to use it. By providing a wide-area network connection in every county office, Extension has enabled lower-level employees to have the same electronic access as those at the top. However, in order for the information pyramid to function seamlessly, clients of Extension would also need to be able to receive materials in this same electronic form. If they cannot, a "translation" or "conversion" function would have to occur. Thus, an organization such as Extension that converts to electronic forms from the top level to the county offices saves tremendously on printing and distribution costs. However, these costs are then shifted to the county level, where local staff must access materials electronically, and then print them or convert them to other forms for their ultimate users. This task can be both expensive and time-consuming.

Assessments of System Value

Three questions were included to gain overall assessments of three aspects of the changes represented by the shift to EXNET-IP: (a) An assessment of the importance of a computer to carry out respondents' jobs; (b) an overall rating of EXNET-IP; and (c) an assessment of the overall importance of rural telecommunications technologies for rural economic development. Results (Table 11.4) show that respondents have become very dependent on computers, place a very high value on EXNET-IP, and believe that telecommunications technologies such as EXNET-IP will greatly strengthen rural communities. These results, especially the ratings of EXNET-IP are striking given the negative assessment of the system when it was first imposed. Because much of the participatory literature suggests that systems imposed from above are doomed to failure, it is worth examining some of the reasons why a staff that opposed a system should embrace it so positively 2 years later.

One factor has to do with the technical characteristics of the new system. Although the original EXNET did enable county offices to communicate via e-mail, 28 percent of staff rated it as either very difficult or somewhat difficult to use. Furthermore, as has been noted, staff often had to work through a secretary or wait for an available computer to send or receive messages. When the new system arrived with Eudora software, each person's messages came to his or her own computer, and message authors and titles could quickly be scanned. The system was integrated into the computer's word processing system, so that files could be created with the word processing program and then e-mailed. In short, the system was technically far superior to its predecessor. In addition, it was combined with a change to provide almost all employees with their own computer—another major

Table 11.4
Three Measures of the Value of New Telecommunications Technologies, 1997 Survey of
Iowa Extension County and Area Staff

How important is your computer for you to successfully carry out your job?	
My computer has become an essential tool for doing my job	87%
My computer is important to my job, but it is not essential	8
My computer is used in my job, but it is not very important	2
I utilize my computer very little in my job	3
Considering your experience with EXNET-IP, how would you rate the system overall?	
Very positively	67
Somewhat positively	24
Neutral	7
Somewhat negatively	1
Very negatively	1
There is a divided opinion about the importance of telecommunication technologies for economic development in Iowa's rural communities. Listed below are three statements about possible impacts. Check the one that best reflects your view.	
Recent advances in telecommunications will serve to strengthen the economies of rural communities by providing them access to the information highway.	79
Recent advances in telecommunications will have little impact upon the economies of rural communities.	11
Recent advances in telecommunications will serve to widen the economic gap between rural and urban communities because rural areas will be comparatively information poor	3
Other	6

change that has been shown to greatly increase staff use of computers. A third important factor was the introduction of MOSAIC, the first graphically oriented Web browser. This presented new and exciting possibilities that had been unknown before the new system was installed. Finally, access to the new system permitted staff to see how the USDA and other agencies were now making materials available in electronic form. The conclusion: when an innovation meets the perceived needs of its users in a way that is clearly superior to the old way, it will be adopted and embraced even if originally opposed. A corollary might be: participatory methods of change work best when those contemplating the changes have a clear idea of their alternatives and the potential impacts of the changes. When the changes are less well understood—and the rapid transformation of information into electronic form was not well understood by county level Extension staff during the early 1990s—local participation is less effective. That said, there have been several areas in which the system has not been used as effectively as it might have been that relate to its hierarchical origins. These are discussed later.

Extension staff are very optimistic about the potential of telecommunications technologies for rural economic development—more positive than responses received from surveys of either librarians or newspapers, and similar to farm computer adopters. While this might represent a positive spirit about change that is often found in the Extension system and its concern about economic development, it is also possible that it results at least in part from a positive experience with such a technology. This positive view might cause Extension staff to be more active than some librarians and newspaper editors in efforts to integrate such technologies into rural communities.

Use of EXNET-IP to Communicate with Different Audiences

In late 1994, when in-depth interviews were held with Extension staff, several respondents laughed at the idea of communicating with farmers and other Extension clients via e-mail. Farmers and other clients simply weren't connected, they said, and didn't seem very interested in getting connected. While 4-H specialists noted that youth were being exposed to computers at school, the 4-H program specialists often were the last ones at the county office to get a computer. Three years later, although the majority of rural residents still lack Internet access, things have changed considerably. Slightly less than half of the respondents said they now receive e-mail from their clients, a dramatic increase from the 2 percent figure 2 years earlier (Table 11.5). In 1994, several respondents also thought it was ridiculous to think of sending e-mail to persons working in the same county Extension office. "I can just lean around the corner and yell down the hall," said one. However, by 1997, 87 percent reported receiving e-mail from someone in their own office. Within the state Extension system itself, communication via e-mail was high in 1995 and continued to be high in 1997. However, e-mail with nonExtension persons in Iowa and outside the state increased dramatically from 1995 to 1997.

Table 11.5
Changes in Reported Audiences from Whom E-mail Messages Are Received, 1995 and 1997 Iowa Surveys of Extension County and Area Staff

Audience	1995	1997
Clients	1.9%	45.4%
Own extension office	9.0	86.5
Another county office	90.1	97.6
Area extension office	82.1	89.4
Iowa State University	92.0	97.4
Other Iowans	21.7	65.4
Persons outside Iowa	15.6	59.1

The potential use of e-mail, Web pages, and other means of communication by Extension staff to reach target audiences are limited by the access of those audiences as well as the extent to which audience members rely on those channels for information. In both respects there have been increases between 1995 and 1997. To examine the change in perception about how best to communicate with various audiences, respondents were asked to provide their first and second choices of "best" ways to reach the same seven audiences listed in Table 11.5. While Table 11.5 shows the percentages reporting *any* communication activity, Tables 11.6A and 11.6B show the "best" two ways of communicating with each audience in 1995 and 1997. This is a much more powerful indicator of how Extension staff have changed their communication channels. Results show that while first choices remain fairly traditional, especially for reaching clients, the second choice for communicating within the office has shifted dramatically to e-mail. Communication with other persons in Iowa, and with persons outside the state, also have shown dramatic shifts to e-mail as the "best" channel. E-mail is by far the most popular choice for messages within the Extension system itself, and FAXing has declined. For clients, one additional question in 1997 asked respondents to rate the readiness of "your clients now" to receive information in computerized forms. A total of 24 percent said clients were ready "now." Another 15 percent said clients would be ready in a year, 31 percent in 2 to 3 years, 16 percent in 3 to 5 years, and 14 percent more than 5 years or never.

Table 11.6A
First and Second "Best" Ways to Communicate with Extension Audiences, 1995 Survey of Iowa Extension County and Area Staff

1995 Survey	Face to Face	Phone	U.S. Mail	FAX	E-mail	Other
Clients 1st	92%	7%	0%	0%	0%	1%
Clients 2nd	5	85	8	0	0	2
Own office 1st	98	1	0	0	0	1
Own office 2nd	4	40	4	3	30	19
Another Ext. county 1st	17	53	2	1	25	1
Another Ext. county 2nd	1	26	14	28	28	3
Area Ext. office 1st	16	41	6	1	34	2
Area Ext. office 2nd	1	25	15	30	27	2
Iowa State 1st	11	39	9	1	38	2
Iowa State 2nd	0	25	17	30	25	3
Other Iowans 1st	11	45	15	5	24	1
Other Iowans 2nd	0	26	31	28	13	2
Outside Iowa 1st	8	39	22	7	23	1
Outside Iowa 2nd	0	26	32	27	13	2

Table 11.6B
First and Second "Best" Ways to Communicate with Extension Audiences, 1997 Survey
of Iowa Extension County and Area Staff

1997 Survey	Face to Face	Phone	U.S. Mail	FAX	E-mail	Other
Clients 1st	77%	14%	4%	1%	2%	2%
Clients 2nd	10	70	11	1	3	5
Own office 1st	89	3	0	0	8	< 1
Own office 2nd	7	16	1	1	74	2
Another Ext. county 1st	6	20	1	< 1	72	< 1
Another Ext. county 2nd	3	64	2	26	22	< 1
Area Ext. office 1st	4	21	2	< 1	73	< 1
Area Ext. office 2nd	3	58	7	9	23	< 1
Iowa State 1st	3	20	3	1	72	< 1
Iowa State 2nd	2	60	6	8	22	2
Other Iowans 1st	4	31	6	1	57	2
Other Iowans 2nd	1	48	19	10	19	3
Outside Iowa 1st	4	28	10	1	55	2
Outside Iowa 2nd	1	39	25	11	21	4

Changes in Horizontal or Vertical Communication Patterns

Three questions concerned whether or not patterns of horizontal and vertical communication have changed because of the use of EXNET-IP technology. Respondents were asked "Do you believe you have increased or decreased your communication with: (a) persons in similar job positions to yours [horizontal communication] as a result of EXNET-IP; (b) campus Extension staff [vertical communication]; or (c) persons in similar job positions but outside the state [horizontal communication]." Results in Table 11.7 show an *increase* in all three categories, indicating respondents believe both horizontal and vertical communications have increased. The percentages of change are highest for the first category—those in similar job positions in Iowa, with 83 percent saying it had increased at least somewhat. But 75 percent also said communication with campus Extension staff had increased at least somewhat. Slightly less than 40 percent believe communication has increased with persons in similar positions outside the state.

Effect of Organizational Position on Use of EXNET-IP

Not all staff benefit equally from installation of computers. Nor do they all value them equally. When Extension county offices were studied in 1994–95, a signifi-

Table 11.7
Change in Patterns of Horizontal and Vertical Communication, 1997 Survey of Iowa
Extension County and Area Staff

	Similar Job Positions in Iowa	Campus Extension Staff	Similar Job Position Outside Iowa
Increased greatly	39%	27%	11%
Increased somewhat	44	48	28
Stayed the same	16	23	59
Decreased somewhat	2	3	1
Decreased greatly	0	0	1

cant gap was found between use of EXNET and position in the organization. In a county office, there are four categories of employees: directors, field specialists, program assistants, and office assistants. Directors supervise office staff and work with local Extension councils. They are typically very oriented toward the counties they serve. Field specialists in crops, livestock, horticulture, the family, and other areas have a home office in one county but serve a number of counties. Field specialists were all given laptop computers with modems several years ago. Program assistants are typically involved in youth and 4-H work, although some work in nutrition, the family, and other areas. They are often part-time employees and tend to focus on the needs of their specific counties. Office assistants handle secretarial work and are usually full-time employees. They were key individuals in most offices uploading and downloading EXNET messages.

The 1995 results found that program assistants were least likely to have either a computer or an EXNET account. For this reason, the 1997 results were also broken down by the four positions. Even though the percentage of adopters and users of EXNET-IP is very high, the group most likely not to have a computer and not to send or receive many messages is program assistants. A total of 27 percent of program assistants still lack a computer of their own. Ninety-eight percent of other types of employees have one. Program assistants also are slightly less positive in their rating of EXNET-IP (although they are still very positive) (Table 11.8). In addition, program assistants are significantly less likely to rate a computer as essential to their jobs. One-fourth of program assistants don't use the World Wide Web at all, and as a group they score significantly lower in Web uses than the other groups. Program assistants also indicated that their patterns of communication had changed the least, either horizontally or vertically, since the installation of EXNET-IP.

There were significant differences between program assistants and directors/field specialists in the overall number of e-mails sent and received. While directors said they usually receive an average of 117 messages per week, program assistants said they receive 40 (field specialists said 94, and office assistants said 56). Program assistants and office assistants initiate significantly fewer messages (an average of 14 each) than directors (39) or field specialists (35) in an average week. Finally,

Table 11.8

Relationship between Organizational Position and Use of EXNET-IP, 1997 Survey of
Iowa Extension County and Area Staff

	Directors	Field Specialists	Program Assistant	Office Assistant	F value & Probability
Overall rating of EXNET-IP	3.75	3.66	3.40	3.58	F = 3.57
4 = very positive					p < .014
0 = very negative					
Importance of computers to job	2.92	2.96	2.48	2.88	F = 14.1
3 = essential					p < .000
0 = very little					
Mastery of WWW	2.17	2.33	1.53	2.09	F = 15.6
3 = large number of uses					p < .000
0 = not a user					
Has peer communication increased?	3.17	3.33	3.01	3.19	F = 2.72
4 = great increase					p < .05
0 = decreased greatly					
Has Iowa State communication increased?	2.88	3.11	2.92	2.98	F = 1.49
(same as peer)					p < n.s.
Has outside communication increased?	2.58	2.80	2.24	2.29	F = 12.9
(same as peer)					p < .000
Rating score of applications used	9.06	9.78	4.80	7.50	F = 41.6
(16 possible)					p < .000

Note: Values represent mean scores. Respondents rated items on 0–3 or 0–4 scales for the first four variables. For the rating score, one point was awarded for each of the 16 applications used.

program assistants scored lowest on the number of specific computer applications they had mastered—only half as many as field specialists. Regular access to and use of a computer on one's own desk is closely linked to acquiring more skills and uses.

Community Spread Effects

The original EXNET-IP project coordinator envisioned a system that would provide citizens with the ability to dial in to Extension databases. However, a number of decisions to date by technologists working with the system and concerns about competition with private providers have largely prevented direct public access to the system. In most communities, citizens are generally unaware that the most advanced Internet system in their area can be found at the local Extension office. When the system was installed, technologists expressed serious concerns about the capacity of the servers and other equipment to carry the volume of messages that direct citizen access could generate. In addition, the Iowa Network Services (INS) organization serving the state's private rural telephone cooperatives expressed strong concern about Extension providing access to citizens. INS insisted that citizens pay it to gain Internet access. Although 15 of the state's county offices were given Public Access terminals that could provide citizens with direct linkages to Iowa State University, the great majority of offices do not provide public access of any kind. Clients request materials, and office staff use their computers to provide responses.

It is important to note that this technology could still be in an early transitional stage. The recommendations of the Future Technology Committee of Iowa State University Extension still list as Goal 1 that "[c]ounty Extension offices will have public 'learning centers' with various educational technologies—public access computers, video-conferencing equipment, satellite, etc." Specifically, the committee recommended that 90 percent of county offices have public access computers by the year 2000 (Future Technology Committee, 1997). Besides direct access, there are a number of other ways in which Extension might use its wide-area network to benefit communities. It could, for example, begin a local Web site for businesses, education, or community groups. The site could not only provide useful local information but could also link to other sites and databases, including those useful for community economic development.

Another important way in which Extension's experience with a wide-area network might be useful to a community is by assisting that community in organizing itself to assess and take advantage of information technologies. In fact, there is high demand for such services by communities. In February 1998, the two Iowa State University Extension employees who work with communities in this way reported that they could not keep up with the demand for assistance. They also noted that it takes a long time to help a community think through its information-age resources and make decisions. What becomes apparent is that most Extension staff do *not* work with communities in this area and do not see themselves as qualified to assist communities in making these types of decisions. Thus, the fact that Extension staff know how to use computers to access databases and link to outside resources continues to be useful to clients only indirectly, as Extension staff provide them with answers to their questions. By and large, the existence of the system is not having much direct effect in terms of demonstrating to the community the potential of these systems or in assisting communities to adopt. In fact, the decision by Extension to go it alone and install its own wide-area network has in some ways hindered community adoption by draining off demand for Internet services that could have been aggregated by Extension, schools, libraries, hospitals, and local government. Thus, it has reduced the interest of private firms in providing such services.

CONCLUSIONS

Although the long-term impacts of the imposition of a high-speed, wide-area network on Iowa State University Extension will not be known for several years, it is possible to make some observations at this point about organizational adoption of new information technologies in rural areas. Extension is somewhat unique in both its structure and mission, but its experiences may provide lessons for other organizations considering how to use new information technologies.

Support for the Positions of Dertouzos and Kling

Dertouzos. It was the enlightened management of Extension, for example, that saw the potential for a wide-area network to speed delivery of text, graphics, and even video material. To Dertouzos, good management is the key to understanding and using the "electronic bulldozers." Supporting Dertouzos is the fact that after installation, staff praise the system, use it heavily, and are beginning the process of transferring the knowledge of the organization into digital forms. From the case study, we conclude that hierarchical imposition of a uniform Internet system can be successful. In addition to innovative management, several key factors were responsible for its success. First, it was technically a good system, much better and more user friendly than its predecessor. As the diffusion literature points out (Rogers, 1995), innovations that offer a clear advantage are much more likely to be adopted. Second, extensive training was provided to all employees on how to use the system. Although the training did not always coincide exactly with the time when the system was received, survey results show that virtually all employees received such training, and that it did cover materials they found useful. Third, the movement of Extension materials—from memos and e-mail to newsletters, publications, and databases to the Internet and Web sites—strongly encouraged employees to utilize the system's new features. Fourth, near universal imposition of the system meant that a critical mass was reached, and the system became quite useful as a means of reaching all Extension employees. Dertouzos might be expected to cite this case study as a good example of how an organization that can benefit from new information technologies reinvented itself for the next century.

Kling. Two of Kling's concerns were also supported. He maintains that computers tend to be pushed by advocates who then become the major beneficiaries both in terms of organizational power and technology uses. In this case, central Extension administration was in a position to benefit from adoption of this technology in several ways. Not only did it improve its political position by becoming "modern," but also it became able to join the U.S. Department of Agriculture initiatives and strengthen its role as an information provider. In broader terms, we need to consider the extent to which local organization or community individuals should be involved in the selection and design of these systems. The disadvantage of involving them is that they generally have low levels of experience and knowledge about such systems. Extension employees interviewed before the installation of EXNET-IP had only a fragmentary understanding of what it involved and how it worked. Several thought of it as a system in which clients would come in, sit down at a terminal, type in a request, and the answer would magically appear. On the other hand, the fact that the system was imposed hierarchically has meant that except for persons directly employed by or connected to Extension, few have benefited from its adoption. Had the communities themselves played a more active role in system design, one might expect that they would have been much more interested in public access, e-mail, and the use of the system to aggregate total local demand in order to negotiate lower provider fees.

A second area in which Kling's concerns seem appropriate is in translation costs at the county office level. While the state and federal level systems save considerably by being able to transmit information such as publications electronically instead of having to print and mail them, the county offices find themselves having to serve clients who for the most part do not have computers or the Internet and in any event might prefer to receive information in printed form. As a result, county office staffs now find themselves in the position of "translators," taking material in electronic forms and then converting it into a form the client can use. There are significant time and production costs associated with this translation process. In general, it is much more expensive to print material on demand, one copy at a time, than it is to print a large number of copies (assuming one can estimate demand correctly). County office laser printers are becoming the printing presses of the new electronic age, and county offices either must charge for this service or find other ways to pay for it. Some maintain that this translation cost will be temporary, and as soon as clients adopt computers in large numbers, they will be sent information directly in electronic form. Others wonder if the ultimate consumers of information will ever want to receive it on a disk or via e-mail. Kling's point, however, is not about its ultimate form. It is the fact that had county offices played a more active role in the design of the overall system, there might have been more attention paid to the problem of translation costs.

Changes in Organizational Communication Patterns

Within two years, a properly designed wide-area network can change organizational communication patterns in important ways. Certain kinds of changes, including the overall volume of messages, intraoffice communication patterns and reliance on the system to reach other connected groups, can be expected to significantly alter the communication structure of an organization within two years. The selection of the Internet and e-mail as a "best source" or "second-best source" is a powerful indication of how important the EXNET-IP system has become to the organization. A second area in which powerful changes have occurred is in the area of electronic archiving of newsletters, reports, publications, and other materials. Although these changes have not been as sweeping, important proportions of materials are now stored electronically. King (1996) noted that a confluence of three factors was necessary to gain productivity from computers (adoption of the technologies, the know-how needed to apply them successfully, and the wholesale substitution of the new technologies and new methods for older ways of doing things).

Extension results show that wholesale changes have taken place in terms of e-mail and communication. Changes in storage and retrieval of materials are beginning to change. A third area of impact concerns horizontal and vertical communication patterns. The development of new networks of individuals who converse virtually could be expected to take more time than the other changes we have been discussing. There is some evidence that both horizontal and vertical communication have increased as a result of EXNET-IP. Thus, it may be that the effect of these systems is not necessarily to "flatten" the hierarchy but to increase communication of all types. Although horizontal communication message flows have increased, vertical ones still dominate the system.

The technical capabilities of EXNET-IP could permit a radical restructuring of some of the communication functions in the organization. Administrative functions that were centralized because of an inability to meet the communication demands of a decentralized structure could be reshaped. One could imagine clusters of county directors communicating and forming alliances in ways that would have been very expensive and time-consuming before the system. An office assistant who believes a new work policy is unfair could quickly e-mail counterparts at the other 99 county offices to find out what they think. Specialists with content interests could develop their own databases. One might even imagine alliances and information sharing across state lines, a practice that in some past time periods was forbidden by administrators. Is raising a hog or growing a strawberry or supporting a 4-H group really that different when one crosses a state line? These types of changes in administrative and content communication areas, if they occur, will not come quickly, because the first tendency when adopting a new technology is to make it fit existing organizational practice. However, these longer-term changes may represent the technology's most significant impacts.

Use and Satisfaction

Direct experience with a wide-area network system including having one's own computer leads to high levels of use and satisfaction. Although a number of studies have cited "techno-fear" and other attitudinal concerns they detected among Extension employees prior to installation of these systems (Smith & Kotrlik, 1990; Babbitt, Desmarais, Koehler, Lacy, & Merchant, 1988), our interviews among individuals after adoption very seldom found fear or anxiety to be important. The overwhelming conclusion from our 1997 data is that direct access, including having a computer, one's own e-mail account, and software led to high levels of use and satisfaction. The only individuals in the Extension system who have reservations about the importance or value of computers and the Internet are those who lack them—program assistants who don't have their own computer. Said another way, employees who must get up and go somewhere to use the Internet and a computer will use this technology less and are less satisfied with it.

12

TELECOMMUNICATIONS AND ECONOMIC DEVELOPMENT: CHASING SMOKESTACKS WITH THE INTERNET

Brent Hales, Joy Gieseke, and Delfino Vargas-Chanes

INTRODUCTION

Rural communities are finding it increasingly difficult to compete with larger, more technologically advanced communities in an era of rural disinvestment. To keep up, rural communities must participate in the transition to an information-intensive society. Furthermore, their future vitality depends at least partly on maintaining technological currency. If communities lack the physical infrastructure to embrace technological developments, it is very unlikely that these communities will be able to capitalize on current or future technological advancements. Rapid changes in the telecommunications industry intensify the need for rural communities to include plans to address telecommunications needs in their development programs. Instant global communication capabilities can reduce the isolation of rural communities and enhance their economic strength by providing technology driven jobs and by increasing the availability of outside influence and expertise.

But in order for telecommunications technologies to be an effective economic development tool there must be strong local community capacity and open lines of communication for cooperation in delivery of services and utilization of the technologies. "Capital and information alone will not ensure the emergence of local economic development unless residents are able to work together and motivate others to create an environment favorable to successful economic activities" (Ryan, 1988, p. 367). Successful local development requires coordinated efforts involving the whole community (Kelley, 1993) and frequent, open lines of communication that cross community sectors (Wilkinson, 1986; Tilly, 1974; Shaffer, 1990; Ryan, 1994). Luloff and Swanson (1995) refer to this as *community agency*, or the community's capacity for collective action. It is an important, but often neglected, component of development efforts. McDowell (1995) asserts that communities can gain or lose opportunities depending upon their preparedness, willingness, and capacity to act.

The purpose of this chapter is to examine the role of telecommunications technologies in community economic development from the perspective of the local economic development professional. Economic development professionals obviously are central to any decisions on strategies communities use for economic development. If they believe that a technique will work, they are more likely to encourage the techniques' integration into local development strategies. The use of telecommunications technologies for economic development is a strategy increasingly employed by economic development professionals and communities, but it is not widely used as a tool for economic development in Iowa. To illustrate the adoption process, we provide both a descriptive account and empirical assessment of the impact of the economic development professionals' adoption or non-adoption of various strategies for economic development.

Telecommunications technologies are providing new economic development options for rural communities. One application of these technologies that has increasing utility for rural communities is the Internet. The Internet, a network of networks, has more than 5 million host computers located in at least 60 countries connecting more than 50,000 networks with over 27 million users (Hudson, 1997). When combined with the World Wide Web and the convenience of Web browsers, the information and communication potentials for the Internet become almost limitless.[1] Community and economic development can be facilitated by receiving timely information about the availability of financial and other resources from state and federal governments (Slechta & Marner, 1997; Blechman & Levinson, 1991). Of course, telecommunications, including the Internet, are not a panacea for solving rural America's problems and unenlightened implementation may prove detrimental to rural areas (e.g., Dillman & Beck, 1988; Dillman, Beck, & Allen, 1989; Glasmeier & Howland, 1995; National Association of Development Organizations, 1994; Tweeten, 1987; Wilson, 1992). State-of-the-art telecommunications infrastructure to support the needs of business and industry however is fundamental to a community's sustainable economic development (Allen, Johnson, Leistritz, Olsen, Sell, & Spilker, 1995; Fox, 1988; Hudson, 1997; Schmandt, Williams, Wilson, & Strover, 1991; Steinnes, 1990).

TELECOMMUNICATIONS AND ECONOMIC DEVELOPMENT

Economic development is any effort to increase local capacity to generate income and employment to maintain, if not improve, the community's economic position. The five basic strategies of economic development are attracting new basic employers, improving the efficiency of existing firms, improving the ability to capture dollars, encouraging new business formation and increasing aids and/or transfers received (Shaffer, 1989). Economic development includes job creation, income enhancement, and development of infrastructure to achieve these ends (Summers, 1986; Shaffer, 1989).

The development of infrastructure has long been a standard component of economic development strategies. From the 1960s to the 1980s rural communities were encouraged to organize Community Development Corporations and to construct industrial parks to attract new businesses and industries (Gunn & Gunn, 1991). Industrial parks offered ready access to the infrastructure needed to establish a firm's operations (land, electrical power, water, sewet, communications, transportation, etc.). In the 1980s business incubators became popular as a strategy to provide the infrastructure to facilitate the initiation and survival of new entrepreneurial ventures (Weinberg, 1987). Business incubators provided not only access to the necessary physical infrastructure, but also access to services (secretarial, legal, accounting), and interaction with the other tenants of the incubator that fostered the sharing of knowledge and experience. Beginning with the late 1980s into the 1990s a predominant component of infrastructure necessary for the viability of a large share of business organizations, whether in an industrial park, incubator, research park, main street, or the rural hinterlands, has been telecommunications.

The economic viability of rural communities is consistently challenged from without. Sustained economic growth in these communities is tied to their willingness and ability to invest in adequate infrastructures. Infrastructure to support telecommunications technologies can significantly contribute to a community's overall economic development. Telecommunications infrastructure investments, notably in technologies that involve the use of computers to access databases or other online services, provide entry into information exchange networks, and permit participation in video educational programs, classes, or conferences are critically important to the economic vitality of rural communities (Fox, 1988).

As we move into the 21st century, holistic community development strategies will become increasingly reliant on advances in telecommunications technologies as an asset and tool to attract new businesses, encourage local entrepreneurship, and foster growth from within the local economy. Economic forces, especially globalization, volatile employment opportunities in manufacturing, and growth of the service sector, reaffirm the necessity of telecommunications-based development strategies. By exploring and exploiting telecommunications opportunities, new markets can be created where previously there were none. Chapter 1, for example, discussed home office workers, or "lone-eagles," whose numbers have increased dramatically as a result of telecommuting. According to the U.S. Department of Labor, "more than 21 million persons did some work at home as part of their primary job in May 1997" (1998, n.p.). About 60 percent used a computer for this work.

Although the number of persons doing job-related work at home overall did not grow dramatically between 1991 and 1997, the number of wage and salary workers doing paid work at home did. Nearly 90 percent of workers doing paid work at home were in white-collar occupations. Seven out of ten persons who did some work at home in 1997 were married. The work-at-home rate for married parents was about the same as the rate for married persons without children. Men were as likely as women to work at home. Significantly, Whites were more than twice as

likely to be engaged in some form of home-based work than either Blacks or His-
panics (U.S. Department of Labor, 1998, n.p.).

The trend in corporate down-sizing, along with further growth expected in the
number of home workers and continued expansion of the service sector, could
translate into an increasing number of jobs dependent on adequate telecommunications infrastructure in rural communities. Unquestionably, the new telecommunications technologies hold profound implications for the restructuring and
success of local economies. Some areas already are prospering from a rapid growth
of technology-related industries (e.g., California's Silicon Valley) and the siting of
information-dependent companies. Firms adopting the new technologies often
gain an economic advantage over their competitors as a result of increased operating efficiency and lower capital costs associated with inventory reduction. In fact,
the economic benefits of these technologies are sufficiently apparent that about
half of the states are proposing or constructing telecommunications networks as
an economic development strategy to attract and retain firms (Read & Youtie,
1996).

A much discussed merit of telecommunications technologies is their seeming
utility in resuscitating stagnant rural economies and decayed social institutions. By
overcoming costs of physical space, including pressures on rural firms to relocate
to larger places and by loosening the ties of information-intensive companies to
urban areas, the new technologies are purported by many analysts as offering an
important vehicle for rural development. The majority of economic development
professionals in Iowa report that as telecommunications technologies become
more popular in rural places, the gap between rural and urban communities will
decline (Gieseke, Korsching, Abbott, Bultena, & Gregg, 1995). But there is not
agreement that these technologies will prove beneficial for rural areas. Indeed,
some critics feel that these technologies ultimately will widen, rather than narrow
the economic gap between rural and urban communities.

Iowa is considered an innovator in the implementation of advanced telecommunications technologies (see discussions in Chapters 2 and 3). As an innovator, it
provides a useful case to examine both the possibilities and the problems that may
exist for telecommunications-based economic development for rural areas.

AN IOWA STUDY

In a study conducted by the Rural Development Initiative in 1995 (Gieseke et al.,
1995), telephone surveys of local Iowa economic development professionals were
conducted to determine their use of telecommunications technologies in promoting economic development. The respondents were selected from a directory of
community economic development contacts provided by Iowa's Department of
Economic Development. A random sample of 229 economic development professionals from communities of less than 25,000 population located outside Census
Bureau standard metropolitan statistical areas (SMSAs) were selected and 207 interviews were completed. The survey focused on four types of innovative telecommunications technologies: computers with modems, enhanced telephone services,

videoconferencing, and teletext services (Gieseke, Korsching & Bultena, 1996, p. 1). These communities ranged from 159 to 23,936 in population. The respondents ranged in age from 23 to 80 years old and had worked in economic development from one to 45 years. Eleven percent of the development professionals had advanced degrees and over 80 percent had at least some post-high school education.

At the time of this study local telephone service was provided by 154 small, locally owned independent companies and three large, absentee-owned corporations. The state owns and operates a statewide fiber optic system, the Iowa Communications Network (ICN), which has at least one point of presence in each of the 99 counties. The ICN is primarily for educational purposes with legal users limited to educational institutions (K–12 schools, public and private colleges and universities), libraries, hospitals, and state and federal government offices. In the private sector, the three large telephone companies provide fiber optic cable to the metropolitan areas and larger cities and the Iowa Network Services, a consortium of 130 small, independent companies, operates a fiber optic network that serves many rural communities.

When asked if the telecommunications services provided by their local telephone companies were an asset, a liability, or irrelevant to the community, 70 percent of the economic developers characterized the services as assets. While 20 percent saw them as irrelevant to the economic development in their communities, only 10 percent viewed the services provided as a liability.

Nearly all of the respondents believed that telecommunications technologies were useful for economic development and over half perceived them as essential (Figure 12.1). About three-fifths of those surveyed rated both the ICN and the Internet as being very or somewhat beneficial for economic development. It is interesting that although economic development professionals reported that they considered telecommunications technologies important for economic development activities, they reported that a lack of telecommunications technologies would not seriously affect the likelihood of business recruitment.

Respondents rated the importance of telecommunications technologies for several economic development and quality of life issues in their communities (Figure 12.2). Each item could be rated very important, somewhat important or not important. Telecommunications technologies were deemed most important for educational services, followed by medical services and good paying jobs. Telecommunications technologies also ranked high in importance in recruiting new businesses and industries, developing home-based businesses, growth of the local economy, job creation, establishment of new businesses, library services, business profitability, and business retention and expansion. Areas for which telecommunications technologies were seen as having less importance are the quality of working conditions, retail trade, and entertainment and recreation.

To further examine this relationship, respondents were asked to rate the importance of eight community infrastructure and service components for the work of local economic development. Each item was rated on a 10-point scale with 1 being "not important" to 10 being "very important." The items were: local schools, transportation system, medical facilities, telecommunications, water and sewer,

Figure 12.1
Economic Development Professionals' Perceptions of the Importance of
Telecommunications for Economic Development

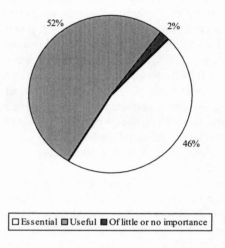

Figure 12.2
Importance of Telecommunications for Economic Development and Quality of Life in
Iowa Municipalities: Proportion of Respondents Who Deemed Telecommunications
"Very Important" or "Important"

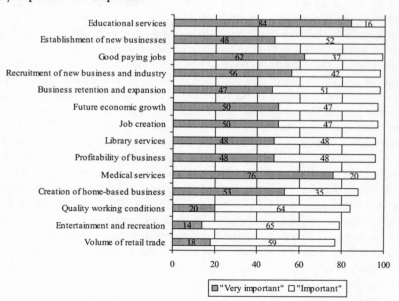

recreational facilities, retail and service, and housing stock. Although no item was rated unimportant, water and sewer, schools, and housing were of greatest importance. When asked which one component was most important for economic development in their communities, 33 percent mentioned housing stock. Schools and public utilities followed with 20 percent. Telecommunications technologies ranked near the bottom in being selected as the most important component for economic development. It is interesting to note that although the respondents had predominantly rated telecommunications technologies as useful for economic development, they failed to see their importance in the infrastructure of their communities.

Respondents rated the adequacy of telecommunications services in their local areas for economic development on a scale of 1 to 5, with 1 being poor and 5 excellent. On average, the perceived adequacy of telecommunications services was rated 3.23. When asked to predict the adequacy of telecommunications services in their local area by the year 2000, the average rating increased to 4.09.

Although the ratings on the adequacy of local telecommunications services for economic development are not strong, most of the economic development professionals in the communities surveyed did not believe this discouraged businesses from locating in their communities. Only 8 percent (17) think that inadequate telecommunications service is a factor in failing to attract businesses to the community. Again, although the economic development professionals reported that telecommunications technologies were important for overall community economic development and quality of life, they failed to make the connection with business recruitment.

Local Use of Telecommunications Technologies

Respondents were asked to rate the present use of telecommunications technologies by 11 different sectors of their communities. Each sector was rated on a 4-point scale, from nonexistent (1) to high (4). The average ratings exclude respondents who were unfamiliar with use in a particular sector. The highest use is seen in county cooperative Extension offices, hospitals and schools. Importantly for local development, the lowest use is in economic development, local government, and chambers of commerce (Figure 12.3).

Economic development professionals in Iowa are using these technologies on a limited basis. While four-fifths of the respondent's reported using personal computers in their economic development activities, very few were using electronic mail or the Internet. The most common use of the computers was word processing. Of those respondents who had a personal computer, one-fourth of the professionals accessed electronic databases, 15 percent used electronic mail, and 15 percent accessed an online Internet service. While use of the Internet and Iowa's ICN are growing, only 21 percent of the professionals believed that they were very well or well informed about the Internet. Those respondents who were informed

Figure 12.3
Economic Development Professionals' Perceptions of Community Use of
Telecommunications Reported as the Average (Mean) Based on a Four-Point Scale,
(1 = None Existent, 2 = Low, 3 = Medium, 4 = High)

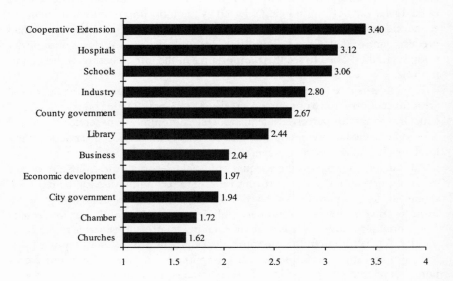

used it infrequently. Only 1 in 10 reported using the Internet to acquire economic development information.

Although economic development professionals saw benefits from using tele-communications technologies, there were several likely reasons these technologies were not being used often. For example, respondents were asked if the following reasons kept them from using the ICN: restrictions on who is eligible to use it, distance to an access point, scheduling problems, or that the ICN is not a useful tool for economic development. The most frequently mentioned (33%) deterrent to using the ICN was distance to an access point. Access points to the ICN are limited in many of Iowa's counties to one point in the entire county. These access points are often located in libraries or schools with minimal hours of access time.

Barriers to using the Internet are perceived as greater than barriers to using the ICN. The respondents were asked which of the following reasons kept them from using the Internet in their economic development activities: lack of access to a computer, financial costs, lack of knowledge, or not seeing the Internet as a useful economic development tool. Lack of knowledge was the largest barrier to using the Internet. Sixty-three percent of the respondents who answered this question felt that they did not have the knowledge required to use the Internet. One-half felt that the financial cost of access was keeping them from using the Internet.

Telecommunications and Economic Development

Among the relevant actors in the process of economic development, telephone companies can play a major role in providing the infrastructure for enhancing telecommunications. Active participation of telephone companies is an important factor in the economic development plans and activities of rural communities. A substantial majority of the economic development directors (70%) recognized their telephone company as an asset to their communities. On the other hand, most of the economic development leaders in this study rated as average the adequacy of telecommunications services provided by the telephone company. Despite this low rating, most economic development professionals did not believe that poor quality telecommunications service caused any failure in new business recruitment. Interestingly, over half of the economic development professionals had little or no contact with their telephone companies. Overall, the economic developers perceived the telephone companies as contributing little to their economic development efforts (Figure 12.4).

This was related to the ownership structure of the telephone company (Gieseke & Korsching, 1998) Interaction between the telephone company and the economic development professional was four times higher in communities served by locally owned independent telephone companies. Also, leadership in economic development related activities by telephone companies and their financial support of community projects were more than twice as high in these communities as compared to communities served by large, absentee-owned telephone corporations.

Important to a discussion of telecommunications technologies are their likely effects on rural communities. Despite their much heralded benefits, it remains unclear whether these technologies will offset powerful historical forces, especially geographical isolation and sparse population that have severely eroded rural economies and institutions. When asked how telecommunications technologies would impact the rural-urban gap, 66 percent of the economic development professionals reported that they would strengthen local economies, while 16 percent feared they would widen the gap, and another 16 percent felt these technologies would have little or no impact on the gap (Figure 12.5).

Our data suggest that economic development professionals in Iowa seem to have some understanding of the potential telecommunications technologies have for economic development, so it is interesting, if not frustrating, that these technologies currently are not used more readily for economic development in rural communities. It may be that economic developers lack vision or are unwilling to stray from tried methods of traditional economic development.

Although nothing can guarantee accrual of benefits from telecommunications, leaders with insight and vision for innovative telecommunications applications within their local communities can provide the impetus and direction to heighten potentials of success. "Ultimately, rural development (through telecommunications) is a community process. . . . Development depends on local leadership, local initiative and local cooperation" (Parker, Hudson, Dillman, Strover, & Williams, 1992, p. 8). Within rural communities a key person for providing the vision, impe-

Figure 12.4
Economic Development Professionals' Contact with Local Telephone Company

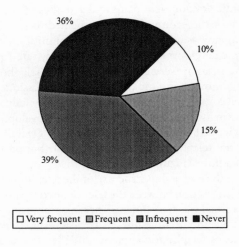

Figure 12.5
Economic Development Professionals' Perceptions of the Impact of
Telecommunications Technologies on the Local Community

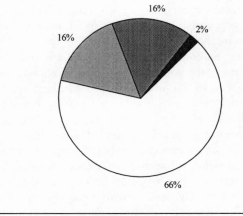

tus, and direction for telecommunications-based economic development is the professional economic developer, often the local economic development director. This person provides economic development leadership through his or her position as a community *change agent*.

Role and Importance of the Change Agent

A change agent's basic function is to establish a link between the perceived need of the client and the means of satisfying that need. As a change agent, the local economic development professional works to bring about economic change in the community that will create jobs and income for the local population. The skills, knowledge, and interests of the developer will affect the specific development goals and the trajectory and means to reach those goals (Zaltman & Duncan, 1977). The trajectory and means include the development and utilization of appropriate infrastructure, and, in the current economy, the emphasis is on innovative telecommunications infrastructure. Many innovations have attributes that facilitate their adoption and use and therefore little assistance and few incentives are necessary for their implementation. Other innovations, however, are not readily adopted unless they are actively promoted and assistance and incentives are provided for their use, particularly if the potential adopters are not familiar with the technology. Telecommunications technologies are among the group of innovations that may not be readily adopted in rural areas. Attributes that inhibit quick adoption and rapid diffusion of telecommunication technologies are relatively high initial cost and slow return on investment; incompatibility with existing values, beliefs, and methods for accomplishing tasks; and perceived complexity because of lack of knowledge about its use and potentials. The latter issue is of particular salience with rural populations lacking experience with and exposure to innovative applications of telecommunications technologies (Dillman, 1991b).

When potential users of an innovation are not inclined to adopt even though it may offer benefits or advantages over existing technologies, the adoption and diffusion process may be facilitated through a change agent. Change agents vary in their effectiveness in producing the needed changes or impacts that satisfy client needs, depending on their professional qualifications (Lackey & Pratuckchai, 1991). For diffusion of technical innovations, technical qualifications are notably important, that is, the change agent must be knowledgeable and competent in the field to have credibility with the clients (Rogers, 1995; Zaltman & Duncan, 1977). A good change agent also will have the capability to increase the clients' ability to evaluate new ideas and technologies (Zaltman & Duncan, 1977). Both qualifications are critical to expanding telecommunications-based economic development within a community. Furthermore, the degree to which the developer promotes the use of telecommunications technologies and is seen as competent in that role will be related to the developers own orientation toward the technologies, his or her perceptions of their usefulness for economic development, and whether or not they are personally used by the developer. As Rogers (1995) states, the change agent must be accepted and seen as credible before the clients will accept the innovations that he or she promotes: "The innovations are judged on the basis of how the change agent is perceived" (p. 337).

The change agent also must have a vision of the importance of telecommunications technologies for economic development. Without this vision, the incorporation of telecommunications technologies into holistic economic development strategies will be less likely. Garkovich (1989) states that an important function of community leaders is that they have an image or vision of what their communities can become through productively channeled activities. Ayres and Potter (1989) state that the attitudes of community leaders toward change play an important role in determining the types of social action initiated in the community. In the communities with innovative applications of telecommunications technologies studied by Schmandt et al. (1991), "the key to successful community development was the presence of visionary community leaders who were willing to spend much time and energy in planning, promoting and implementing community development" (p. 209).

THE ECONOMIC DEVELOPMENT PROFESSIONAL AS AGENT OF CHANGE

As change agent, the economic development professional can play a pivotal role in determining what methods are used in promoting economic development. Adoption and use of telecommunications technologies by the change agent can and does affect the desire and ability of rural communities to harness the potential of these technological advancements. This takes place in several ways. First, foundations for adoption must be laid. By this, we mean the extent of the interaction between local telephone service providers and local economic development leaders can significantly impact the likelihood that viable partnerships are formed to encourage telecommunications development. Second, if the economic development professional does not promote these technologies as indispensable for economic development, it may be difficult for rural communities to acquire the infrastructure necessary to support telecommunications-based development. Finally, without the infrastructure base, and without the change agent's use and support of these technologies, it is not possible for others to adopt these technologies or employ them for wider community benefit.

There are factors in addition to the economic developers' vision, leadership, and commitment that affect his/her ability to promote telecommunications-based economic development. Education is important in that those with more years of formal education and advanced degrees are more likely to use telecommunications technologies such as the Internet and e-mail (Wresch, 1996). Therefore, they can serve as role models in telecommunications use for others in the community. Also, change agents with higher levels of education are viewed as better technically qualified to address development issues (Rogers, 1995; Zaltman & Duncan, 1977).

Projecting the impact of other personal characteristics such as age of the developer and years of professional experience is more problematic. One would expect older professionals with more years of experience to have higher levels of perceived competence and credibility. On the other hand, research has shown that the use of computers and related technologies is negatively correlated with age, that is, those who are older are less likely to use these technologies (Binstock & George, 1995). Finally, peoples' perceptions of new technologies are, in part, influenced by existing knowledge and past experiences (Rogers, 1995). Socialization continues over a lifetime and past experiences help define future realities and expectations (Ritzer, 1992). Economic developers learn directly from their own experiences and vicariously from the experiences of others, and they make future decisions based on that knowledge. As new technologies emerge, perceptions of the compatibility of the technologies with existing values and practices, the ease or difficulty in using the technologies, and the benefits that might accrue are influenced by the developer's cumulative fund of knowledge. If the use of innovative technologies are not part of the developer's repertoire of mastered skills, he or she is less likely to use them. And older developers are less likely to have had exposure in their formative education and training to the innovative telecommunications technologies than younger developers.

To examine further the role of Iowa economic development professionals as change agents in the adoption of the telecommunications technologies by the wider community, a conceptual model, as diagramed in Figure 12.6, was developed to guide analysis. The analysis examines the impact of economic development professionals' personal characteristics, age, education, and number of years worked in economic development on the likelihood of the their adoption of tele-

Figure 12.6
Conceptual Model of Variables that Influence Community Use of Telecommunications Technologies and the Paths of Hypothesized Significant Influence

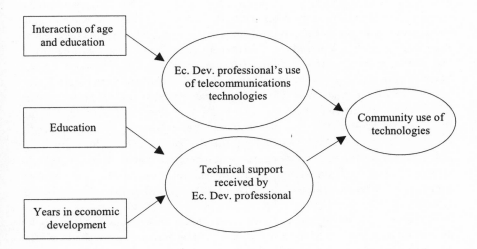

communications technologies for personal use in economic development efforts. It also examines whether these characteristics affect the likelihood of their receiving support in the form of workshops, conferences, and consultation in their economic development efforts. Finally, the analysis examines whether the economic development professionals' personal use of the technologies and their having received support affect the communities' likelihood of using these technologies. For a comprehensive examination of relationships in the model, it is necessary and appropriate to use a statistical technique that allows us to examine these relationships from a cause and effect perspective. Therefore, path analysis with structural equations were used for the analysis, allowing us to determine the variables that influence community use of telecommunications technologies as well as their paths of significant influence.

Hypotheses

Several hypotheses are incorporated in the model. Community adoption of telecommunications technologies (*community*) is hypothesized to be influenced by several factors. First, the personal characteristics of the development professional such as *age, education,* and *years in economic development* are expected to be significantly correlated to the endogenous variables (use of telecommunications technologies and technical support received). Years in economic development are expected to be positively associated with technical support received due to the changes in technology and need for further education (Rogers, 1995; Lackey & Pratuckchai, 1991). Education is expected to be negatively associated with technical support received due to the economic development professional's base of knowledge. It is less likely that the professionals need technical support services if their knowledge of telecommunications is already sufficient (Wresch, 1996; Rogers, 1995; Zaltman & Duncan, 1977). The *interaction of age and education* is expected to be negatively related to the level of personal *use of telecommunications technologies* by the economic development professional. As people age, their knowledge base derived from formal education many years past begins to lose pace with changes in technology (Reshetyuk, 1992). Older workers are less likely to adopt new technologies because of market confusion, inadequate dissemination of information, and failure of delivery systems (Enders, 1995). Other reasons cited for the lack of adoption are that information on these technologies in its general orientation is either overly descriptive or overly futuristic (Lesnoff-Caravaglia, 1989). Others yet claim that the failure to adopt new technologies may be due to a self-categorization wherein the aging workers begin to remove themselves from the rapid pace of technological advancement as they prepare for retirement and withdrawal from the economic/work sphere (Giles & Condor, 1988).

The economic developers' personal use of telecommunications technologies and technical support received is hypothesized to be positively associated with the communities' adoption of these technologies (Dillman, 1991b). This hypothesis remains consistent with the Rogers's (1995) conception of the change agent as a social and technological gatekeeper. As noted earlier, the degree to which the developer promotes the use of telecommunications technologies and is seen as competent in that role will be related to the developer's own orientation toward and personal use of these technologies, as well as his or her perceptions of their usefulness for economic development.

Variable Measurement

The model includes three latent variables (variables not observed or directly measured, but generating the pattern or structure of variables that are observed and measured) each composed of different observable measures. The developers' *personal use of telecommunications technologies* is the first latent variable. It consists of two observable variables, *computer use* and *Internet use*. Computer use asked the economic development professionals whether they were using computers in their professional capacity. Internet use inquired about their familiarity and use of the Internet in their economic development efforts.

Technical support received in telecommunications technologies uses four indicators to measure this concept. Two of the measures consist of summed scores of six items. The first measure inquired if development professionals had received *written materials, attended workshops,* or had been offered *personal consulting* on the use of telecommunications. The second inquired whether they had received or been offered *personal consulting* services by universities or other information sources. The third variable asked whether they had attended *workshops on the ICN or received relevant material from the ICN* on telecommunications. Written materials were the most common source of information about telecommunications technologies. More respondents obtained information from the Iowa Department of Economic Development (IDED), Iowa State University, and community colleges than from other sources.

The final latent construct was the dependent variable, *community use* of telecommunications technologies. Community use was constructed from of two scaled indicators that measured the overall use of various technologies by the community—*telecommunications technologies* and *economic development groups*. Telecommunications technology inquired as to the communities' use of these technologies in various community and civic settings, with responses of nonexistent, low, medium, or high. Economic development groups inquired as to whether economic development groups were currently using telecommunications technologies in their economic development efforts or planning.

Results

Table 12.1 contains descriptive statistics of the observed variables in this study.[2] Respondents have an average of 15 years of education and have worked about 10 years in economic development. A majority were familiar with and used computers and the Internet. In addition, the means of the variables *seminars and conferences* and *personal consulting* indicate that the respondents generally show a high

Table 12.1
Descriptive Statistics

Variable	Mean	Std. Dev.	Minimum	Maximum
Years in economic development	10.35	8.77	1	45
Education	15.23	1.94	12	18
Computer use	1.80	.30	1	2
Internet use	1.86	.30	1	2
Seminars and conferences attended	1.87	.18	1	2
Personal consulting received	1.82	.19	1	2
Workshops attended	1.54	.24	1	2
Telecommunications technologies	2.37	.64	1	4
Economic development	3.15	.65	2	4

Note: n = 111

level of participation in activities that provide technical support for computer use. The last two variables make up the latent construct used in the variable, community use of telecommunications technologies. By comparing the mean score of the variables, we can see that economic development groups are using telecommunications technologies to a greater extent than the rest of the community (as measured by telecommunications technologies).

Figure 12.7 shows the model with standardized estimates.[3] The paths for *education, years in economic development* and the *interaction of age and education* are statistically significant, indicating that they affect *technical support received* and *use of telecommunications technologies* by economic development professionals. The path between *education* and *support received* in telecommunications is negative indicating that more educated development professionals are less likely to require support or information on telecommunications. More years worked in economic development increased the likelihood that the economic developer received support and/or information on telecommunications. The *interaction between age and education* is negatively related with *technology use* indicating that those who have a higher education are more likely to use telecommunications technology in their economic development efforts, but that age becomes more important in the relationships as the professional's education becomes dated and he or she fails to keep current on technological advances.

The economic development professionals' *use of telecommunications technology* in economic development is positively related with *community use* of telecommunications. *Support received* in telecommunications is also positively related with *community use* of telecommunications. This implies that the use and knowledge of telecommunications technologies by the economic development professionals is a predictive tool for communities' use of these technologies.

Figure 12.7
Conceptual Model of Variables that Influence Community Use of Telecommunications
Technologies with Standard Estimates (Path Coefficients) for Significant Paths of
Influence Using Asymptotic Distribution Free Estimation

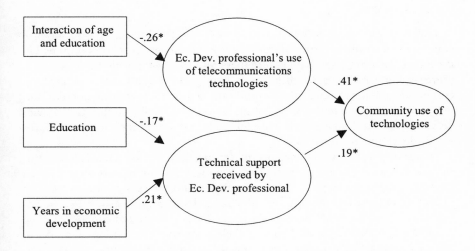

Note: * Significant at 0.05 level; $\chi^2 = 49.616$ with 31 degrees of freedom

Discussion

Figure 12.7 demonstrates that all hypotheses were supported. The *interaction of age and education* was negatively and significantly related to the *use of telecommunications technologies* by the economic development professionals. *Education* was negatively and significantly related to the *technical support received*. As expected, *years in economic development* were found to be positively associated with the *support received. Use of telecommunications technologies* and *technical support received* were found to be positively associated with the *community use* of the technologies.

Telecommunications technologies are vital assets in community economic development strategies that should be explored for rural community economic development (Gillette, 1996; Dillman, 1991b; Hudson & Parker, 1990; National Telecommunications and Information Administration, 1995; Parker, Hudson, Dillman, & Roscoe, 1989; Parker et al., 1992; Read & Youtie, 1996; Salant et al., 1996; Schuler, 1996; Office of Technology Assessment, 1991). To determine what facilitates adoption of these technologies, we developed a causal model and tested the model with data from Iowa communities. While it is possible to argue that the industry is pushing the adoption of these technologies into rural areas, this chapter has demonstrated the importance of the economic development professional as a change agent in rural communities' adoption of telecommunications. The model (Figure 12.7) indicates that development professionals' personal characteristics are significantly related to their use of telecommunications technologies and their reception of support in telecommunications. These variables in turn were found to significantly impact the communities' use of telecommunications technologies.

CONCLUSIONS

Telecommunications technologies generally and the Internet specifically will continue to be sources of support for rural community economic development in the future. Communities that tap into these resources will position themselves to more readily attract outside industries to their locale while also encouraging local entrepreneurship. Whether rural communities can compete with larger metropolitan areas in this race remains to be seen. However, if rural communities can and do adopt these technologies, they are more likely to be competitive in their economic development efforts.

There are several important factors influencing the integration of telecommunications technologies with communities' economic development strategies. Local economic development professionals are important to the adoption and effective utilization of these technologies. As this chapter has demonstrated, the economic development professionals' knowledge, personal adoption, and support of these technologies increases the likelihood of their communities adopting them. Without the economic development professionals' leadership and support, rural communities are less likely to include telecommunications technologies in their economic development strategies. Furthermore, by embracing and using telecommunications technologies, the development professional is a role model on state-of-the-art technology use for the larger community.

Unfortunately, the majority of economic development professionals in Iowa are not currently embracing these technologies. As noted, only 15 percent currently use electronic mail, 15 percent access the Internet, and even fewer (10%) access the Internet to acquire information on economic development. It is also significant that 52 percent of the professionals reported telecommunications as essential for economic development, but 92 percent believe that inadequate telecommunications services are not a factor in attracting business to their communities.

As demonstrated by our data, economic development professionals have the ability to significantly impact their communities' adoption of telecommunications technologies. As we have seen however, due to minimal use and a lack of perceived importance of the technologies for economic development, the professionals are not emerging as a driving force for this adoption. It is our recommendation that economic development professionals reexamine the usefulness of telecommunications technologies as viable tools for economic development, especially in rural communities. Failure to do so may widen the gap between their communities and urban areas that have embraced these technologies in their economic development efforts.

NOTES

1. In April 2000, there were nealy 16 million registered Internet domains worldwide (DomainStats.Com, 2000) and estimatees of the number of Internet users by 2001 is over 700 million (Matrix Maps Quarterly, 2000).

2. To reduce the number of cases with missing values, we pared the number of respondents from 229 to 111, ensuring through reliability in and confidence that the remaining sample was representative.

3. Presentation of the model and the data have been simplified for the reader not familiar with path models and structural equations. The complete model with all relevant statistics may be found in Figure 12.8.

Figure 12.8
Full Model of Variables that Influence Community Use of Telecommunications Technologies with Standard Estimates (Path Coefficients) for Significant Paths of Influence Using Asymptotic Distribution Free Estimation

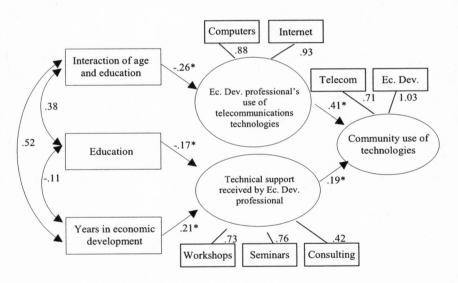

Note: * Significant at 0.05 level; χ^2 = 49.616 with 31 degrees of freedom

13

ON-RAMPS AND ROAD BLOCKS TO THE INFORMATION SUPERHIGHWAY

Peter F. Korsching, Eric A. Abbott, and Patricia C. Hipple

MYTHS AND REALITIES

Popularized images of "having all the right connections" portray people, businesses, and communities re-inventing themselves to take advantage of the global reach and interconnectivity made possible by advanced telephone services, Internet, World Wide Web, cable television, satellite technology, interactive video, and other new telecommunications technologies. These technologies make it possible for school children in small town America to meet and chat with children in Germany, Japan, or anywhere else in the world. They make it possible for a heart specialist at the Mayo Clinic to diagnose and treat an ailing patient stranded aboard ship in the mid-Atlantic. They make it possible for entrepreneurs in Aladdin, Wyoming, to advertise and market their products in Katmandu, and vice versa. Fortune 500 employees and owners of modest home-based businesses can locate wherever the telecommunications infrastructure allows and telecommute to work and market. Electronic communities such as Blacksburg, Virginia, enable citizens to surf the Internet for local shopping, church news, and e-mail, to transact e-commerce, and to fulfill civic duties. A wealth of federal, state, and local government materials that would have taken weeks of effort to acquire before the World Wide Web are now available through a few quick strokes on the keyboard.

Main street businesses, hospitals, libraries, newspapers, and Extension offices are all beginning to find that their futures are in some way connected to networks of computers and the information they provide. The small town pharmacist, with the click of the computer mouse, now verifies insurance coverage, updates warning labels, and cross-checks the patient's total prescription regimen for drug interactions and harmful side effects, all in less than 30 seconds. An isolated rural hospital has the potential to link to the finest doctors in the land should additional consultation be needed for diagnosis or treatment of patients. The online library operating in an old Carnegie building offers access to information databases and self-education possibilities for rural citizens that exceed the wildest dreams of Andrew Carnegie himself. And digitization of information enables the newspaper and county Extension office to obtain mountains of news and information of interest to local citizens.

What we describe here represents the fulfillment of some of the highest hopes of the creators of the Internet, World Wide Web, and other telecommunications systems. These things are truly happening in some parts of America, even rural America. But the popularized, and idealized, images promote technological utopianism, making it easy to assume that the diffusion and implementation of these technologies will have positive impacts for everyone—that the technologies that make possible electronic cottages, electronic commerce, telemedicine, virtual libraries, and cyber citizenry will have universally beneficial results. As we have seen throughout this book, however, the actual impacts of the telecommunications revolution are not always positive for rural communities, and its failure to reach into depressed urban enclaves and deprived rural locations has profound and detrimental implications for areas most in need of opportunity and development. The idealized images of telecommunications innovations and their uses projected by computer ads, magazine articles, and books are neither accurate nor complete. The reality is much more complex and ambiguous than the myths would imply, with likely winners and losers.

LESSONS FROM OUR RESEARCH

Three themes reflecting the differential impacts of telecommunications for rural areas emerged from our research. They include (a) unequal access to technological infrastructure, hardware, and software to deliver or access needed services, (b) patterns of negative as well as positive, intended as well as unintended, consequences of technologies when they can be accessed, and (c) the dominance of vertical influences driving local telecommunications development, adoption, and use. We discuss each in turn before elaborating the characteristics that distinguish communities likely to succeed or fail in applying telecommunications for economic and community development.

Unequal Access to Technological Infrastructure

Having a technological infrastructure that supports the use of innovative telecommunications services such as the Internet, the World Wide Web, videoconferencing, and public packet switching is necessary, though certainly not sufficient, for rural community economic viability. The importance of quality local public infrastructure was addressed in almost all chapters. Respondents from all sectors rated existing telecommunications services deficient although they were optimistic those services would be substantially improved within 3 years. Most telephone companies engaged in continual upgrades and will be offering services not previously available within the next five years. A sizeable number, however, still do not have plans to offer some of the more sophisticated services, such as public packet switching, that many businesses and institutions find indispensable.

Research demonstrates that adoption of telecommunications hardware and software is becoming more widespread, but disparities between rural and urban areas remain. Computer use by rural and urban workers is increasing quickly, but at a faster pace in urban settings; this means the gap between rural and urban use is increasing. Computer use in urban areas grew from 28 percent in 1984 to almost 49 percent in 1993, while in rural areas it grew from about 18 percent to 36 percent (Kusmin, 1996). During this time period, the gap between rural and urban computer use grew from 10 percentage points to almost 13. Kusmin attributes most of the 13-point difference to job needs and personal user characteristics, but about one-fourth of the difference (3%) is attributed exclusively to locale. A study conducted by the National Telecommunications and Information Administration to determine who is and who is not connected to the information superhighway confirmed differences between rural and urban areas in percentage of households with online services (McConnaughey & Lader, 1997). As Table 13.1 shows, rural areas in the Northeast had a higher percentage of households with online services than did urban areas, but the pattern was reversed in the Midwest, South, and West, where rural areas lagged behind by at least 5 to 7 percent. The rural Midwest, West, and South have one-quarter to one-third *fewer* households with online service than their respective urban areas. These results suggest that fewer rural communities (and their residents) have all the right connections. Nor do they have the potential to be connected like urban America.

Access is only one indicator, of course. Some research has found little, if any, difference in *use* of telecommunications technologies between rural and urban areas (LaRose & Mettler, 1989). Some research even suggests that rural people use telecommunications technologies more than their urban counterparts (Johnson, Allen, Olsen, & Leistritz, n.d.). Closer examination of the data, however, reveals that the telecommunications technologies used by rural residents and businesses are the more common technologies such as facsimile (FAX), telephone answering machines, and personal computers, while advanced online applications such as Internet, e-mail, and the World Wide Web are much less common (Johnson et al., n.d.).

Table 13.1
Percent of Rural and Urban Households in the United States with Online Service by Region in 1997

Region	Rural	Urban	Difference
Northeast	19.7%	18.0%	1.7%
Midwest	13.9	18.6	-4.7
South	12.7	19.4	-6.7
West	17.8	23.1	-5.3

Patterns of Negative and Positive, Intended and Unintended Consequences

Our chapter on farmers noted that although innovative agricultural technologies have markedly increased farm production, these technologies have not been universally beneficial to rural communities. In fact, most labor-saving technologies adopted since the mid-1930s by farmers and others in extractive industries such as lumbering, mining, quarrying, and fishing have reduced not just drudgery, but job opportunities as well. By reducing net demand for rural labor, they have helped to empty the rural countryside of its once abundant population. Rural communities have been forced to either develop a new economic base for employment and income opportunities or face their own demise. As a result, some communities prospered by becoming regional trade centers or developing a specialized niche, but most have declined and a few have disappeared altogether. As with agricultural innovations, telecommunications technologies can not be expected to have universally positive impacts for rural communities. We must look beyond the idealized images or myths that serve as packaging for these technologies and examine the likely detrimental (as well as beneficial) consequences of these technologies for different sectors of rural communities.

Perhaps one of the most pervasive myths is that telecommunications development will be the savior of rural communities that have lost their traditional economic base. Just as communities built industrial parks in the 1960s and 1970s to stimulate economic growth, and then sponsored business incubators in the 1980s, they pursue the promises of telecommunications in the 1990s, strong in the belief that "if you build it, they will come." Run a fiber-optic, digitally switched system to town and it will attract business and stimulate economic growth. A growing body of evidence unquestionably indicates a significant positive relationship between investments in telecommunications and economic growth (Garcia & Gorenflo, 1997), and over half of the economic developers in our research felt that telecommunications were essential for economic development. However, our research findings and others demonstrate that telecommunications technologies alone are not sufficient for rural community economic development, and blind pursuit of their benefits could result in significant unintended or unwanted consequences.

Opinions vary on the efficacy of a telecommunications base for rural economic development. Tweeten (1987) contends that the impact of telecommunications on rural areas will be minimal, arguing that although business is generally attracted to the low wages, reduced taxes, fresh air, clean water, little traffic congestion, and financial incentives offered by rural communities, high technology firms are influenced less by these than by the availability of a technically skilled workforce, consultation opportunities with university personnel, good ground transportation and shipping facilities, easy access to air transportation, and high quality medical services. These latter characteristics are usually associated with proximity to a metropolitan area (Tweeten, 1987).

While popular images of professional employees telecommuting from their rural homes abound, the number of professionals with lucrative employment opportunities operating from electronic cottages in rural areas is very small, and contrary to claims that they are proliferating rapidly, they constitute only a tiny minority of the rural labor force. Instead, telecommunications opportunities presented to rural communities often come in the form of telemarketing or data processing firms seeking lower rents, slack telephone capacity, and employees willing to work for $5 to $8 per hour. Leistritz (1993) examined the impact of such telemarketing and data processing firms on North Dakota's rural economy. He concluded that the economic contributions of these telecommunications firms were comparable to new light-manufacturing firms in terms of wages, benefits, and community multiplier effects. They thus represent legitimate opportunities for rural citizens but tend to result in low-skill, low-wage jobs that have few advancement opportunities. Furthermore, these highly mobile firms often abandon the community and relocate to other towns offering additional cost savings. The ultimate effect of telecommunications-induced development for rural communities is often far different from the conjured images of electronic cottages or telecommuting professionals.

One of the fastest growing business sectors in rural areas is producer services, that is, services sold primarily to business and government (Beyers & Lindahl, 1996). The degree to which such local businesses export their services regionally, nationally, or internationally, bringing income from outside the community, contributes to the local economic base. But while telecommunications can aid the export of producer services, a study in Washington State found that telecommunications can make it difficult for rural areas to attract and retain producer service firms (Kirn, Conway, & Beyers, 1990). Producer service firms prefer being close to customers and situate themselves near larger customer bases in urban areas. Telecommunications are more likely to be used to reach into the rural community customer base from outside. Many producer service companies do have back-office operations (such as data processing for credit cards or insurance) that are often located in rural areas, but again, rural communities should be wary of recruiting such operations because the jobs they provide are predominantly low skill and low wage (Kirn et al., 1990). A better solution would be to encourage the start-up of new local producer service businesses.

Farmers provide another example of how new telecommunications technologies can increase rural inequalities rather than provide universal benefit. Farmers with large operations and the knowledge for using computers to increase profits are heavy users of computers and telecommunications technologies such as teletext for farming. But for farmers at the other end of the spectrum—those with small operations—computers and other telecommunications technologies are neither particularly useful for farm management nor cost efficient. With smaller farm scale and a lower farm management knowledge base among these farmers, the financial investment to purchase and operate a computer, coupled with the time and productivity losses in learning how to exploit its capabilities, result in exceedingly small payoffs. Farmers with small operations are adopting computers and telecommunications technologies such as the Internet to enhance nonfarm income opportunities, education, and entertainment, rather than for farm management. Their continued ability to enjoy the country lifestyle may depend on whether they can utilize computers and the Internet to maintain or increase their nonfarm income.

In many cases, the predominant applications of telecommunications, as well as their impacts, are quite different than the popularized images would have us believe. For example, television advertising implies that as soon as school computers are networked to the world, children's education will be enriched by interchange with children from other cultures. But in reality few urban or rural schools are adopting computers for this reason. Rather, the prime purposes of school computers are to provide uniform grading systems and to further development of self-instructional modes of education by replacing paper with electronic worksheets. Proponents of telemedicine paint scenes of highly skilled medical specialists saving lives by remote diagnosis and treatment, in some cases "virtual surgery," afforded by telecommunications innovations. To a high degree, however, telecommunications adoption and use by hospitals is mandated by insurance providers, state health departments, licensing agents, and others. As a result, the most prevalent uses of computer-mediated communications is for electronic billing and ordering, while interactive video is used primarily for teleconferencing among hospital administrators and routine training of health care providers.

Depending on implementation and use, telecommunications may not necessarily be a benefit to the community and may indeed have detrimental effects. Stoll (1995) scorns the "technocratic belief that computers and networks will make a better society" and warns us to be skeptical of the broad claims made by telecommunications proponents of the Internet that "access to information, better communication, and electronic programs can cure social problems" (p. 50). Although the 40 percent of economic developers in our research who failed to see much benefit in the Internet may have been too skeptical, healthy caution is warranted nonetheless.

Dominance of Vertical Influences Driving Telecommunications Development, Adoption, and Use

A key finding of the research in this volume is that forces outside the community are playing a major role in determining whether, when, and how telecommunications technologies are available to and adopted by sectors of the rural community. Libraries provide a clear example of how outside forces shape technology use and impacts. In many states the state library system is the driving force behind the virtual library of the future. The state library's vision is of a single state virtual library system permitting remote access to material from all branch libraries at the touch of a computer mouse. Efforts to achieve this vision have resulted in the expenditure of large amounts of money to digitize the holdings of large state and research libraries, while almost no investment is made in rural libraries themselves. The potential for developing rural libraries as community information hubs has been largely ignored. Thus, the probable impact of telecommunications for local libraries will be to make them mere access points for the rich holdings of central libraries, with control passing from local library directors and boards to directors and boards of large regional and state libraries.

Similarly, the driving force for the creation of a high speed communications network linking Iowa's 100 county Extension offices was Extension administrators who clearly saw that digitization was about to occur in federal and state offices and wanted to position Iowa Extension to take advantage of this trend. By converting Extension publications and other materials to digital form and making them available online, the state Extension office could save large sums of money formerly spent printing and mailing publications. In the county offices, laser printers became the printing presses for these online publications, and the printing costs formerly borne by the state office were effectively shifted to the counties. In addition, although most Iowa county Extension offices now had the most modern, high-speed connections to the Internet and World Wide Web, local citizens were, for the most part, unaware of this technological capability. Nor were local citizens included in the planning and implementation of these technologies. Had they been, they would likely have pushed for network access for a broader spectrum of community members.

Whether for libraries, county Extension offices, local schools, rural hospitals, or small businesses, the impetus behind most telecommunications applications comes from vertical forces outside the community. Thus, to understand the potential impacts of these technologies, we must understand what animates these outside forces. We previously noted that vertical linkages are the relationships community residents, businesses, agencies, organizations, and institutions have with others outside the community, while horizontal linkages are relationships within the community. Vertical linkages are important for introducing new ideas, technologies, and resources to the community, and horizontal linkages spread those ideas, technologies, and resources to others within the community for greater community benefit. We found vertical linkages dominant in almost all the sectors we studied, but we found very little horizontal communication or sharing of telecommunications information, technologies, or training within the community. Successful telecommunications-based economic development in rural communities requires vertical linkages, but to achieve that development and derive communitywide benefits, horizontal linkages must be nurtured simultaneously.

CHARACTERISTICS THAT DISTINGUISH COMMUNITIES LIKELY TO SUCCEED OR FAIL

Examples like Aurora, Nebraska, Blacksburg, Virginia, and Nevada, Missouri, demonstrate that when a community comes together, marshals its resources, and aggregates demand, telecommunications technologies become powerful tools to stimulate both economic and community development. But it is largely the unique characteristics, histories, resource endowments, and organizing experiences of communities that shape their development and use of physical, financial, and human capital for community improvement. Most rural communities have made little progress in organizing themselves or mustering their resources to take productive advantage of new telecommunications technologies. As a result, telecommunications technologies can be expected to benefit some rural communities handsomely, but they may contribute to the decline of others. Unless counteracted by strong community action, the tendencies of these technologies to make it easier for local residents to shop elsewhere, get information elsewhere, be entertained elsewhere, receive services elsewhere, and generally become more cosmopolite in orientation will lead to weakened economic and social bonds that hold rural communities together.

Rowley and Porterfield (1993) argue that rural communities must individually assess their strengths for using telecommunications technologies and then adapt strategies to exploit those strengths: "The diversity of rural America, the need for rural areas to be self-reliant and able to adapt, and the fact that successful strategies for development require local participation preclude a one-size-fits-all policy. Development must be locally driven, and policies to promote rural development through enhanced telecommunications must reflect locally determined goals" (p. 6).

At the same time we recognize this need for unique solutions for each community, our research identifies several generalizations that rural communities might find useful. We discuss them within the conceptual framework that we introduced in Chapter 1—community capital, community leadership, and critical mass—concluding with a comparison between economic development and community development.

Community Capital

In Chapter 1 we stated that a community's resources, assets, and wealth—in short, its capital—are critical to telecommunications-based economic development. We identified four types of capital: physical, financial, human, and social.

Physical Capital. This type of capital consists of the natural, physical, and material resources existing within a community, including its telecommunications infrastructure. Access and service from a local telecommunications provider with state-of-the-art infrastructure is important to the long-term viability of rural communities. Telecommunications capability was rated high across all community sectors in importance for community economic development and quality of life, and a majority of respondents trust telecommunications will strengthen local communities by providing them access to valuable resources found on the information superhighway. Despite improvements in the provision of infrastructure around the state, however, a sizeable proportion of respondents in every sector rated telecommunication infrastructure below par. The extent to which upgrades and extension of infrastructure continue can be a deciding factor in the success of rural communities eager to deploy telecommunications for development.

One key to telecommunications success is broad access not only to the requisite physical infrastructure, but also hardware and software. The perceived relative complexity and cost of networked computer systems precludes participation by many small libraries, newspapers, municipal offices, businesses, and, we might assume, homes. The success of Iowa Extension's implementation of a wide-area network resulted in part from attention to providing direct computer access to all potential users. In contrast, rural businesses cite lack of a local access number as a significant barrier to Internet use. By providing a minimum of one point of presence in each of Iowa's 99 counties, the Iowa Communications Network (ICN) opens the potential for universal access, but lack of local service, compatible hardware and software, expense of connections, and toll charges for Internet and World Wide Web access still stand as deterrents. The efforts of Wise County, Virginia, demonstrate the importance community leaders from various sectors attach to state-of-the-art telecommunications infrastructure, as well as their commitment to securing that infrastructure.

CASE STUDY: MAKING WISE CHOICES ABOUT TELECOMMUNICATIONS TECHNOLOGIES

Wise County, Virginia, located in the heart of the Appalachian Mountains, has a population of about 40,000 and an economy traditionally driven by coal production and related services. To diversify the economy the Wise Community Technology Project seeks to establish information technology (IT) work-ports, call centers, and telecommuters for mid-size and large companies in need of back office services and "knowledge workers." The effort was initiated in November of 1998 by Jack Kennedy, Clerk of the Wise Circuit Court, and is supported by a group of 40 community leaders representing business, education, and government. Participants include the county board of supervisors, the school board, presidents of Mountain Empire Community College and Clinch Valley College, a bank president, a lawyer, and other business interests. Although the local telephone company, Bell Atlantic, is not involved directly in the project, it provides state-of-the-art services to the locale. In addition, the company contributes funds to support development efforts, and the local Bell Atlantic director serves on the regional development authority and has been involved in other development activities.

A primary goal of the Wise Community Technology Project is to build a model for the Certified Technology Community concept. The Certified Technology Community grew out of Virginia's Certified Business Location concept. Communities that are Certified Business Locations display green signs at highway entrances indicating that the community has the necessary infrastructure and supports business development. Based on the same principle, the Certified Technology Community communicates to prospective business and industry that it is technology-based development friendly (Kennedy, 1999).

Financial Capital. Hard currency, savings, revenues, and credit available to the community and to various community sectors is required for initial planning and purchase of telecommunications infrastructure and technology, for maintaining and upgrading equipment, for hiring competent technical expertise and computer knowledgeable staff, and for training current novices and prospective users. The challenges of securing financial capital for acquisition and use of telecommunications technologies are ongoing for most individuals, businesses, and organizations. Rural institutions such as small town libraries are particularly constrained in telecommunications use by the lack of financial capital. Most do not have Internet connections. The local telephone company needs financial capital to develop and maintain the telecommunications infrastructure that serves the community. Deregulation of the telecommunications industry raised concerns for existing local telephone companies but provided new opportunities for communities that considered themselves underserved. If the existing provider in the local exchange is a large, absentee-owned company that refuses to upgrade services, the community can entice another provider to compete in the exchange or alternatively establish its own telephone company to compete with the incumbent. If the existing provider is a small locally owned company with limited resources, it may be possible to negotiate a joint public-private venture to fund infrastructure development. Small, locally owned rural telephone companies understand the symbiotic relationship that exists between their company and the community's welfare.

Human Capital. The technical knowledge and skills relevant to installing and using telecommunications technologies, as well as the management and organizational skills to optimize that use is human capital. Innovative applications of telecommunications technologies often occur when the adopter has the ability to not only use but also to adapt or reinvent the technology. Our findings corroborate other research demonstrating a direct relationship between knowledge, skills, and use of telecommunications technologies. Farmers with stronger managerial skills tend to use telecommunications technologies more for managing their farming operations. Iowa Extension's successful implementation of its wide-area network was bolstered by specific training of Extension staff in the use of that network and the provision of consultants to answer questions and troubleshoot problems. Hospitals, on the other hand, whose computerization is often mandated by insurance companies and others, receive little assistance in determining the type of hardware or software to acquire and even less training in its use. Telephone company managers considered skills and training of employees an important factor in upgrading technologies. Unfortunately for rural communities, rural residents often lack the knowledge or skills for using telecommunications. Well-educated librarians with the knowledge to adapt the Internet to benefit their libraries are attracted primarily to large well-endowed urban libraries, leaving small rural libraries in want of such expertise. Most economic development professionals we surveyed, although current in traditional development expertise, had little knowledge or personal experience using telecommunications technologies. Rural communities have major obstacles to overcome to achieve sufficient levels of technical knowledge and skills in telecommunications development and expertise in computer use. These include securing competent technical assistance, identifying knowledgeable installers and operators, building technological capacity among decision makers and leaders, and promoting computer literacy across all sectors of the community. The goal is not simply use of telecommunications, but creative and productive use. This involves deepening local capacity to create visions of how the community can collectively employ telecommunications. As a first step, incumbent city officials, economic development professionals, community development planners, and others within the community should be trained in telecommunications-based development and use—a veritable telecommunications Chautauqua to spread knowledge communitywide. As a second step, telecommunications expertise should be a criterion for future hires from outside the community.

Social Capital. Networks, norms, cultural resources, relationships, and trust foster cooperation and facilitate coordination within the community to achieve mutual benefits. Social capital is about identifying resources, assembling talents, generating enthusiasm, directing efforts, and creating synergies. It also is about building capacity and investing in community. Evidence of the importance of social capital to the implementation and use of telecommunications is found throughout our research. Locally owned telephone companies with a stake in the welfare of the community are more likely to be involved in community activities and provide leadership and financial support for those activities. Companies that are more involved with their communities, in turn, are more likely to provide innovative services. Local government is more innovative in its use of telecommunications when the community supports that use. Aurora, Nebraska, and Wise, Virginia, reveal the role of social capital in telecommunications development. A major part of Aurora's success in developing and using its telecommunications infrastructure can be attributed to involvement from a broad representation of local citizens. Similarly, the Wise Community Technology Project has leaders that represent broad community interests.

Community Leadership

Community leadership, a component of human and social capital, requires the knowledge, skills, commitment, and vision to mobilize local capacity to collectively meet community needs and solve local problems. Research has shown that a champion, be it one person or a small group of individuals, is needed to initiate the introduction of telecommunications technologies and to promote and guide their implementation. Champions are found in businesses, organizations, and communities and their roles are of great value in all those settings. All eight communities whose successful telecommunications programs were showcased in *Making Wise Choices: Telecommunications for Rural Community Viability* (Abbott, 1997) had an individual or small group of leaders that shared their innovative ideas with the communities and then provided the impetus for community action to achieve the resulting vision.

Critical Mass

Critical mass refers to a minimum threshold necessary to permit or justify telecommunications infrastructure development in rural communities. Businesses, organizations, and communities in rural areas suffer a lack of critical mass in at least three major respects: insufficient financial capital, insufficient human capital, and insufficient scale for feasible implementation.

Insufficient Financial Capital. This was an important individual limitation on adoption of computers and other telecommunications technologies by individuals in the early 1980s, when it commonly cost $5,000 or more to buy a computer, printer, and software. It still is a problem at the community level, where telephone switching equipment, computers, software, and networking can require expenditures of hundreds of thousands of dollars. In some situations a financial threshold can be identified below which telecommunications development is unlikely to take place. For example, rural libraries that are connected to the Internet in Iowa generally have an annual budget of $50,000 or more. Newspapers connected to the Internet and World Wide Web have circulations of at least 800. In other cases, the relationship is more linear, with adoption levels gradually increasing as the budget and other financial resources increase. For example, hospitals with more beds, more admissions, and more revenue tend to make more use of telecommunications technologies. Telephone companies with more employees tend to be more innovative in the technologies they adopt and the services they provide. And local governments in larger communities use more telecommunications technologies than their smaller counterparts and they tend to use them more for economic development.

Insufficient Human Capital, Especially the Requisite Technical Knowledge and Skills. Due to the complexity of telecommunications infrastructure and hardware, the rapidly changing regulatory environment, the confusing service matrix, and fast-paced technological changes, it is difficult for rural communities to establish the needed expertise and maintain currency with the ever changing potential of telecommunications. General computer literacy, as well as knowledge about the Internet and World Wide Web, tends to be concentrated among those who are young, highly educated, or recently trained. Younger, better educated economic development professionals were more likely to report facility with computers than their older counterparts with many years' experience in development but little recent exposure to telecommunications technologies. Editors of weekly small-town newspapers have limited their adoption of new technologies because of the lack of computer-trained employees. When the one person in town who knows how to use a technology moves away, the newspaper is left without the technical skills needed, replacement skills that urban editors locate much more easily. Similarly, businesses with in-house information technology expertise were more likely to have adopted telecommunications technologies than those not having such expertise immediately available.

Insufficient Scale for Feasible Implementation. Small community size and population make it difficult to justify local investment or attract outside service providers. Simple supply and demand dynamics work against small or remote communities because supply is directed where demand is greater and that is usually away from rural communities. Small community size, not just in population, but in the number of potential telecommunications users, can dissuade investors from locating state-of-the-art infrastructure and services in a locale because they fear they will be unable to recoup their investment and realize profits. We were unable to identify the exact community size or critical mass required to make telecommunications investment profitable, but we know that communities under 3,000 residents struggle to secure infrastructure and adopt basic technologies, while communities under 8,000 struggle to optimize their use. Communities under 500 are most vulnerable, and in Iowa, this represents 54 percent of all communities. Unless they are able to join forces with their neighbors to aggregate demand, present a sufficient prospective market, earn price incentives, and pool expertise, it could be impossible for them to counteract the disadvantages of their size and scale to achieve a payoff from telecommunications investments. Similarly, for small rural libraries, hospitals, newspapers, governmental units, or businesses minimum investments needed in infrastructure and training may exceed individual sector payoffs. We have seen that for individual adopters, such as farmers, computers and telecommunications technologies are extremely scale sensitive. They pay off handsomely for larger farms, but the questionable increase in profit for small farmers does not warrant the investment.

The crux for small rural communities that are poorly endowed with any or all physical, financial, human, or social capitals is to maximize external linkages and to aggregate demand and resources. Without outside assistance, many of the smallest communities and rural institutions lack the critical mass of financial and human capital necessary for success. Often it is difficult to secure external resources unless the community can demonstrate sufficient existing capacity (again, critical mass) to effectively utilize those new resources. More rural communities need to develop *linkages with surrounding communities* as a strategy to aggregate demand and resources and achieve critical mass. Variously called multi-community collaborations, multicommunity development organizations, or simply cluster communities, this strategy brings together communities and their resource for collaborative development efforts to achieve certain economies of scale and promote synergies (Korsching & Borich, 1997). Communities using this strategy have higher levels of all four capitals to mobilize local action and attract additional outside resources.

COMMUNITY ECONOMIC DEVELOPMENT

The introductory chapter distinguished between economic development and community development. Whereas economic development is narrowly focused to address economic welfare and growth, with little attention given to broader community concerns or unintended consequences, community development is a more comprehensive process involving local residents participating in planning and implementation to improve their community. Community development is development *of* the community; it not only seeks to achieve a goal but also builds capacity within the community to address problems and issues. *Community development* develops not only the physical and financial capital, but human and social capital as well. Without the capacity building of the community development process, economic development is simply development *in* the community.

The primary economic development strategy used by rural communities in recent decades has been recruitment of new business and industry. When used in isolation or to the exclusion of alternatives, this strategy frequently leads to concessions that leave communities vulnerable to enterprises with little commitment to the community, whose primary motivations are to exploit subsidized infrastructure and amenities, minimal regulation, low taxes, low wages, and a variety of other economic development incentives. Business retention and expansion, along with grow-your-own strategies, are more viable alternatives that also enhance community telecommunications capacity. For example, telecommunications technologies can greatly expand market opportunities for home-grown businesses, and business growth can translate into higher demand for telecommunication capacity and service. Additionally, local families who operate home-grown businesses usually have a stronger commitment to the welfare of the community than absentee owners and managers of branch plants.

Telecommunications technologies such as computers and the Internet are powerful tools that can be used for many purposes. Use of telecommunications technologies, however, is simply not sufficient. More important is telecommunications-induced development, that which Gary Warren, executive vice president of Hamilton Communications in Aurora, Nebraska, was referring to when he stated: "[W]hat I'm interested in is having people [within the community] who are on the Internet who have figured out a way to get somebody from the outside to pay them some money to do something" (Warren & Whitlow, 1997, p. 55). Viable rural communities will be those that identify ways telecommunications can be used productively—to enhance existing capital and grow new capital.

How can the storehouse of this kind of knowledge be increased in rural communities? One solution begins with the community's youth, by integrating into school curricula instruction that goes beyond mere use of the Internet and World Wide Web to instruction that introduces applications of telecommunications technologies to real local issues. Rural students, in turn, can train other community members to use these technologies to more fully exploit their capabilities, a strategy that has been successful in a number of communities. Teaching youth about the local economy, history, and culture can open their eyes to local needs, assets, and opportunities. Providing students information about their own communities, coupled with insights on telecommunications applications, promotes their potential as future entrepreneurs and community leaders (Hobbs, 1997). It can thereby stem the flow of young people leaving the community because they perceive no future opportunities.

Telecommunications-based development is most successful when it builds on existing organization, experience, expertise, and leadership. The success of Aurora, Nebraska's telecommunications program may in part be attributed to the community's long history of local community development efforts. But communities without such social capital should not despair nor be dissuaded from telecommunications-based development. There are many public and private organizations that can provide assistance in this effort.

We recognize that telecommunications are not the sole or sufficient route for communities seeking directions to a revitalized local economy, but they provide an important vehicle to move communities toward that destination. Advanced telecommunications are part of the overall package the community offers to both insiders and outsiders, which also includes a quality educational system, health and welfare facilities and services, cultural and recreational amenities, and retail goods and services. Dholakia and Harlam (1993) found these other factors actually may strengthen the relationship between telecommunications infrastructure and economic development. We found that the community having all the right connections is one connected not only to quality telecommunications infrastructure, but one that invests in itself, developing the economic, social, and cultural infrastructure and connections necessary to create opportunities for those living in the community that also are attractive to others on the outside.

NOTE

The authors gratefully acknowledge the assistance provided by J. Paul Yarbrough of Cornell University in the preparation of this chapter.

GLOSSARY

Access charges: Since the divestiture of AT&T, most telephone customers pay access charges, sometimes called subscriber line charges, for access to the interstate public switched network. The charge theoretically makes up for the subsidies paid by the long-distance companies to the local exchange telephone companies.

Advanced Research Projects Agency (ARPA): An agency of the Department of Defense that created the computer network that evolved into the Internet.

Advanced Research Projects Agency Network (ARPANET): The precursor to the Internet.

Analog: A method of transmission in which information is sent over a medium, such as copper cable or microwave, by changing the voltage of signals.

Analog signal: A continuously varying electrical signal in the shape of waves, whose size and number change as the information source varies. Fluctuations in the signal produce the differences in loudness, voice, or pitch heard by the user.

Analog switch: Telephone exchange that switches signals in analog (as opposed to digital) form.

Asynchronous transfer mode (ATM): A type of switching that is expected to bridge the gap between packet and circuit switching. ATM uses packets called cells that are designed to switch cells so fast that there is no perceptible delay, making it suitable for voice and video. Network communication technique capable of handling high-bandwidth multimedia information applications.

Audio conferencing: Connecting more than two locations into one telephone conversation.

Audiotex: An interactive audio information service available for a fee to users of touch-tone telephones.

Baby Bells: The seven regional telephone companies created as a result of the 1984 divestiture of AT&T's local telephone holdings. The Baby Bells include Ameritech, Bell Atlantic, BellSouth, NYNEX, Pacific Telesis, Southwestern Bell, and U.S. West.

Backbone: A high-capacity communications channel that carries data accumulated from smaller branches of the computer or telecommunications network.

Bandwidth: The width of an electrical transmission path or circuit in terms of the range of frequencies it can pass, bandwidth serves as a measure of the information carrying capacity of a communications channel.

Basic services: A Federal Communications Commission designation for transmission capacity offered by a common carrier (e.g., phone company) to move information between two or more points.

Baud: The baud rate is the standard unit of measure for data transmission speed or capability. (Typical rates are 1200, 2400, 9600, and 14,400 baud.) Baud rates represent the highest number of single information elements (bits) transferred between two devices (such as modems or fax machines) in one second.

Bell Operating Company (BOC): The BOCs are grouped under seven regional holding companies (RHCs or RBOCs). See also Baby Bells.

Bit: A contraction of "binary digit." The smallest unit of digital information that a computing device handles, represented by "on/off," yes/no," or 0/1. All characters, numbers, and symbols are processed as electronic strings of bits, thus, in electrical communications systems, a bit can be represented by the presence or absence of a pulse.

Bits per second (bps): The number of binary digits transmitted per second.

Broadband: Refers to high-speed, high-capacity transmission of signals in a frequency-modulated fashion, over a segment of the total bandwidth available, thereby permitting simultaneous transmission of several messages.

Broadband carrier or broadband channel: A high-capacity telecommunications medium capable of carrying a wide range of frequencies required for multimedia applications, and with bandwidth in excess of that required for high-quality voice transmission.

Broadband communication: Communication system with a bandwidth greater than that of a voiceband.

Bulletin board system: A computer system allowing users of an electronic network to leave messages that can be read by many other users.

Bypass: Telecommunications transmission that avoids part or all of the public switched network.

Byte: A strong or grouping of 8 bits used to represent a character or value.

Bytes per second (Bps): The number of characters transmitted per second.

Cellular (radio) telephone service: Mobile telephone service using a series of transmitters in local areas or cells. The transmission changes frequency as the driver moves between cells. The system allows frequencies to be re-used, thus providing much greater capacity than older mobile systems. Cellular telephone calls are connected into the public switched network.

Centrex: Referring to both a service and a switch, Centrex entails having a switch located in a telephone company's central office that provides witching services to an institutional user; typically that user requires more than 30 internal lines. Centrex facilitates intra-institutional calling and offers various special calling features.

Coaxial cable: One or two transmission wires covered by an insulating layer, a shielding layer, and an outer jacket that can transmit either broadband (several signals) or baseband (one signal). Coaxial cable can carry several channels of telephone and TV signals simultaneously, transmitting data, voice, and video.

Coder/decoder (codec): A series of integrated circuits that perform a specific analog-to-digital conversion, such as the conversion of an analog voice signal to a digital bit stream, or an analog television signal converted to a digital format.

Common carrier: Telecommunications network supplier which carries communications from others. In data communications, a common carrier is a public utility company that is recognized by an appropriate regulatory agency (FCC or various state public utility commissions) as having a vested interest and responsibility in furnishing communications services to the general public. They are subject to federal and state regulations that establish operating rules and tariffs in order to make services available at a fair price and on a nondiscriminatory basis. Telephone companies are common carriers.

Communications Act of 1934: The basic national legislation governing electronic communications. The act mandated a national telephone system—the first such legislation to recognize the importance of telecommunications for social and economic growth.

Compact disk (CD): An optical storage medium used for music and for computer data, among other services.

Compatibility: The ability for computer programs and computer readable data to be transferred from one hardware system to another without losses, changes, or extra programming.

Compressed video: Video images that have been processed or digitized to reduce the amount of bandwidth needed to capture the necessary information so that the information can be sent over a telephone network.

Compression: The reduction of certain parameters of a signal while preserving the basic information content. The result is to improve overall transmission efficiency and to reduce cost. In media operations, the most extensive use of compression techniques is in cable TV coaxial cable, where compression can double or triple the number of available channels.

Computer conferencing: Group communications through computers or the use of shared computer files, remote terminal equipment, and telecommunications channels for two-way, real-time communications.

Convergence: Coming together of all information and communication forms into common underlying approaches based on digital techniques.

Cross-subsidy: Telephone term meaning that funds from one part of the business (e.g., long distance) are used to lower prices in another (local service).

Custom calling features: Special services for telephone customers, e.g., three-way calling, call forwarding, and call waiting.

Cyberspace: Term indicating the virtual universe created by networked information flows.

Data compression: Methods to reduce data volume by encoding it in a more efficient manner, thus reducing image processing, transmission times, and storage space requirements.

Dedicated line: A permanent telephone line owned by or reserved exclusively for one customer, available 24 hours a day, usually to provide certain special services not otherwise available on the public-switched network.

Deregulation: The opening of an industry to competition by legislative or regulatory action. In the telephone industry, deregulation has removed subsidies from certain services and separated the telephone industry into several competitive areas, including equipment manufacturing and sales, long-distance service, and specialized transmission services.

Dial-up access: Connecting to a network by dialing a number, rather than being connected to it permanently.

Digital: Digital communication is the transmission of information using discontinuous, discrete electrical or electromagnetic signals (represented by strings of 0's and 1's) that change in frequency, polarity, or amplitude, allowing the simultaneous transmission of audio, data, and video (as bits of information) over the same line.

Digital compression: Techniques which enable large volumes of information to be sent using fewer bits.

Digital image: An image composed of discrete pixels, each of which is characterized by a digitally represented luminance level. For example, a common screen size for digital images is a 1,024 by 1,024 matrix of pixels x 8 bits, representing 256 luminance levels.

Digital switch: Computerized telephone exchange that processes voice and data in digital (as opposed to analog) form.

Digital switching: A process in which connections are established by processing digital signals without converting them to analog signals.

Digitize: The process by which analog (continuous) information is converted into digital (discrete) information.

Direct Broadcasting by Satellite (DBS): Satellite system designed with sufficient power so that inexpensive earth stations can be used for direct residential or community reception, thus reducing the need for a local loop by allowing use of a receiving antennae with a diameter that is less than one meter.

Disc or disk: A flat, circular, rotating platter that can store and replay various types of information, both analog and digital. "Disk" is often used when describing magnetic storage media. "Disc" usually refers to optical storage media.

Divestiture Agreement: The plan, finalized in 1984, in which AT&T and the federal government set the terms for AT&T's relinquishment of control over its local phone companies in exchange for permission to enter the information services market. The resulting document is also called the Modified Final Judgment (MFJ).

Dots per inch (Dpi): Film resolution for digitized images, commonly expressed as dots (pixels) per inch.

Downlink: An earth station (e.g., antenna) used to receive signals from a satellite; the signal transmitted for the satellite to earth. Also, the path from a satellite to the earth stations that receive its signals.

Download: To send an electronic document, software, or other computer file across a network.

DS0, DS1, DS3: Digital telecommunications channels, capable of carrying high volume voice, data, or compressed video signals. DSI and DS3 are also known as T1 and T3 carriers.

Enhanced 911 (E911): Emergency service; the caller's telephone number, location, and other important information are stored in a computer and automatically displayed for the dispatcher when a 911 call is received.

Electronic data interchange (EDI): The transmission of data without paper or human intervention between two devices or applications, using a standard data format. The ability to exchange documents and other information, such as orders and invoices, between organizations electronically.

Electronic mail (E-mail): A store and forward service for the transmission of textual messages from a computer terminal, computer system, word processing, and facsimile equipment. A message sent from one computer user to another is stored in the recipient's "mailbox" until that person logs into the system. The system then can deliver the message.

Encoding: The process of transforming an analog signal into a digital signal, or a digital signal into another digital format.

Encryption: The rearrangement of the "bit" stream of a previously digitally encoded signal in a systematic fashion to make it unrecognizable until restored by the necessary authorization key. This technique is used for securing information transmitted over communication channel with the intent of excluding all other than the authorized receivers from interpreting the message.

Enhanced services: A telecommunications category established by the FCC to describe services that result in additional, different, or restructured transmitted information or that involve user interaction with stored information, whether voice or data. See Basic services.

Equal access: Under the AT&T divestiture agreement, subscribers must be able to reach the long-distance carrier of their choice by dialing a "1" plus 10 digits (1+ dialing). The requirement that local telephone companies must provide all long distance companies access equal in type, quality, and price to that provided to the dominant carrier.

Ethernet: A communications protocol that runs on different types of cable at a rate of 10 Mbps.

Exchange area: A local geographic area serviced by one or more telephone companies' central offices or switching centers.

Facsimile (FAX): Equipment that transmits and receives documents over telephone lines and reproduces them at the receiving end. Also, the black-and-white reproduction of a document resulting from a fax transmission.

Federal Communications Commission (FCC): The federal independent regulatory board of five members (commissioners) appointed by the president under the Communications Act of 1934 with the authority to regulate all interstate telecommunications originating in the United States. The FCC licenses and sets standards for telecommunications and electronic media.

Fiber optics or fiber optic cable: Insulated strands of hair-thin, flexible glass-core cable that uses light pulses instead of electricity to transmit audio, video, and data signals. Fiber optics allow high capacity transmission at very high speeds, e.g., billions of bits per second, with low error rates.

Fiber to the home (FTTH): High bandwidth optical fiber links to individual households.

Gateway: A connection between networks using different protocols. In computer networking, a gateway is the hardware and software used to interconnect different networks, for example two or more local area networks (LANs). In the context of telephone policy, gateway also means the hardware and software used by a telephone company to connect a subscriber to an information provider.

General Telephone and Electronics (GTE): The largest "independent" (non-Bell) local telephone operator in the United States.

Geostationary satellite: Satellite, with a circular orbit 22,400 miles in space, that lies in the satellite plane of the earth's equator and turns about the polar axis of the earth in the same direction and with the same period as that of the earth's rotation. Thus, the satellite is stationary when it is viewed from the earth. Also referred to as satellite or communications satellite.

Gigabits per second (Gbps): A measure of bandwidth and rate of information flow in digital transmission.

Gigabyte (Gbyte): A measure of computer memory and storage capacity. One gigabyte equals 1.074 billion bytes or 1,000 Mbytes.

Hacking: Accessing a computer-based system illegally.

Hardware: Electrical and mechanical equipment used in telecommunications and computers systems. See Software.

Hertz: A measure of the number of complete cycles made by an analog signal in a given time period.

High-definition television (HDTV): A television system or format with 1,125 lines of horizontal resolution, capable of producing high quality video images.

Hypertext: Computer software that allows users to link information together through a variety of paths or connections. HyperCard programs ("stacks") are made up of "cards," which when activated, allow users to move to another part of the material they are working with.

Icon: In computer operations, a symbolic, pictorial representation of any function or task.

Image technology: A general category of computer applications that convert documents, illustrations, photographs, and other images into data that can be stored, distributed, accessed, and processed by computers and special-purpose workstations.

Independent telephone company: A local exchange carrier that is not part of the BOCs (Bell Operating Companies), often cooperative in rural areas.

Infomatics or informatics: The application of computer science and information technologies to the management and processing of data, information, and knowledge.

Information economy: An economy in which the processing and transmission of information is a prime activity.

Information infrastructure: Provision of underlying network capabilities to support a variety of services based on computing and telecommunications capabilities.

Information service provider: Organization, group, or individual who creates and packages information content carried by electronic networks.

Information society: Refers to the increasing centrality of ICTs to all forms of social and economic activity.

Information superhighway: Term coined by the Clinton administration for an advanced information infrastructure accessible to all individuals, groups, and firms.

Information Technologies (IT): Computer-based techniques for storing, processing, managing, and transmitting information.

Information utility: The concept of a national, and eventually global, electronic network, which will supply a full range of media and other information resources to all locations (comparable to water and electricity utility networks).

Instructional television fixed service (ITFS): A microwave frequency allocated by the FCC for educational use.

Integrated services digital network (ISDN): A long-term plan for the transition of the world's telecommunications systems from analog to digital technology and a software standard that will eliminate present technical incompatibilities between telecommunications systems and allow uninterrupted transfer of traffic between them.

Interactive media or services: Media resources that involve the user in providing the content and duration of a message, permitting individualized program material. Also used to describe media production operations that take maximum advantage of random access, computer-controlled videotape, and videodisc players.

Interactive television: Networked service allowing TV sets to be used for two-way communication with various services, such as for teleshopping.

Interactive video: A combination of video and computer technology in which programs run in tandem under the control of the user. The user's choices and decisions directly affect the ways in which the program unfolds.

Interexchange carriers (IXC): Telephone companies that are authorized by federal or state regulatory agencies to provide connections between service areas. Telephone companies (e.g., AT&T, MCI, US Sprint) that connect local exchanges and local access and transport areas (LATAs) to one another (i.e., they provide interLATA service).

Interface: The point or boundary at which hardware or software systems interact (e.g., the connection between a computer and a terminal) and across which data are transferred.

Internet: The worldwide network of interconnected computer networks offering electronic mail and database services.

Just-in-time system (JIT): Use of information and computer technologies to coordinate deliveries from suppliers to ensure a minimum of locally stored inventory is needed to support production processes. A production system in which parts are delivered from suppliers to the manufacturers as needed, rather than being stored on site.

K: 1,000 in general terms; 1,024 for measurements specific to computing.

Kilobits per second (Kbps): A measure of bandwidth and rate of information flow in digital transmission. One Kbps is 1,024 kilobits per second.

Kilobyte (Kbyte): A measure of computer storage and memory capacity. One Kbyte equals 1,024 bytes.

Laser: Technically, light amplification by simulated emission of radiation. An intense light beam that can be modulated for communications. Lasers amplify and generate energy in the optical, or light, region of the spectrum above the radio frequencies. In a typical media application, lasers are used to read the micropits on a videodisc, which contain video or sound signals.

Local Access and Transport Area (LATA): The geographical area or boundaries within which Bell operating companies are permitted to carry long-distance telecommunications traffic without violating the terms of the MFJ barring them from long distance services.

Local area network (LAN): A small private network of computers with limited reach, such as for a building or campus, linked to allow access and sharing of information and computer resources by users.

Local exchange carrier (LEC): Companies that provide connections to individual telephone users and process calls within a serving area. LECs are responsible for traffic within LATAs These carriers, consisting of local telephone companies, also provide connections to interexchange carriers.

Local loop: The link, typically twisted copper wire pair, between the subscriber's premises and the local exchange.

Megabits (Mbits): Millions of bits of data.

Megabits per second (Mbps): A measure of bandwidth and rate of information flow in digital transmission. One Mbps equals one million bits per second.

Megabyte (Mbyte): A measure of computer storage and memory capacity. One Mbyte is equal to 1.024 million bytes, 1,024 thousand bytes or 1,024 kbytes.

Metropolitan area network (MAN): A network linking computers at several sites in an urban area.

Microprocessor: An electronic circuit, usually on a single chip, which performs arithmetic, logic, and control operations, customarily with the assistance of internal memory.

Microwave link: A communications system using particularly high frequency radio signals (above 800 megahertz) for audio, video, and data transmission. Microwave links require line of sight connection between transmission antennas.

Modem: The acronym for modulator/demodulator, a modem is the device that converts digital signals to pulse tone (analog) signals for transmission over telephone lines and reconverts them to digital form at the point of reception to allow computers to be connected via nondigital networks.

Modified Final Judgment (MFJ): The Consent Decree that ended the antitrust case against American Telephone & Telegraph Company and broke up the Bell system.

Monitor (video): Usually refers to the video screen on a computer but has more technical meanings as well.

MOSAIC: Simple graphical user interface developed for the World Wide Web which has influenced interface designs of many other tools for exploring information networks.

Multimedia: Information delivery systems that combine or integrate different content formats (e.g., text, video, sound) and storage facilities (e.g., video tape, audio tape, magnetic disks, optical disks) in computer and telecommunication systems.

Multiplexing: Process of combining two or more signals from separate sources into a single signal for sending on a transmission system from which the original signal may be recovered.

Multipoint control unit (MCU): A centrally located service offered by switched network providers to which three or more users can be connected, permitting audio and video teleconferencing.

Narrowband, or narrowband channel: A telecommunications medium that uses (relatively) low frequency signals, generally up to 1.544 Mbps, so it handles only relatively small volumes of data. A communication channel, such as copper wire or part of a coaxial cable channel, that transmits voice, facsimile, or data at rates of kilobits per second, but not high speed data or video.

Narrowcasting: Targeting communication media at specific segments of the audience.

National information infrastructure (NII): The integration of hardware, software, and skills that will make it easy and affordable to connect people with each other, with computers, and with a vast array of services and information resources. See NTIA.

National Telecommunications and Information Administration (NTIA): The federal agency in the Department of Commerce responsible for the National Information Infrastructure initiative. See NII.

Network: A set of nodes, points, or locations connected by means of data, voice, and video communications for the purpose of exchange, or interconnected telecommunications equipment used for data and information exchange. Three common network types are local area networks (LANs), metropolitan area networks (MANs), and Wide-area networks (WANs). See also Rural area network (RAN).

Off-line: Equipment not connected to a telecommunications system or an operating computer system.

Online: Activity involving direct interaction with a computer-based system via a telecommunications link. Also, a device normally connected to a microcomputer that permits it to run various programs and handle scheduling, control of printers, terminals, memory devices, and so forth.

Online computer services: Information services that are accessible via telephone lines from personal computers and computer terminals.

Operating system: Software, such as Microsoft DOS and Windows, Unix, and IBM MVS, which manages the computer's basic functions so they can be exploited effectively by users.

Packet: A basic message unit or group of bits that includes data, call control signals, and error control information (referring to its source, content, and destination) that are arranged in a specified format and switched and transmitted as a composite whole in a packet-switched network.

Packet switching or packet-switching network (PSN): These terms refer to the method of coding and transmitting digital information using addressed packets of information that travel along different routes in a network and are reassembled at their destination. Packet switching is more efficient than modem transmission because the channel is occupied only during packet transmission rather than throughout the transmission.

PANS: Jargon for "pretty amazing new stuff." Compare with POTS, plain old telephone service.

Party line telephone service: Telephone service which provides for two or more telephones to share the same loop circuit.

Peripherals or peripheral equipment: In a data processing system, any equipment, distinct from the central processing unit, that may provide the system with outside communication or additional facilities (e.g., printer, scanner, etc.)

Personal communications network (PCN): A proposed network composed of a variety of wireless services including cordless telephones, wireless private branch exchanges, and wireless local area networks.

Pixel: Stands for picture element, the smallest piece of information that can be displayed on a CRT. Represented by a numerical code in the computer (dots per square inch), pixels appear on the monitor as dots of specific color or intensity. Images are composed of many, many pixels.

Point of presence (POP): The point at which an interexchange carrier's circuits connect with local circuits for transmission and reception of long distance calls.

Pooling or revenue pooling: Telephone industry term meaning setting up special collections of funds for intended cross-subsidy, as in averaging rates between high-cost rural services and less-expensive urban ones. Intended to offset cost disparities.

POTS: Jargon for "plain old telephone service," basic voice-only telephone services.

Price cap: A regulation that sets the maximum price telephone companies can charge for a designated group of services. Regulators may change the price cap over time based on inflation and targets for improvements in productivity.

Protocol: Detailed definition of the procedures and rules required to transmit information across a telecommunications link or system.

Random access memory (RAM): The most commonly used method of defining computer capability (e.g., 64K RAM), random access memory is the computer's temporary memory space where programs are run, images are processed, and information is stored. Information stored in the RAM is lost when the power is shut off.

Rate of return: A method of regulation that defines the total revenue a telephone company requires to provide services. The revenue requirement includes operating expense, depreciation and taxes, and a "fair" return on its capital investment ("rate base").

Read-only memory (ROM): The permanent memory space of a computer. A computer chip stores data and instructions in a form that cannot be altered, thus, programs and information stored in ROM are not lost when the power is shutdown. Compare RAM.

Real time: The capture, processing, and presentation of data, audio, and/or video signals at the time that data are originated on one end and received on the other end. When frames containing such data are transmitted at rates of 30 per second, real time is achieved.

Regional Bell Operating Company (RBOC): Local US telecommunication companies created after the break-up of AT&T in 1984. One of seven companies formed by the AT&T divestiture, including Ameritech, Bell Atlantic, BellSouth, NYNEX, Pacific Telesis, Southwestern Bell, and US West.

Regional holding company (RHC): More commonly called RBOC.

Rural Area Network (RAN): Shared-usage networks, configured to include a wide range of users in rural communities, such as educational, health, and business entities.

Satellite connections: A communications relay device or system that uses radio signals sent to and from a satellite orbiting the Earth. A satellite receives a signal from an earth station, amplifies it, and retransmits it to one or more receiving stations. This mode of transmission permits connections between points at great distance from each other on the Earth's surface, between which direct transmission is difficult, as well as to remote areas lacking cables for telephone lines.

Serving areas: Geographic areas in which local telephone companies provide local exchange service and connection to interexchange carriers.

Signaling System 7 (SS7): A control system for the public telephone network that allows telephone company computers to communicate directly with each other for routing calls on signaling circuits separate from the circuits used for the telephone calls themselves, making telephone call processing faster and more efficient and making more services available to consumers.

Software: A set of computer programs, procedures, rules, and associated documentation concerned with the operation of network computers (e.g., compilers, monitors, editors, utility programs). The written detailed instructions or programs that tell the computer what to do, byte by byte.

Store-and-forward (SAF): A type of telemedicine interaction that produces a multimedia electronic medical record. Data and images are captured and stored for later transmission, consultation, or downloading, obviating the need for the simultaneous availability of the consulting parties and reducing transmission costs due to low bandwidth requirement.

Switch: A mechanical or solid-state device that opens or closes circuits, changes operating parameters, or selects paths or circuits on a space or time division basis.

Switched line: Communication link for which the physical path, established by dialing, may vary with each use.

Switched network: A type of telecommunications system where each user has a unique address and any two points can be connected directly, using any combination of available routes in the network.

Switched service: A telecommunications service, usually based on telephone technology, that changes (switches) circuits to connect two or more points.

Synchronous transmission: The process by which bits are transmitted at a fixed rate with the transmitter and receiver synchronized, eliminating the need for start/stop elements, thus providing greater efficiency.

T1, T3: A digital transmission system for high-volume voice, data, or compressed video traffic. T1 and T3 transmission rates are 1.544 Mbps and 44.736 Mbps, respectively. Also known as DS1, DS3.

Tariffs: Price structures for communications facilities set forth by federal or local governments, intended to allow telephone companies (LATAs) a fair rate of return on their capital investments.

T-carrier: A family of high speed, digital transmission systems using pulse code modulation technology at various channel capacities and bit rates to send digital information over telephone lines or other transmission media. A T1 carrier has an operating capacity of 1.544 megabits per second and higher.

Telco: Jargon for telecommunications company or telephone company.

Telecommunications: The use of wire, radio, optical, or other electromagnetic channels to transmit or receive signals for voice, data, and video communications.

Telecommuting: Using telecommunications to perform work at home or a work center that would otherwise involve commuting physically to a more distant place of work.

Teleconferencing: A meeting of three or more people at two or more locations links by telecommunications, communicating simultaneously through electronic media.

Teleconsultation: Geographic separation between two or more providers who use telecommunications during a consultation.

Telediagnosis: The detection of a disease by evaluating data transmitted to a receiving station from instruments monitoring a distant patient.

Telehealth: The electronic transfer of health information from one location to another for purposes of preventive medicine, health promotion, diagnosis, consultation, education, and/or therapy.

Telemarketing: Method of marketing that emphasizes the creative use of the telephone and other telecommunications systems.

Telemedicine: The use of audio, video, and other telecommunications and electronic information processing technologies to provide health services and health education, support patient care and patient-related activities, and assist health care personnel at distant sites.

Telemonitoring: The use of audio, video, and other telecommunications and electronic information processing technologies to monitor patient status at a distance.

Teleport: Site-specific telecommunication infrastructure, such as a land-station link to a satellite, associated with related land and building development.

Teletext: A broadcasting service transmitted by television signal which allows users to call up a wide variety of information from a central database on a TV screen. Teletext is the generic name for a set of systems that transmit alpha-numeric and simple graphical information over the broadcast (or one-way cable) signal, using spare line capacity (vertical blanking intervals) in the signal for display on a suitably modified TV receiver.

Telework: Use of an electronic network to enable individuals to work from home or a decentralized work center.

Telex: Dial-up telegraph service; a public switched network connecting teletypewriters or other devices transmitting at 50 bits per second for transmission of text messages.

Terabits per second (Tbps): A measure of bandwidth and rate of information flow in digital transmission. One Tbps equals one trillion bits per second.

Terminal: Point at which a communication can either leave or enter a communication network.

Throughput: The amount of data actually transmitted over a network in a given period of time, expressed in bits per second. Throughput rates are related to baud rates, but are usually a little lower because transmission conditions are never ideal.

Transmission control protocol/Internet protocol (TCP/IP): The underlying communications protocols or standard that permit computers to interact with each other on the Internet yet governs data exchanges.

Transmission speed: The speed at which information passes over the line, defined in either bits per second (bps) or baud.

Twisted pair: Term given to the two wires that connect local telephone circuits to the telephone central office. The most prevalent type of medium in public switched telephone networks' local loops, twisted-pair is insulated copper wires wrapped around each other to cancel the effects of electrical noise. It can transmit voice, data, and low-grade video.

U.S. West: The Bell regional holding company serving the Northwest, Rocky Mountain states, parts of the Southwest and Midwest, including Iowa.

Universal service: Originally, the concept put forth by the first Bell System chairman that residential telephone service be priced low enough so everyone could afford it. Now it refers to the goal of providing a minimum set of telecommunications services at affordable rates to virtually every household.

Uplink: Communications path or link from a transmitting earth station to the satellite.

Upload: To transfer information out of the memory or disc file of your computer to another computer.

USENET: International collection of electronic discussion groups on a multiplicity of topics, accessible through the Internet.

Videoconferencing: Teleconferencing involving video communication; that is, real-time, usually two-way transmission of digitized video images between two or more locations.

Videotext: The transmission of information by television channels, FM frequencies, or telephone circuits to a TV or computer monitor. Videotext is the generic name for a computer-based network service which delivers textual and graphical information, typically as pages of information stored on remote computers over the ordinary telephone lines for display on a video monitor.

Videotext publishing: A market for supplying videotext materials from large databases to home computers and other terminals. Many newspapers and magazines now offer their printed products electronically in this manner.

Virtual organization: Operation involving many individuals, groups, and firms in different locations using electronic networks to act as if they were a single organization at one site.

Virtual reality (VR): A computer-based technology for simulating visual, auditory, and other sensory aspects of complex environments. Computer software that produces multidimensional visual images that can create "realities," which are manipulated in many different formats by a user wearing computerized gloves or helmet.

Voice mail: A voice messaging system in which spoken messages are recorded and stored in electronic "mailboxes" for later playback or transfer to others.

Wide-area network (WAN): A data communication network that links together distinct networks and their computers over large geographical areas. WANs are typically larger than metropolitan area networks. See also MAN, LAN, and RAN.

Wireless: Communications via radio waves rather than wire or cable.

Workstation: A functional grouping of computer hardware and software (e.g., monitor, keyboard, hard drive) for individual uses such as word, information, and image processing.

World Wide Web (WWW or the "Web"): An internet information resource through which information-producing sites offer hyperlinked multimedia information to users. System which allows information sites around the world to be accessed via the Internet through the MOSAIC interface.

Sources for these definitions include: Bashshur et al., 1997; Dizard, 1994; Dutton, 1996; Field, 1996; Hudson, 1997; Parker et al., 1989; Parker et al., 1992; Schmandt, 1991; and Zetzman, 1995, among others.

REFERENCES

Abbott, Barbara Ann Bean. (1986). Electronic gatekeeping: How Iowa Extension home and family news releases are affected by electronic distribution. Unpublished master's thesis, Iowa State University, Ames, Iowa.

Abbott, Eric A. (1988, July). The volunteer newspaper: A communication solution for small rural communities? Paper presented to Newspaper Division, Association for Education in Journalism and Mass Communication, Portland, OR. Also available as an ERIC document.

Abbott, Eric A. (1989, Summer). The electronic farmers' marketplace: New technologies and agricultural information. *Journal of Communication, 39*(3), 124–136.

Abbott, Eric A. (1997). *Telecommunications for rural community viability: Making wise choices*, workshop proceedings, Kansas City, MO, February 25–27. Ames, IA: Department of Journalism and Mass Communications, Iowa State University.

Abbott, Eric A., & April Eichmeier. (1998, August 7). The hoopla effect: Toward a theory of regular patterns of mass media coverage of innovations. Paper presented to Communication Theory and Methodology Division, Association for Education in Journalism and Mass Communication Annual Convention, Baltimore MD.

Abbott, Eric A., & J. Paul Yarbrough. (1992, Spring). Inequalities in the information age: Farmers' differential adoption and use of four information technologies. *Agriculture and Human Values, 9*(2), 67–79.

Abbott, Eric A., & J. Paul Yarbrough. (1993, May 24–28). The unequal impacts of microcomputer adoption and use on farmers. *Proceedings of the International Conference on Information Technology and People* (n.p.). Moscow, Russia: ICITP.

Adams, Paula J., & Robert Stephens. (1991). Introduction. In Jurgen Schmandt, Frederick Williams, Robert H. Wilson, & Sharon Strover (Eds.), *Telecommunications and rural development: A study of private and public sector innovation* (pp. 1–17). New York, NY: Praeger Publishers.

Adams, Paula J., Scott J. Lewis, and Robert Stephens. (1991). Public service delivery. In Jurgen Schmandt, Frederick Williams, Robert H. Wilson, & Sharon Strover (Eds.),

Telecommunications and rural development: A study of private and public sector innovation (pp. 61–94). New York, NY: Praeger Publishers

Ahuja, Manju K., & Kathleen M. Carley. (1998, June). Network structure in virtual organizations. *Journal of Computer Mediated Communication, 3*(4), n.p. Available online at http://www.ascusc.org/jcmc/vol3/issue4/ahuja.html.

Allen, John C., & Bruce B. Johnson. (1995). *Rural American home-based businesses and the information age* (Community Economics Report No. 220). Madison, WI: Department of Agricultural Economics, University of Wisconsin.

Allen, John C., Bruce B. Johnson, & Larry F. Leistritz. (1993, Fall). Rural economic development using information technology: Some directions for practitioners. *Economic Development Review,* 30–33.

Allen, John C., Bruce B. Johnson, F. Larry Leistritz, Duane Olsen, Randy Sell, & Matt Spilker. (1995). *Telecommunications and economic development in rural communities: Policy directions and questions.* Columbia, MO: Rural Policy Institute, University of Missouri.

Amabile, Teresa. (1988). From individual creativity to organizational innovation. In Kjell Gronhaug & Geir Kaufmann (Eds.), *Innovation: A cross-disciplinary perspective* (pp. 139–166). New York, NY: Norwegian University Press.

American Telemedicine Association. (1998, December 11). State Medical Licensure Committee, draft report to ATA Board of Directors. Available online at www.atmeda.org.

Anderson, James E. (1966). *Politics and the economy.* Boston, MA: Little, Brown and Company.

Anderson, Robert M., Jr. (1993). Wide-area network connection to field Extension offices. Ames, IA: Extension Council Exchange, Iowa State University.

Anyanwu, Alphonsus Chijioke. (1982). Communication behavior of professional disseminators of scientific agricultural knowledge: A study of the Iowa Cooperative Extension Service. Unpublished doctoral dissertation, Iowa State University, Ames, IA.

AT&T v Iowa Utilities Board. (1999). 119 Iowa Supreme Court 721.

Ayres, Janet S., & Harry R. Potter. (1989). Attitudes toward community change: A comparison between rural leaders and residents. *Journal of the Community Development Society, 20*(1), 1–18.

Babbitt, Elizabeth, Jill Desmarais, Christine Koehler, Barbara Lacy, & Marjorie Merchant. (1988, Winter). Overcoming "techno-fear." *Journal of Extension, 26*(4), n.p.

Bashshur, Rashid L. (1997). Telemedicine and the health care system. In Rashid L. Bashshur, Jay H. Sanders, & Gary W. Shannon (Eds.), *Telemedicine: Theory and practice* (pp. 5–36). Springfield, IL: C. C. Thomas.

Bashshur, Rashid L., Jay H. Sanders, & Gary W. Shannon (Eds.). (1997). *Telemedicine: Theory and practice.* Springfield, IL: C. C. Thomas.

Batte, Marvin T., Eugene Jones, & Gary D. Schnitkey. (1990, November). Computer use by Ohio commercial farmers. *American Journal of Agricultural Economics, 72*(4), 935–945.

Beath, Cynthia. M. (1991). Supporting the information technology champion. *MIS Quarterly, 15*(3), 355–374.

Becker, F., C. M. Tennessen, & D. Young. (1995). Information technology for workplace communication. Ithaca, NY: International Workplace Studies Program, Cornell University.

Bell Atlantic. (1998, December 1). Bell Atlantic launches Physician Linkage, the virtual medical practice: Information network allows even doctors to telecommute. Available online at http://www.ba.com/nr/1998/Dec/19981201004.html.

Bell, Daniel. (1973). *The coming of post-industrial society: A venture in social forecasting.* New York, NY: Basic Books.

Bergson, Tom. (1994). Corporate executives rate site selection factors. In William H. Read, & Jan L. Youtie, *Telecommunications strategy for economic development* (p. 11). Westport, CT: Praeger Publishers.

Bertot, John Carlo, & Charles R. McClure. (1997, December). Impacts of public access to the Internet through Pennsylvania public libraries. *Information Technology and Libraries, 16*(4), 151–164.

Bertot, John Carlo, Charles R. McClure, & Douglas L. Zweizig. (1996, July). The 1996 National Survey of Public Libraries and the Internet. Final Report. Washington, DC: National Commission on Libraries and Information Sciences.

Beyers, William B., & David P. Lindahl. (1996). Lone eagles and high fliers in rural producers services. *Rural Development Perspectives. 11*(3), 2–10.

Binstock, Robert H., & Linda K. George. (1995). *Handbook of aging and the social sciences.* San Diego, CA: Academic Press.

Bleakley, Fred R. (1996, January 4). Rural county balks at joining global village. *Wall Street Journal,* pp. B1, B4.

Blechman, Bruce, & Jay Conrad Levinson. (1991). *Guerrilla financing: Alternative techniques to finance any small business.* Boston, MA: Houghton Mifflin Company.

Bliss, Ralph K. (1960). *Extension in Iowa: The first 50 years.* Ames, IA: Iowa State University.

Bogart, Leo. (1993, Spring). Newspaper of the future: Our look at the next century. *Newspaper Research Journal, 14*(2), 2–10.

Bollier, David (Ed.). (1988, July 24–27). *The importance of communications and information systems to rural development in the United States.* Report of an Aspen Institute Conference, Truro, MA.

Bonnett, Thomas. (1997). *The twenty-one most frequently asked questions about state telecommunications policy.* Washington, DC: Council of Governors' Policy Advisors.

Boyce, Judith I., & Bert R. Boyce. (1995, June 22). Library outreach programs in rural areas. *Library Trends, 44*(1), 112–128.

Brown, Lawrence. (1981). *Innovation diffusion: A new perspective.* New York, NY: Methuen.

Brunner, Michael E. (1998, November/December). Community anchors. *Rural Telecommunications, 66.*

Buehler, Phillip. (1997, April 2). Advertising and the Internet. Sound recording, Iowa State University Lecture Series, Iowa State University, Ames, IA.

Bultena, Gordon L., & Peter F. Korsching. (1996, Aug. 15–18). Telecommunications technologies as a community economic development strategy: Views of Iowa rural development professionals. Paper presented at the annual meeting of the Rural Sociological Society, Des Moines, IA.

Bundy, M. L. (1960). The attitudes and opinions of farm families in Illinois toward matters related to rural library development. *Dissertation Abstracts International, 21*(09). (University Microfilms No. AAD61-00092.)

Burstein, Daniel, & David Kline. (1995). *Road warriors: Dreams and nightmares along the information superhighway.* New York, NY: Dutton Books.

Business Wire. (1998, October 13). Eight California community coalitions and Pacific Bell to launch $50 million in community technology grants. *Business Wire.*

Byerly, Greg. (1996). Ohio: Library and information networks. *Library Hi-Tech, 14*(2–3), 245–254.

Byerly, Kenneth R. (1961). *Community journalism.* Philadelphia, PA: Chilton Company.

Byerly, Tom. (1998, September). Privatizing public networks. *Government Technology,* 18.

Cable Telecommunications Association (CATA). (1998). Municipal ownership: An ongo-
    ing review of the status of municipal cable television systems. Available online at:
    http://www.catabet,org/general/wpmuni.html.
CanWest Agricultural Research. (1984, April). *A user evaluation of the Grassroots System.*
    Report prepared for Infomart.
Carey, James. (1969, January). The communications revolution and the professional com-
    municator. *Sociological Review Monograph, 13,* 23–38.
Carey, James. (1998, Spring). The Internet and the end of the national communication sys-
    tem: Uncertain predictions of an uncertain future. *Journalism and Mass Communica-
    tion Quarterly, 75*(1), 28–34.
Caristi, Dom. (1997, September). The Iowa Communications Network: The policy implica-
    tions of publicly funded infrastructure. In *Proceedings of the Twenty-fifth Telecommu-
    nications Policy Research Conference,* Alexandria, VA, September 27–29, 1997 (pp.
    10–18). Ann Arbor, MI: TPRC, Inc.
Caristi, Dom. (1998, September). The Iowa communications network: The policy implica-
    tions of publicly funded infrastructure. *Telecommunications Policy, 22*(7), 617–627.
Case, Donald, Milton Chen, Hugh Daley, Joung-Im Kim, Nalini Mishra, William Paisley,
    Ronald E. Rice, & Everett M. Rogers. (1981, December). *Stanford evaluation of the
    Green Thumb box: Experimental videotext project for agricultural Extension information
    delivery in Shelby and Todd counties, Kentucky* (Final Report, USDA Contract
    53-3K06-1-63). Stanford, CA: Stanford University.
Census Bureau. (1996). *Statistical Abstract of the United States* (116th Edition). Washington,
    DC: U.S. Department of Commerce.
Center for Technology in Government. (1997). *1997 annual report.* Washington, DC: Cen-
    ter for Technology in Government.
Center for Technology in Government. (1998). *Celebrating five years of collaboration and in-
    novation: 1998 annual report.* Washington, DC: Center for Technology in Govern-
    ment.
Christensen, Kathleen E. (1989). Home-based clerical work: No simple truth, no single real-
    ity. In Eileen Boris & Cynthia R. Daniels (Eds.), *Homework: Historical and contempo-
    rary perspectives on paid labor at home* (pp. 183–197). Urbana, IL: University of Illinois
    Press.
Christianson, James A., & Jerry W. Robinson, Jr. (1980). *Community development in Amer-
    ica.* Ames, IA: Iowa State University Press.
Cisler, Steve. (1995, June 22). The library and wired communities in rural areas. *Library
    Trends, 44*(1), 176–189.
Clearfield, Frank, & Paul D. Warner. (1984). An agricultural videotex system: The Green
    Thumb pilot study. *Rural Sociology, 49*(2), 284–297.
Cleveland, Harlan. (1985). The twilight of hierarchy: Speculations on the global informa-
    tion society. *Public Administration Review 45,* 185.
Coen, Robert. (1999, December). Robert Coen's presentation on advertisers' expenditures.
    McCann Erickson World Group News and Views. Available online at: http://
    www.mccann.com.
Cohn, Gary. (1998, February 12). Interview with author. Senior Information Services Coor-
    dinator, City of Iowa City; Iowa City, Iowa.
Cole, Barry. (1995). State policy laboratories. In P. Teske (Ed.), *American regulatory federal-
    ism and telecommunications infrastructure* (pp. 35–45). Hillsdale, NJ: Lawrence
    Erlbaum Associates.

Coleman, James. (1988). Social capital in the creation of human capital. *American Journal of Sociology, 94,* S95-S120.

Cordes, Donald. L., & LaVonne A. Straub. (1992). Changing the paradigm of rural health-care delivery. In LaVonne A. Straub & Norman Walzer (Eds.), *Rural health care: Innovation in a changing environment* (pp. 105–116). Westport, CT: Praeger Publishers.

Cox, Robert. (1997). Telemedicine: Hays, Kansas. In Eric A. Abbott (Ed.), *Making wise choices: Telecommunications for rural community viability: Proceedings, February 25–27, 1997* (pp. 35–37). Kansas City, MO, NCRCRD. (Available from the Rural Development Initiative, Department of Sociology, Iowa State University, Ames, 50011).

Curtis, Terry, & Jorge Reina Schement. (1995). Communication rights. In Charles M. Firestone & Jorge Reina Schement (Eds.), *Toward an information bill of rights and responsibilities* (pp. 39–60). Washington, DC: Aspen Institute.

Cutler, Richard, & Harmeet S. Sawhney. (1991). Conclusions. In Jurgen Schmandt, Frederick Williams, Robert H. Wilson, & Sharon Strover (Eds.), *Telecommunications and rural development: A study of private and public sector innovation* (pp. 223–239). New York, NY: Praeger Publishers.

Daft, Richard L., & Robert H. Lengel. (1986). Organizational information requirements, media richness, and structural design. *Management Science, 32*(5), 554–571.

Davenport, Lucinda, Frederick Fico, & David Weinstock. (1996, Summer/Fall). Computers in newsrooms of Michigan's newspapers. *Newspaper Research Journal, 17*(3–4), 14–28.

Davidson, Jeffrey P. (1991). *Marketing to home-based businesses.* Homewood, IL: Business One Irwin.

Davis, Stan. (1998, January 20). Interview with author. Manager of Information Services, City of Ames, Iowa.

DeGruyter, M. L. (1982). *Patterns of rural adult public library use.* Chicago, IL: The University of Chicago.

Dertouzos, Michael. (1997). *What will be.* New York, NY: Harper Edge.

Des Moines Register. (1999, January 5). More cities in Iowa want right to operate own phone system. *Des Moines Register,* p. 3M

Des Moines Register. (1998, January 20). Big telecom bill doesn't live up to expectations. *Des Moines Register,* p. 1M.

Dewey, John. (1916). *Democracy and education.* New York, NY: Macmillan.

Dholakia, Ruby Roy, & Bari Harlam. (1993). Telecommunications and economic development: Econometric analysis of the U.S. experience. *Telecommunications Policy. 18*(6), 470–477.

Dicken-Garcia, Hazel. (1998, Spring). The Internet and continuing historical discourse. *Journalism and Mass Communication Quarterly, 75*(1), 19–27.

Dillman, Don A. (1991a, May). Community needs and the rural public library. *Wilson Library Bulletin, 65*(9), 31–33, 155–156.

Dillman, Don A. (1991b). Telematics and rural development. In Cornelia Butler Flora & James A. Christianson (Eds.), *Rural policy for the 1990s* (pp. 292–306). Boulder, CO: Westview Press.

Dillman, Don A., & Donald M. Beck. (1988, January/February). Information technologies and rural development in the 1990s. *Journal of State Government, 61*(1), 29–38.

Dillman, Don A., Donald M. Beck, & John C. Allen. (1989). Rural barriers to job creation remain, even in today's information age. *Rural Development Perspectives, 5*(2), 21–26.

DomainStats.com. (2000). Available online at http://www.domainstats.com/

Dotinga, Randy. (1998, January). Newsroom tech: Newspapers with little or no Web news. Available online at mediainfo.com, 36–37.

Drumm, John E., & Frank M. Groom. (1997, January). The cybermobile: A gateway for public access to network-based information. *Computers in Libraries, 17*(1), 29–33.

Dubrovsky, V. J., S. Kiesler, & B. N. Sethna. (1991). The equalization phenomenon: Status effects in computer-mediated and face-to-face decision making groups. *Human-Computer Interaction, 6*, 119–146.

Duesterberg, Thomas, & Kenneth Gordon. (1997). *Competition and deregulation in telecommunications: The case for a new paradigm.* Indianapolis, IN: Hudson Institute.

Durand, Philippe. (1983, June). The public service potential of videotex and teletext. *Telecommunications Policy, 7*(2), 149–262.

Dutton, W. H., Everett M. Rogers, & Suk-Ho Jun. (1987). Diffusion and social impacts of personal computers. *Communication Research, 14*(2), 219–250.

Eckhoff, Jeff. (1997, December 14). TCI faces criticism at talks with city. *Des Moines Register*, pp. 1A, 4A.

Economic Research Service. (1997, July). *Forces shaping U.S. agriculture: A briefing book.* Washington, DC: U.S. Department of Agriculture.

Economic Research Service. (1996). 1) U.S. farm workers by type: 1910–1995, 2) U.S. number of farmers and all farm workers 1910–1995. Washington, DC: U.S. Department of Agriculture. Available online at: http://www.usda.gov/nass/aggraphs/fl_typwk.htm.

Editor & Publisher. (1996, December 28). Weekly runs profitable Web site. *Editor & Publisher*, p. 26.

Edwards, Elaine Harvey. (1994). Iowa's early agricultural press: A content analysis of the Iowa Farmer and Horticulturist, 1853–1856. Unpublished master's thesis, Iowa State University, Ames, IA.

Ellis, Leslie. (1998, May 13). La Grange mulls launching its own cable TV system. *Louisville Courier-Journal*, p. 01N.

Emmons, Willis. (1993). Franklin D. Roosevelt, electric utilities, and the power of competition. *Journal of Economic History, 53*(4), 880–907.

Enders, Alexandra. (1995, Spring). The role of technology in the lives of older people. *Generations, 19*(1), 7–12.

Estabrook, L. (1991). National opinion poll on library issues. Urbana-Champaign, IL: Graduate School of Library and Information Science, Library Research Center, University of Illinois.

Ettema, James S. (1983, August). Information equity and information technology: Some preliminary findings from a videotex field trial. Paper presented to the Communication Theory and Methodology Division, Association for Education in Journalism and Mass Communication, Corvallis, OR.

Ettema, James S. (1984a). Communication research needs—The researcher's perspective. In North Central Regional Communication Research Committee (Eds.), *Proceedings of the NCR-90 Research Conference on the Application of Computer Technology to Communication Processes* (pp. 10–18). Minneapolis, MN: North Central Regional Communication Research Committee.

Ettema, James S. (1984b, Fall). Three phases in the creation of information inequities: An empirical assessment of a prototype videotex system. *Journal of Broadcasting, 28*(4), 383–395.

Farrell, Joseph. (1997, September). Prospects for deregulation in telecommunications. In *Proceedings of the Twenty-fifth Telecommunications Policy Research Conference*, Alexandria, VA, September 27–29, 1997. Ann Arbor, MI: TPRC, Inc.

FCC Hearing. (1997, December). Federal Communications Commission—FCC televised hearings.

Federal Communications Commission (FCC). (1997). In the Matter of the Public Utility Commission of Texas, 13 FCC Rcd 3460.

Fett, John. (1993, October 7–29). Using public libraries to reach clientele with Cooperative Extension information. Paper presented at the North Central Regional Agricultural Communication research meetings, West Lafayette, IN, Purdue University.

Fidler, Roger. (1997). *Mediamorphosis: Understanding new media*. Thousand Oaks, CA: Pine Forge Press.

Field, Marilyn (Ed.). (1996). *Telemedicine: A guide to assessing telecommunications in health care*. Committee on Evaluating Clinical Applications of Telemedicine, Division of Health Care Services, Institute of Medicine. Washington, DC: National Academy Press.

Flora, Cornelia Butler. (1995). Vital communities: Combining environmental and social capital. Paper presented at the Rural America: A Living Tapestry Conference, Knoxville, TN.

Flora, Cornelia Butler, & Jan L. Flora. (1993). Entrepreneurial social infrastructure: A necessary ingredient. *Annals 529*, 48–58.

Flora, Cornelia Butler, & Jan L. Flora. (1995). The past and the future: Social contract, social policy and social capital. In S. A. Hallbrook & C. E. Merry (Eds.), *Increasing understanding of public problems and policies* (pp. 53–64). Oak Brook, IL: Farm Foundation.

Flora, Jan, Jeff Sharp, Bonnie L. Newlon, & Cornelia Butler Flora. (1997). Entrepreneurial social infrastructure and locally initiated economic development in the nonmetropolitan United States. *The Sociological Quarterly, 38*(4), 623–645.

Fox, William F. (1988). Public infrastructure and economic development. In David L. Brown, J. Norman Reid, Herman Bluestone, David A. McGranahan, & Sara M. Mazie (Eds.), *Rural economic development in the 1980s: Prospects for the future* (pp. 281–306). Washington, DC: Economic Research Service, U.S. Department of Agriculture.

Future Technology Committee Recommendations. (1997, November 11). Ames, IA: Iowa State University Extension.

Gamble-Risley, Michelle. (1999, January). TEACH—Wisconsin's answer to e-rate problems. *Converge, 2*(1), 34.

Gapenski, Louis C. (1993). *Understanding health care financial management: Text, cases, and models*. Ann Arbor, MI: AUPHA Press and Health Administration Press.

Garcia, Linda D., & Neal R. Gorenflo. (1997, September). Best practices for rural Internet deployment: The implications for universal service policy. In *Proceedings of the Twenty-fifth Telecommunications Policy Research Conference*, Alexandria, VA, September 27–29, 1997, (n.p.). Ann Arbor, MI: TPRC, Inc.

Garkovich, Lorraine E. (1989). Local organizations and leadership in community development. In James A Christenson & Jerry W. Robinson, Jr. (Eds.), *Community development in perspective* (pp. 196–218). Ames, IA: Iowa State University Press.

Garofalo, Denise A. (1995, March). Rural public libraries' use of the Internet: Assistance or aggravation? *Computers in Libraries, 15*(3), 61.

Garrison, Bruce. (1983, April 5–8). Technological developments in journalism: The impact of the computer in the newsroom. Paper presented at the Annual Meeting of the Southern Speech Communication, Orlando, FL.

Garrison, Bruce. (1997, Summer/Fall). Online services, Internet in 1995 newsrooms. *Newspaper Research Journal, 19*(3–4), 79–93.

Gasman, Lawrence. (1994). *Telecompetition*. Washington, DC: CATO Institute.

Gatignon, Hubert, & Thomas Robertson. (1989). Technology diffusion: An empirical test of competitive effects. *Journal of Marketing, 53*(1), 35–49.

Gieseke, Joy A., & Peter F. Korsching. (1998). Effects of local versus absentee telephone company ownership on participation in rural community economic development. *Journal of the Community Development Society, 29*(2), 256–275.

Gieseke, Joy A., Peter F. Korsching, Eric A. Abbott, Gordon Bultena, & Jennifer Gregg. (1995). *Telecommunications for rural development: Iowa telephone and local economic development survey results.* Ames, IA: Department of Sociology, Iowa State University.

Gieseke, Joy A., Peter F. Korsching, & Gordon Bultena. (1996, April 8–11). Social capital and the successful utilization of telecommunications technologies in rural economic development. Paper presented at the annual meeting of the Rural Sociological Society, Des Moines, Iowa.

Giles, Howard, & Susan Condor. (1988, June). Ageing, technology, and society: An introduction and future priorities. *Social Behaviour, 3*(2), 59–69.

Gillette, Jay E. (1996). The information renaissance: Toward an end to rural information colonialism. *Pacific Telecommunications Review, 18*(2), 29–37.

Glasmeier, Amy K., & Marie Howland. (1995). *From combines to computers: Rural services and development in the age of information technology.* Albany, NY: State University of New York Press.

Goes, James B., & Seung Ho Park. (1997). Interorganizational links and innovations: The case of hospital services. *Academy of Management Journal, 40*(3), 673–696.

Goldberg, Carey. (1999, January 3). Rural town takes on obstacles to Internet ties. *The New York Times,* p. 14.

Gonzalez, Hernando. (1988). How much information is enough? Comparing agricultural teletext with other media. *Journal of Applied Communications, 71*(1), 3–11.

Goodrich, Carter (Ed.). (1967). *The government and the economy: 1783–1861.* Indianapolis, IN: Bobbs-Merrill.

Government Accounting Office, U.S. (1989). *Profiles of existing government corporations.* Washington, DC: U.S. Government Printing Office.

Government Accounting Office, U.S. (1999). *Rural health clinics: Rising program expenditures not focused on improving care in isolated areas* (Publication No. GAO-HEHS-97-24.) Washington, DC: U.S. Government Printing Office.

Government Technology. (1997, December). Iowa governor Terry Branstad. *Government Technology, 10*(6), 26, 28.

Government Technology. (1999, April). Arizona may require physician info online. *Government Technology, 12*(5), 14.

Granovetter, Mark. (1973). The strength of weak ties. *American Journal of Sociology, 78,* 1360–1380.

Granovetter, Mark. (1982). The strength of weak ties: A network theory revisited. In Peter V. Marsden & Nan Lin (Eds.), *Social structure and network analysis* (pp. 105–130). Beverly Hills, CA: Sage.

Gregg, Jennifer L. (1997). Revolution or evolution in organizations: Impact of a computer-mediated communication system on the Iowa State University Extension Service. Unpublished master's thesis, Iowa State University, Ames, IA.

Gregg, Jennifer L., Eric A. Abbott, & Peter F. Korsching. (1996). *Telecommunications use in the heartland: Profile of Iowa communities.* Ames, IA: Iowa State University, Department of Sociology.

Grigsby, Jim. (1995). Current status of domestic telemedicine. *Journal of Medical Systems, 19,* 19–27.

Grigsby, R., L. N. Adams, & J. Sanders. (1995). Telemedicine: Is there a doctor on the channel? *HealthScope*, 13–18.

Grossman, Lawrence K. (1995). *The electronic republic: The transformation of American democracy*. New York, NY: Viking.

Groves, Marjorie Pfister. (1978). Social systems, roles and information source use in two adult education programs. Unpublished doctoral dissertation, Iowa State University, Ames, IA.

Gruley, Bryan. (1999, February 10). As phone wars move to rural towns, tactics are growing rougher. *Wall Street Journal*, pp. A1, A8.

Gunn, Christopher Eaton, & Hazel Dayton Gunn. (1991). *Reclaiming capital: Democratic initiatives and community development*. Ithaca, NY: Cornell University Press.

Hacker, Kenneth L. (1996). Missing links in the evolution of electronic democratization. *Media, Culture, and Society, 18*(2), 213–232.

Handfield, Robert B., & Ernest L. Nichols. (1998). *Supply chain management*. Upper Saddle River, NJ: Prentice Hall.

Harper, Christopher. (1996, Summer/Fall). Online newspapers: Going somewhere or going nowhere? *Newspaper Research Journal, 17*(3–4), 2–13.

Harris, Blake. (1997, December). Economic development: The race for bandwidth. *Government Technology, 10*(13) 1, 124–125.

Harris, Blake. (1998a, November). Jane Jacobs: Cities and Web economies. *Government Technology, 11*(15), 16–21.

Harris, Blake. (1998b, March). Telecom wars. *Government Technology, 11*(3), 1, 38–39, 71.

Harris, Scott Blake. (1998, December 20). Satellites and subsidies. *Satellite Communications, 22*(12), 20.

Haug, Bob. (1997, December 8). Cities can beat cable TV: Municipal telecommunications utilities work for consumers. *Des Moines Register*, p. 9A.

Hausman, William J., & John L. Neufeld. (1991). Property rights versus public spirit: Ownership and efficiency of U.S. electric utilities prior to rate-of-return regulation. *The Review of Economics and Statistics, 73*(3), 414–423.

Hellman, Richard. (1972). *Government competition in the electric industry*. New York, NY: Praeger Publishers.

Hirschman, Albert O. (1967). *Development projects observed*. Washington, DC: The Brookings Institution.

Hobbs, Daryl. (1997). Community social capital and leadership: Keys to the technological future of rural communities. In Eric Abbott (Ed.) *Telecommunications for rural community viability: Making wise choices* (pp. 25–29). Ames, IA: Department of Journalism and Mass Communication, Iowa State University.

Holt, Glen. (1995, June 22). Pathways to tomorrow's service: The future of rural libraries. *Library Trends, 44*(1), 190–215.

Houlahan, John. (1991, May). Looking at rural libraries through rose-colored glasses. *Wilson Library Bulletin, 65*(9), 36–38, 156.

Hudson, Heather E. (1984). *When telephones reach the village: The role of telecommunications in rural development*. Norwood, NJ: Ablex Publishing.

Hudson, Heather E. (1997). *Global connections: International telecommunications infrastructure and policy*. New York, NY: Van Nostrand Reinhold.

Hudson, Heather E., & Edwin B. Parker. (1990). Information gaps in rural America. *Telecommunications Policy, 14*, 193–205.

Hughes, John, Steinar Kristoffersen, Jon O'Brien, & Mark Rouncefield. (1996). The organisational politics of meetings and their technology: Two case studies of video-supported communication. In Karlheinz Kautz and Jan Pries-Heje (Eds.), *Diffusion and adoption of information technology* (pp. 52–64). London: Chapman & Hall.

Hughes, Jonathan. (1991). *The government habit redux.* Princeton: Princeton University Press.

Iacono, Suzanne, & Rob Kling. (1996). Computerization movements and tales of technological utopianism. In Rob Kling (Ed.), *Computerization and controversy: Value conflicts and controversy* (2nd Ed., pp. 85–107). San Diego, CA: Academic Press.

Iddings, R. Keith, & Jerold W. Apps. (1990, Spring). The enthusiasm and frustration computers bring. *Journal of Extension, 28*(1), n.p.

Iddings, R. Keith, & Jerold W. Apps. (1992, Fall). Learning preferences and farm computer use: Implications for Extension programming. *Journal of Extension, 30*(3), n.p. Available online at: http://www.joe.org/joe/1992fall/a4.html.

Iowa Cable News. (1998, December 4). Muscatine group seeks "clean" cable programming. *Iowa Cable News,* 4.

Iowa Department of Public Health. (1999a). Bureau of Rural Health and primary care shortage maps. Des Moines, IA: State of Iowa. Available online at http://www.dia-hfd.state.ia.us/publicatons/books/index.htm.

Iowa Department of Public Health. (1999b). Primary care physicians: Private practice, all ages. Des Moines, IA: State of Iowa. Available online at http://www.dia-hfd.state.ia.us/publicatons/books/index.htm.

Iowa Hospital Association. (1995). *Iowa hospitals: A profile of service to the people.* Des Moines, IA: Iowa Hospital Association.

Iowa Methodist Medical Center. (1995). *HCFA Telemedicine Grant: Final Report* (Grant No. 18-C-90254/7-01). Des Moines, IA: Iowa Methodist Medical Center.

Irving, Larry. (1997, October 21). The 1996 Telecommunications Act's impact on rural America. Keynote address presented to RuralTeleCon '97, the 1997 Rural Telecommunications Conference, Aspen, CO.

Jaakkola, Hannu. (1996). Comparison and analysis of adoption models. In Karlheinz Kautz and Jan Pries-Heje (Eds.), *Diffusion and adoption of information technology* (pp. 65–82). London: Chapman & Hall.

Janowitz, Morris. (1952). *The community press in an urban setting.* Glencoe, IL: Free Press.

Jarvis, Anne Marie. (1990). Computer adoption decisions—implications for research and Extension: The case of Texas rice producers. *American Journal of Agricultural Economics, 72*(5), 1388–1394.

Jerome, Kent J. (1998). Interviews with the author. Executive Vice President, Iowa Telecommunications Association, West Des Moines, IA.

John, Patricia LaCaille. (1995, June 22). The Rural Information Center assists local communities. *Library Trends, 44*(1), 152–175.

Johnson, Bruce B., John C. Allen, Duane A. Olsen, & F. Larry Leistritz. (n.d.). *Telecommunications in rural communities: Patterns and perceptions* (Report No. RB 323). Lincoln, NE: University of Nebraska, Department of Agricultural Economics.

Johnson, Curtis. (1998, February 12). Interview with author. Manager of Electric and Communications Engineering, Cedar Falls Utilities; Cedar Falls, Iowa.

Johnson, J. T. (1995, September). Newspapers slow to embrace advances in computer world. *Quill,* 20.

Kaufman, Harold F. (1985). Action approach to community development. In Frank A. Fear & Harry Schwarzweller (Eds.), *Rural sociology and development* (pp. 53–65). Greenwich, CT: JAI Press, Inc.

Kautz, Karlheinz. (1996). Information technology transfer and implementation: The intro-
    duction of an electronic mail system in a public service organization. In Karlheinz
    Kautz and Jan Pries-Heje (Eds.), *Diffusion and adoption of information technology* (pp.
    83–95). London: Chapman & Hall.
Kelley, Brian. (1993, September). Rebuilding rural America: The Southern Development
    Bancorporation. *Annals, AAPSS, 529,* 113–127.
Kelley, Doris. (1998, February 12). Interview with author. Marketing Coordinator, Cedar
    Falls Utilities, Cedar Falls, Iowa.
Kennedy, Jack Jr., (1999). Interview with author. Clerk, Wise Circuit Court; Wise, VA.
Kenyon, Alan. (1997). Nevada, Missouri: The TeleCommunity Project. In Eric Abbott
    (Ed.), *Telecommunications for Rural Community Viability: Making Wise Choices* (pp.
    45–48). Ames, IA: Department of Journalism and Mass Communication, Iowa State
    University.
Kerr, John, & Walter E. Niebauer Jr. (1986). Use of full text, database retrieval systems by
    editorial page writers. *Newspaper Research Journal, 8*(2), 21–32.
Keyworth, G. A., Jeffrey Eisenach, Thomas Leonard, & David Colton. (1995). *The telecom
    revolution: An American opportunity.* Washington, DC: Progress and Freedom Foun-
    dation.
Kiel, J. M. (1993). How state policy affects rural hospital consortia: The rural health care de-
    livery system. *The Milbank Quarterly, 71,* 625–643.
Kielbowicz, Richard. (1987). *The role of communication in building communities and mar-
    kets: An historical overview.* Report prepared for the Office of Technology Assessment,
    U.S. Congress, as part of its project Communication Systems for an Information Age.
    Seattle, WA: University of Washington, School of Communication.
Kielbowicz, Richard. (1988). *Societal values that have guided the U.S. communication system:
    A short history.* Report prepared for the Office of Technology Assessment as part of its
    project Communication Systems for an Information Age. Seattle, WA: University of
    Washington, School of Communication.
Kiesler, S., & L. Sproull. (1992). Group decision making and communication technology.
    Organization Behavior and Human Decision Processes, 52, 96–123.
Kimberley, John R. (1981). Managing innovation. In Paul C. Nystrom & Wilham H.
    Starbuck (Eds.), *Handbook of organizational design: Volume 1* (pp. 84–104). New
    York, NY: Oxford University Press.
King, John Leslie. (1996). Where are the payoffs from computerization? Technology, learn-
    ing, and organizational change. In Rob Kling (Ed.), *Computerization and controversy:
    Value conflicts and controversy* (2nd ed., pp. 239–260). San Diego, CA: Academic Press.
Kirn, Thomas J., Richard S. Conway, Jr., & William B. Beyers. (1990, Fall). Producer ser-
    vices development and the role of telecommunications: A case study in rural Wash-
    ington. *Growth and Change. 21,* 33–50.
Klinck, Patricia E. (1996). Vermont: Library and information technology. *Library Hi-Tech,
    14*(2–3), 309–316.
Kling, Rob. (1996a). The centrality of organizations in the computerization of society. In
    Rob Kling (Ed.), *Computerization and controversy: Value conflicts and controversy* (2nd
    ed., pp. 108–132). San Diego, CA: Academic Press.
Kling, Rob (Ed.). (1996b). *Computerization and controversy: Value conflicts and social choices*
    (2nd ed.). San Diego, CA: Academic Press.
Korsching, Peter F., & Timothy O. Borich. (1997). Facilitating cluster communities: Lessons
    from the Iowa experience. *Community Development Journal, 32*(4), 342–353.

Kusmin, Lorin D. (1996). Computer use by rural workers is rapidly increasing. *Rural Development Perspectives, 11*(3), 11–16.

Kusserow, R. P. (1989). *Hospital closure: 1987* (Publication No. 0A1-04-89-00740) Washington, DC: Office of the Inspector General.

Kwoka, John E., Jr. (1996). *The origins of public enterprise* (Discussion Paper No. 96-01). Washington, DC: Center for Economic Research, George Washington University.

Kwon, Tae H., & Robert W. Zmud. (1987). Unifying the fragmented models of information systems implementation. In Richard J. Boland & Rudy A. Hirschheim (Eds.), *Critical issues in information systems research* (pp. 252–257). New York, NY: John Wiley.

Lackey, Alvin S., & Worawit Pratuckchai. (1991). Knowledge and skills required by community development professionals. *Journal of the Community Development Society, 22*(1), 1–20.

LaRose, Robert, & Jennifer Mettler. (1989). Who uses information technologies in rural America? *Journal of Communication, 39*(3), 48–60.

Lasley, Paul. (1998, July). *Iowa farm and rural life poll: 1998 summary report.* Ames, IA: Iowa State University Extension.

Lawson, Linda, & Richard Kielbowicz. (1988). Library materials in the mail: A policy history. *Library Quarterly, 58*(1), 29–51.

Lazarus, W. F., & T. R. Smith. (1988). Adoption of computers and consultant services by New York dairy farmers. *Journal of Dairy Science, 71*, 1667–1675.

Lee, Hsiou-Zong. (1987). A supply-side case study approach to explain diffusion of a computer network–EXNET–in the Iowa Cooperative Extension Service. Unpublished master's thesis, Iowa State University, Ames, IA.

Leistritz, F. Larry. (1993). Telecommunications spur North Dakota's rural economy. *Rural Development Perspectives. 8*(2), 7–11.

Leistritz, F. Larry, John C. Allen, Bruce B. Johnson, Duane Olsen, & Randy Sell. (1997). Advanced telecommunications technologies in rural communities: Factors affecting use. *Journal of the Community Development Society, 28*(2), 257–276.

Leon, Gonzalo. (1996). On the adoption of software technology: Technological frameworks and adoption profiles. In Karlheinz Kautz and Jan Pries-Heje (Eds.), *Diffusion and adoption of information technology* (pp. 96–116). London: Chapman & Hall.

Lesnoff-Caravaglia, Gari. (1989, March). Old age, ancient attitudes, and advanced technology. *Ageing and Society, 9*(1), 73–77.

Levinson, Anne. (1998, August 18). Keeping phone rates affordable for all. *Seattle Times*, p. B5.

Library Hi-Tech. (1996). Special state reports on libraries and the Internet. *Library Hi-Tech, 14*(2–3).

Linkous, Jonathan D. (1999). Toward a rapidly evolving definition of telemedicine. Available online at www.atmeda.org.

Little, Arthur D., Inc. (1992). *Study of the role of the telecommunications industry in Iowa's economic development.* Cambridge, MA: Arthur D. Little, Inc.

Loveridge, Scott. (1996). On the continuing popularity of industrial recruitment. *Economic Development Quarterly, 10*(2), 151–158.

Luloff, A. E., & Lewis E. Swanson. (1995). Community agency and disaffection: Enhancing collective resources. In Lionel J. Beaulieu & David Mulkey (Eds.), *Investing in people: The human capital needs of rural America* (pp. 351–372). Boulder, CO: Westview Press Inc.

Lynch, T. (1989). Cooperation between libraries and Extension. *Rural Libraries, 9*(2), 97–103.

Marcum, Deanna B. (1998). Redefining community through the public library. In Stephen Graubard & Paul LeClerc (Eds.), *Books, bricks and bytes: Libraries in the twenty-first century* (pp. 191–205). New Brunswick, NJ: Transaction Publishers.

Marien, Michael. (1991, May). The small library in an era of multiple transformations. *Wilson Library Bulletin, 65*(9), 27–29.

Markus, M. Lynne. (1987). Toward a "critical mass" theory of interactive media. *Communication Research, 14*(5), 491–511.

Markus, M. Lynne. (1990). Toward a "critical mass" theory of interactive media. In Janet Frelk & Charles Steinfield (Eds.), *Organization and community technology* (pp. 194–218). Newbury Park, CA: Sage.

Martin, Robert S. (1996). Texas: Library automation and connectivity: A land of contrast and diversity. *Library Hi-Tech, 14*(2–3), 291–302.

Marwell, Gerald, Pamela Oliver, & Ralph Prahl. (1988, November). Social networks and collective action: A theory of the critical mass. *American Journal of Sociology, 94,* 502–534.

Masuo, Diane M., Rosemary Walker, & Marilyn M. Furry. (1992, Fall). Home-based workers: Worker and work characteristics. *Journal of Family and Economic Issues, 13*(3), 245–262.

Matrix Maps Quarterly. (2000). *The Internet World, January 1998.* Available online at http://mids.org/mmq/501/pages.html.

Maynard, Nancy Hicks. (1995, September). Managing the future: Evolving technologies put traditional publishers at the crossroads of survival, excellence. *Quill,* 24–26.

Mazie, Sara Mills, & Linda M. Ghelfi. (1995, June 22). Challenges of the rural environment in a global economy: Rural libraries can provide their communities access to the information superhighway. *Library Trends, 44*(1), 7–20.

McCaughan, William T. (1995). Telecommunications issues impacting rural health care in the United States and Texas. *Texas Journal of Rural Health, XIV,* 5–21.

McClure, Charles R., John Carlo Bertot, & John C. Beachboard. (1995, June). Internet costs and cost models for public libraries. National Commission on Libraries and Information Science. Final Report. Washington, DC: NCLIS.

McConnaughey, James W., & Wendy Lader. (1997). *Falling through the Net II: New data on the digital divide.* Washington, DC: National Telecommunications and Information Administration.

McDowell, George R. (1995). Some communities are successful, others are not: Toward an institutional framework for understanding reasons why. In David W. Sears & J. Norman Reid (Eds.), *Rural development strategies* (pp. 269–281). Chicago, IL: Nelson Hall Publishers.

McGranahan, D. A. (1984). Local growth and the outside contacts of influentials: An alternative test of the "growth machine" hypothesis. *Rural Sociology 49,* 530–540.

McLaughlin, M. L., K. K. Osborne, & N. B. Ellison. (1997). Virtual community in a telepresence environment. In S. G. Jones (Ed.), *Virtual culture: Identity and communication in cybersociety* (pp. 146–168). Thousand Oaks, CA: Sage Publications.

McMahon, Kathleen. (1998). Creating the technology user. *Government Technology, 11*(7), 32.

Milne, Claire. (1997, September). Universal service for users: Recent research results. In *Proceedings of the Twenty-fifth Telecommunications Policy Research Conference,* Alexandria, VA, September 27–29, 1997 (pp. 74–81). Ann Arbor, MI: TPRC, Inc.

Missouri Statutes, 392.410.7 RSMo (Supp 1997).

Molotch, Harvey. (1976). The city as a growth machine: Toward a political economy of place. *American Journal of Sociology, 82*(2), 309–332.

Moore, Gary C., & Izak Benbasat. (1996). Integrating adoption of innovations and theory of reasoned action models to predict utilization of information technology by end-users. In Karlheinz Kautz and Jan Pries-Heje (Eds.), *Diffusion and adoption of information technology* (pp. 132–146). London: Chapman & Hall.

Moulder, Evelina, & Gwen Hall. (1995). Local government telecommunications initiatives. *ICMA Special Data Issue, 1*, n.p. Washington, DC: International City/County Management Association.

Mueller, Jennifer, & David Kamerer. (1995, Summer). Reader preference for electronic newspapers. *Newspaper Research Journal*, 2–13.

Mullner, Ross. M., Robert. J. Rydman, David. G. Whiteis, & Robert. F. Rich. (1989). Rural community hospitals and factors connected with their risk of closing. *Public Health Reports, 104*, 315–325.

Mumma, P. S. (1991). Technology and the rural library. *Rural Libraries, 11*(2), 7–13.

Narigon, Edward James. (1992). Diffusion of Iowa State University research information through mass media: The Cooperative Extension Service, knowledge gaps, and information equity. Unpublished master's thesis, Iowa State University, Ames, IA.

National Association of Development Organizations. (1994). *Telecommunications and its impact on rural America*. Washington, DC: NADO Research Foundation.

National Research Council Committee on the Future of Land Grant Colleges of Agriculture. (1966). *College of agriculture at the land grant university: Public service and public policy*. Washington, DC: National Academic Press.

National Rural Electric Cooperative. (1994). *Health care needs, resources, and access in rural America*. Alexandria, VA: Analytical Service.

National Telecommunications and Information Administration (NTIA). (1988). *Telecom 2000: Charting the course for a new century*. Washington, DC: National Telecommunications and Information Administration.

National Telecommunications and Information Administration. (1995). *Falling through the Net: A survey of "have-nots" in rural and urban America*. Washington, DC: U.S. Department of Commerce, NTIA. Available online at: www.Ntia.doc.gov/ntiahome/fallingthru.html.

Nelson, James A. (1996). Kentucky: Status of technology deployment. *Library Hi-Tech, 14*(2–3), 131–139.

Newcombe, Todd. (1997). Cities become telecom owners. *Government Technology, 10*(6). 1, 47, 49.

Nua. (1999). Internet surveys: How many online? Available online at: http:www.nua.ie/surveys/how_many_online/index.html.

Odasz, Frank. (1991, May). Humanizing the Internet: Librarians, citizens, and community networking. *Wilson Library Bulletin, 65*(9), 85–96.

Office of Technology Assessment, U.S. Congress. (1990). *Health care in rural America* (Publication No. OTA-H-434). Washington, DC: U.S. Government Printing Office.

Office of Technology Assessment, U.S. Congress. (1991). *Rural America at the crossroads: Networking for the future* (OTA-TCA-471). Washington, DC: U.S. Government Printing Office.

Office of Technology Assessment, U.S. Congress. (1995). *Bringing health care online: The role of information technologies* (Publication No. OTA-ITC-624). Washington, DC: U.S. Government Printing Office.

Oliver, Pam, & Gerald Marwell. (1988, February). A theory of critical mass: The paradox of group size in collective action. *American Sociological Review, 53,* 1–8.

Oliver, Pamela, Gerald Marwell, & Ruy Teireira. (1985, November). A theory of critical mass. *American Journal of Sociology, 91,* 522–556.

Olson, Chris. (1998, January 26). Benefits of Internet aren't readily apparent to Nebraskans. *Omaha World-Herald.*

Olson, Mancur. (1965). *The logic of collective action: Public goods and the theory of groups.* Cambridge, MA: Harvard University Press.

Orangeburg v Sheppard, 314 (S.C. 240 1994).

Orr, James R. (1985). The role of microcomputers in the Extension Service: Results from an Iowa study. Paper presented to Agricultural Communicators in Education, Fairbanks, AK.

Palvia, Prashant, Dwight B. Means, & Wade M. Jackson. (1994). Determinants of computing in very small businesses. *Information & Management, 27,* 161–174.

Park, Robert E. (1929, January). Urbanization as measured by newspaper circulation. *American Journal of Sociology, 25,* 60–79.

Parker, Edwin B. (1998). Unrestricted access in the local context. *Telecommunications Policy, 22*(7), 629.

Parker, Edwin B., & Heather E. Hudson, with Don A. Dillman, Sharon Strover, & Frederick Williams. (1992). *Electronic byways: State policies for rural development through telecommunications.* Boulder, CO: Westview Press.

Parker, Edwin B., Heather E. Hudson, Don A. Dillman, & Andrew D. Roscoe. (1989). *Rural America in the information age: Telecommunications policy for rural development.* Lanham, MD: University Press of America, Inc.

Pavlik, John V. (1998, November/December). Finally, a peek at profits: Some news sites find the formula. *Columbia Journalism Review,* 14–15.

Pellerin, Bridget Lee Moser. (1997). Adoption and diffusion of online technologies in Iowa's public libraries. Unpublished master's thesis, Iowa State University, Ames, IA.

Pellerin, Bridget, Jennifer Gregg, Melody Ramsey, Eric Abbott, & Peter Korsching. (1996). *Online technologies in public libraries: Survey results from Iowa.* Ames, IA: Iowa State University, Department of Sociology.

Perednia, Douglas A., & Ace Allen. (1995). Telemedicine technology and clinical applications. *Journal of the American Medical Association, 276*(6), 483–488.

Pfannkuch, Merlin L. (1988). Farmers' use of Iowa State's agricultural teletext information system. Unpublished master's thesis, Iowa State University, Ames, Iowa.

Pins, Kenneth, & Charles Bullard. (1997, August 10). A phenomenon of growth. *The Des Moines Register,* p. 1.

Pletcher, Judy (1998). Interviews with the author. Executive Secretary, Rural Iowa Independent Telephone Association, Des Moines, IA.

Poulos, Christine. (1997, December). Humble beginnings bring excellence: NYC provides superior customer service over the Web. *Government Technology, 10*(13), 46, 131.

Premkumar, G., & K. Ramamurthy. (1995). The role of interorganizational and organizational factors on the decision mode for adoption of interorganizational systems. *Decision Sciences, 26*(3), 303–336.

Premkumar, G., & Margaret Roberts. (1999). Adoption of new information technologies by small businesses. *Omega, The International Journal of Management Science, 27,* 467–484.

Premkumar, G., K. Ramamurthy, & Sree Nilakanta. (1994). Implementation of electronic data interchange: An innovation diffusion perspective. *Journal of Management Information Systems, 11*(2), 157–186.

PR Newswire. (1998a, June 4). Iowa's telecommunications network—set to reach 800 sites by 1999—is lauded by national group. *PR Newswire.*

PR Newswire. (1998b, May 13). Olman announces legislation leveling playing field for cable TV. *PR Newswire.*

Public Power Weekly. (1998, January 5). Over three years, more than two dozen cities in Iowa vote to set up their own telecommunications systems. *Public Power Weekly,* 4–5.

Puskin, Dena. S. (1995). Opportunities and challenges to telemedicine in rural America. *Journal of Medical Systems, 19,* 59–67.

Putler, Daniel S., & David Zilberman. (1988, November). Computer use in agriculture: Evidence from Tulare County, California. *American Journal of Agricultural Economics, 70*(4), 790–802.

Putnam, Robert D. (1993). The prosperous community: Social capital and public life. *The American Prospect, 13,* 35–42.

Putnam, Robert D. (1995, January). Bowling alone: America's declining social capital. *Journal of Democracy, 6*(1), 65–78.

Quinn, Tom. (1993). *Extension Council Exchange, 1.* Ames, IA: Iowa State University.

Rahm, Dianne, Erin Schreck, & Montgomery Van Wart. (1997). Telecommunications technology and local governments: Survey results from Iowa. Unpublished paper, Department of Political Science, Iowa State University, Ames, IA.

Rainey, Hal G. (1997). *Understanding and managing public organizations.* San Francisco, CA: Jossey-Bass.

Ramey, Carl. (1993). The Cable Act and municipal ownership: A growing first amendment confrontation. *Federal Communications Law Journal, 46*(1), 147–182.

Ramsey, Melody. (1996). The role of external and internal information seeking in the diffusion of telecommunications technologies in Iowa's rural hospitals. Unpublished master's thesis, Iowa State University, Ames, IA.

Ramsey, Melody, Joy Gieseke, & Eric A. Abbott. (1996). *Telecommunications for rural development: Rural Iowa hospitals and the adoption of telecommunications technologies.* Ames, IA: Iowa State University. (Available from the Rural Development Initiative, Department of Sociology, Iowa State University, Ames, 50011).

Raymond, Louis, & François Bergeron. (1996). EDI success in small- and medium-sized enterprises: A field study. *Journal of Organizational Computing and Electronic Commerce, 6*(2), 161–172.

Read, William H., & Jan L. Youtie. (1996). *Telecommunications strategy for economic development.* Westport, CT: Praeger Publishers.

Recommendation to restructure the Iowa State University Extension field operation. (1992, January). Ames, IA: Iowa State University Extension, Internal Document.

Redford, Emmette, & Charles B. Hagan. (1965). *American government and the economy.* New York, NY: Macmillan.

Reid, Jim. (1999, January 13). Interview with the author. Director, Midwest Rural Telemedicine Consortium, Des Moines, IA.

Reiss, Tammy. (1998, August 17). The Web for free? Not so fast. *Business Week,* p. 6.

Reshetyuk, A. L. (1992). The working ability of aging workers. *Gerontology and Geriatrics Education, 13*(1–2), 91–102.

Rice, Ronald E., & William Paisley. (1982, September). The Green Thumb Videotex experiment: Evaluation and policy implications. *Telecommunications Policy, 6*(3), 223–235.

Richards, Robert O. (1984). When even bad news is not so bad: Local control over outside forces in community development. *Journal of the Community Development Society, 15,* 75–85.

Riley, Patricia, Colleen M. Keough, Thora Christiansen, Ofer Meilich, & Jillian Pierson. (1998, September). Community or colony: The case of online newspapers and the Web. *Journal of Computer Mediated Communication, 4*(1), n.p. Available online at: http://www.ascusc.org/jcmc/vol4/issue1/keough.html.

Ritzer, George. (1992). *Modern sociological theory.* New York, NY: McGraw-Hill.

Rizzuto, Ronald, & Michael Wirth. (1998, May 2). Municipal cable overbuilds: A case study analysis of Paragould (Arkansas) City Cable. Paper presented at the 1998 National Cable Television Association Convention, Atlanta, GA. Available: http://www.catnet.org/general/MUNOVER.htm.

Roach, Stephen S. (1989). The case of the missing technology payback: Economic perspectives. New York, NY: Morgan Stanley and Company.

Rodas, Leon. (1998, April 28). Interview with author. Assistant General Manager, Spencer Municipal Utilities; Spencer, Iowa.

Rogers, Everett M. (1962). *Adoption of innovations.* New York, NY: The Free Press of Glencoe.

Rogers, Everett M. (1986). *Communication technology: The new media in society.* New York, NY: The Free Press.

Rogers, Everett M. (1989, February 9–13). Networks and "critical mass" in the diffusion of interactive innovations. Paper presented at the Sunbelt Network Conference, Tampa, FL.

Rogers, Everett M. (1995). *Diffusion of innovations* (4th Ed.). New York, NY: The Free Press

Rogers, Everett M., & D. Case. (1984, Aug. 22–26). Information technology in society. Paper presented at the annual meeting of the Rural Sociological Society, College Station, TX.

Roos, Jonathan. (1998, July 26). Wired but not inspired. *Des Moines Register,* p. A1.

Rosenberg, Ronald. (1998, April 29). Town to offer phone service. *Boston Globe,* p. C1.

Rowley, Thomas D., & Shirley L. Porterfield. (1993). Can telecommunications help rural areas overcome obstacles to development? *Rural Development Perspectives, 8*(2), 2–6.

Rursch, Julie A. (1988). Personal computer use by Iowa farmers and their families: A case study. Unpublished master's thesis, Iowa State University, Ames, IA.

Ryan, Bryce, & Neal C. Gross. (1943). The diffusion of hybrid seed corn in two Iowa communities. *Rural Sociology, 8:*15–24.

Ryan, Daniel J. (Ed.). (1997). *Privatization and competition in telecommunications: International developments.* Westport, CT: Praeger Publishers.

Ryan Vernon D. (1988). The significance of community development to rural economic development initiatives. In David J. Brown, J. Norman Reid, Herman Bluestone, David A. McGranahan, & Sara M. Mazie (Eds.), *Rural development perspectives in the 1980s: Prospects for the future* (pp. 359–376). Washington, DC: U.S. Department of Agriculture, Economic Research Service.

Ryan, Vernon D. (1994). Community development and the ever elusive "collectivity." *Journal of the Community Development Society, 25*(1), 5–19.

Ryan, Vernon D., Andy L. Terry, & Terry L. Besser. (1995). Rural communities, structural capacity, and the importance of social capital. Paper presented at the annual meeting of the Rural Sociological Society, Washington, DC

Salant, Priscilla, Lisa R. Carley, & Don A. Dillman. (1996). *Estimating the contribution of lone eagles to metro and nonmetro in-migration* (Technical Report No. 96-19). Pullman, WA: Washington State University, Social and Economic Sciences Research Center.

Sanders, G. L., & J. F. Courtney. (1985). A field study of organizational factors influencing DSS success. *MIS Quarterly, 9*(1), 77–93.

Saunders, C. S., D. Robey, & K. A. Vaverek. (1994). The persistence of status differentials in computer conferencing. *Human Communication Research, 20*(4), 443–472.

Sawhney, Harmeet, Jill Ehrlich, Sangjae Hwang, Dale Phillips, & Liching Sung. (1991). Small rural telephone companies, cooperatives, and regional alliances. In Jurgen Schmandt, Frederick Williams, Robert H. Wilson, & Sharon Strover (Eds.), *Telecommunications and rural development: A study of private and public sector innovation* (pp. 95–171). New York, NY: Praeger Publishers.

Schap, David. (1986). *Municipal ownership in the electric utility industry: A centennial view.* New York, NY: Praeger Publishers.

Scherer, Clifford (1987). Databook: Getting information for daily living. Ithaca, NY: Cornell University, Department of Communication.

Schiesel, Seth. (1998, June 6). Cuts are urged in Internet fund. *New York Times*, p. D1.

Schlesinger, Arthur, Jr. (1987). *The cycles of American history.* New York, NY: Houghton Mifflin.

Schmandt, Jurgen, Frederick Williams, Robert H. Wilson, & Sharon Strover (Eds.). (1991). *Telecommunications and rural development: A study of private and public sector innovation.* New York, NY: Praeger Publishers.

Schmidt, David; S. Kay Rockwell, Larry Bitney, & Elizabeth A. Sarno. (1994, June). Farmers adopt microcomputers in the 1980s: Educational needs surface for the 1990s. *Journal of Extension, 32*(1), n.p.

Schmidt, Steffen. (1998). Fear and loathing in cyberspace. *Policy Studies Journal, 26*(2), 341–345.

Schramm, Wilbur, & Merritt Ludwig. (1951, Summer). The weekly newspaper and its readers. *Journalism Quarterly, 28*(30), 301–314.

Schuler, Douglas. (1994, January). Community networks: Building a new participatory medium. *Communications of the ACM, 37*(1), 39–51.

Schuler, Douglas. (1996). *New community networks: Wired for change.* Reading, MA: Addison-Wesley Publishing.

Seattle Times. (1998, May 14). Unhappy with TCI, city to study forming own cable company. *Seattle Times*, p. B3.

Selwyn, Lee L. (1996). Market failure in "open" telecommunications networks: Defining the new "natural monopoly." In Werner Sichel & Donald L. Alexander (Eds.), *Networks, infrastructure, and the new task for regulation* (pp. 73–87). Ann Arbor, MI: University of Michigan Press.

Shaffer, Ron. (1989). *Community economics: Economic structure and change in smaller communities.* Ames, IA: Iowa State University Press.

Shaffer, Ron. (1990). Building economically viable communities: A role for community developers. *Journal of the Community Development Society, 21*(2), 74–85.

Shaw, James. (1998). *Telecommunications deregulation.* Boston, MA: Artech House.

Sherman, R. (1980). *Perspectives on postal service issues.* Washington, DC: American Enterprise Institute for Public Policy Research.

Shipley, C. L. (1985). *The local telephone company and deregulation: Implications for information access in rural America* (M.P.S. Special Project). Ithaca. NY: Cornell University, Department of Communication.

Sidak, J. Gregory, & Daniel F. Spulber. (1997). *Deregulatory takings and the regulatory contract.* New York, NY: Cambridge University Press.

Slechta, Ronald, & Ray Marner. (1997). Fiber to every home: Kalona, Iowa. In Eric A. Abbott (Ed.), *Telecommunications for rural viability: Making wise choices* (pp. 38–41). Ames, IA: Iowa State University, Department of Journalism and Mass Communication.

Smith, Barbara G. (1996). Maryland: Sailing into the electronic future. *Library Hi-Tech, 14*(2–3), 155–161.

Smith, Kevin. (1998, March 24). Interview with author. Director of Information Systems, City of West Des Moines, Iowa.

Smith, Myra N., & Joe W. Kotrlik. (1990, Winter). Computer anxiety among Extension agents a barrier that can be reduced. *Journal of Extension, 28*(4), n.p.

Soffin, Stan, Carrie Heeter, & Pamela Deiter. (1987, August). Online databases and newspapers: An assessment of utilization and attitudes. Paper presented to Newspaper Division, Association for Education in Journalism and Mass Communication, San Antonio, TX.

Sommer, Judith E., Robert A. Hoppe, Robert C. Green, & Penelope J. Korb. (1995). *Structural and financial characteristics of U.S. farms* (20th Annual Family Farm Report to the Congress, Agriculture Information Bulletin No. 746). Washington, DC: Economic Research Service, U.S. Department of Agriculture.

Steinnes, Donald N. (1990). An analysis of infrastructure provision and local economic development policy. *Journal of the Community Development Society, 21*(1), 33–53.

Stempel, Guido H., III. (1991, Fall). Where people really get most of their news. *Newspaper Research Journal, 12*(4), 2–9.

Sterling, Christopher H., & Timothy R. Haight (Eds.). (1978). *The mass media: Aspen Institute guide to communication industry trends.* New York, NY: Praeger Publishers.

Stoll, Clifford. (1995). *Silicon snake oil: Second thoughts on the information highway.* New York, NY: Anchor Books.

Strang, David, & Sarah A. Soule. (1998). Diffusion in organizations and social movements: From hybrid corn to poison pills. In Alejandro Portes (Ed.), *Annual Review of Sociology, 1998* (pp. 265–290). Palo Alto, CA: Annual Reviews.

Straub La Vonne A., & Norman Walzer (Eds.). (1992) *Rural health care: Innovation in a changing environment.* Westport, CT: Praeger Publishers.

Summers, Gene F. (1986). Rural community development. *Annual Review of Sociology, 12,* 347–371.

Sussman, Gerald. (1997). *Community, technology, and politics in the information age.* Thousand Oaks, CA: Sage.

Swindell, David, & Mark Rosentraub. (1998, January/February). Who benefits from the presence of professional sports teams? The implications for public funding of stadiums and arenas. *Public Administration Review, 58*(1), 11–20.

Sword, Doug. (1998, June 24). Local phone competitors are making few inroads. *Indianapolis Star,* p. C1.

Taschdjian, Martin. (1997, September). Alternative models of telecommunications policy: Services competition versus infrastructure competition. In *Proceedings of the Twenty-fifth Telecommunications Policy Research Conference,* Alexandria, VA, September 27–29, 1997 (pp. 44–48). Ann Arbor, MI: TPRC, Inc.

Tatchell, G. M. (1987). Agricultural Viewdata services in Britain and the role of Rothamsted as an information provider. *Computers and Electronics in Agriculture, 2,* 163–170.

Telemedicine: An Information Highway to Save Lives. Hearing before the Subcommittee on Investigations and Oversight of the Committee on Science, Space, and Technology. (1994). U.S. House of Representatives, 103rd Congress, May 2, 1994, No. 132, 1994. Washington, DC: U.S. Government Printing Office.

Tennessen, Daniel J., Steven PonTell, Van Romine, & Suzanne Motheral. (1997, October). Opportunities for Cooperative Extension and local communities in the information age. *Journal of Extension, 35*(5), n.p.

Terbovich, Vicki. (1996). Montana: Pioneer spirits and library networking. *Library Hi-Tech, 14*(2–3), 197–200.

Thompson, Howard E. (1991). *Regulatory finance: Financial foundations of rate of return regulation.* Boston, MA: Kluwer Academic.

Thong, James Y. L., & Chee Sing Yap. (1996). Information technology adoption by small business: An empirical study. In Karlheinz Kautz and Jan Pries-Heje (Eds.), *Diffusion and adoption of information technology* (pp. 160–175). London: Chapman & Hall.

Thong, James Y. L., Chee Sing Yap, & K. S. Raman. (1996). Top management support, external expertise, and information systems implementation in small business. *Information Systems Research, 7*(2), 248–267.

Thornton, Kenneth K. (1997, December). Networking, computing, and citizen services. *Government Technology, 10*(13), 52–53.

Tichenor, Phillip J., George A. Donohue, & Clarice N. Olien. (1980). *Community conflict and the press.* Beverly Hills, CA: Sage Publications.

Tilly, Charles. (1974). Do communities act? In Marcia Pelly Effrat (Ed.), *Approaches to community: Conflicts and complementaries* (pp. 209–240). New York, NY: The Free Press.

Tornatzky, Louis G., & Katherine Klein. (1982). Innovation characteristics and innovation adoption-implementation: A meta-analysis of findings. *IEEE Transactions on Engineering Management, 29*(11), 28–45.

Trainor, Brenda. (1998, August 24). Municipal networks: An interesting option. *Multichannel News.*

Tunstall, W. (1985). *Disconnecting parties.* New York, NY: McGraw-Hill.

Tweeten, Luther. (1984). *Causes and consequences of structural change in the farming industry.* Washington, DC: The National Planning Association.

Tweeten, Luther. (1987). No great impact on rural areas expected from computers and telecommunications. *Rural Development Perspectives. 3*(3), 7–10.

U.S. Code. (1996). Volume 47, secs. 151, 157.

U.S. Department of Labor Statistics. (1998). *Work at home in 1997.* Available online at: http://stats.bls.gov/newsrels.html.

U.S. House Subcommittee on Technology, Environment and Aviation of the Committee on Science, Space, and Technology. (1993, October). *Information technology and government efficiency.* Washington, DC: U.S. House of Representatives.

U.S. Senate Subcommittee on Science, Technology, and Space of the Committee on Commerce, Science, and Transportation. (1995, October). *The uses of the national information infrastructure in providing services to small industry, state and local governments, and education in rural areas.* Washington, DC: U.S. Senate.

Universitel Field Trial Survey Results (Grassroots). (1984, April 10). Brief prepared by Deloitte, Haskins and Sells Associates.

Van Wart, Montgomery, Diane Rahm, & Scott Sanders. (1997, December). When public leadership outperforms private leadership: The case of rural telecommunications. Unpublished paper, Department of Political Science, Iowa State University, Ames, IA.

Vavrek, Bernard. (1995, June 22). Rural information needs and the role of the public library. *Library Trends, 44*(1), 21–48.

Vavrek, Bernard. (1997, December). A national crisis no one really cares about. *American Libraries, 28*(11), 37–38.

Vedro, Steven R. (1983, May 2–5). Infotext: An agricultural electronic text service from WHA-TV. In *The Use of Computers in Agricultual Information* (pp. 37–41). Chicago, IL: North Central Computer Institute.

Vennard, Edwin. (1968). *Government in the power business.* New York, NY: McGraw Hill.

Vincent, Gary. (1987, March). Ag computerization steams on. *Successful Farming,* pp. 18AR–18AS.

Vogel, Morris. J. (1989). Managing medicine: Creating a profession of hospital administration in the United States, 1985–1915. In Lindsay Granshaw & Roy Porter (Eds.), *The hospital in history* (pp. 243–260). New York, NY: Routledge.

Vogelsang, Ingo, & Bridger M. Mitchell. (1997). *Telecommunications competition: The last ten miles.* Cambridge, MA: MIT Press.

Wall, Ellen, Gabriele Ferrazzi, & Frans Schryer. (1998). Getting the goods on social capital. *Rural Sociology, 63*(2), 300–322.

Warner, Paul D., & Frank Clearfield. (1982, January). An evaluation of a computer-based videotext information delivery system for farmers: The Green Thumb project. Lexington, KY: University of Kentucky Cooperative Extension Service, Department of Sociology.

Warren, Gary, & Dixie Whitlow. (1997). Telecommunications builds on long-term community development: Aurora, Nebraska. In Eric Abbott (Ed.) *Telecommunications for rural community viability: Making wise choices* (pp. 52–55). Ames, IA: Department of Journalism and Mass Communication, Iowa State University.

Warren, Roland L. (1963). *The community in America.* Chicago, IL: Rand McNally.

Warren, Roland L. (1978). *The community in America* (3rd Ed.). Chicago, IL: Rand McNally.

Watad, Mahmoud, & Sonia Ospina. (1996). Information technology and organizational change: The role of context in moderating change enabled by technology. In Karlheinz Kautz and Jan Pries-Heje (Eds.), *Diffusion and adoption of information technology* (pp. 202–220). London: Chapman & Hall.

Watkins, B., & D. Brimm. (1985). The adoption and use of microcomputers in homes and elementary schools. In Milton Chen & William Paisley (Eds.), *Children and microcomputers* (pp. 129–150). Beverly Hills, CA: Sage.

Weinberg, Mark. (1987). Business incubators give new firms in rural areas a head start. *Rural Development Perspectives, 3*(2), 6–10.

Weisfeld, V. D. (Ed.). (1993). *Rural health challenges in the 1990s: Strategies from the hospital-based rural health care program.* Princeton, NJ: Robert Wood Johnson Foundation.

Westin, A. F., & A. L. Finger. (1991). *Using the public library in the computer age: Present patterns, future possibilities.* Chicago, IL: American Library Association.

Wettenhall, Roger. (1998). Public enterprise. In Jay M. Shafritz (Ed.) *The international encyclopedia of public policy and administration, Vol. 3* (pp. 1824–1833). New York, NY: Westview Press.

Wheeler, M. B. (1985). Farm computer diffusion: Factors associated with its process among New York State farm operators. Unpublished master's thesis, Cornell University, Ithaca, NY.

Whiting, Larry Robert. (1981). Communications technologies in the land grant university setting: A focus on computer-based innovations for information dissemination to external audiences. Unpublished doctoral dissertation, Iowa State University, Ames, IA.

Wilcox, Claire, & William G. Sheperd. (1975). *Public policies toward business* (5th Ed.). Homewood, IL: Irwin.

Wilhelmson, L. Dan. (1999). Interviews with the author. General Manager, Consolidated Telephone Cooperative, Dickinson, ND.

Wilkins, Barratt. (1996). Florida: Library networking and technology development. *Library Hi-Tech, 14*(2–3), 85–92.

Wilkinson, Kenneth P. (1972). A field-theory perspective for community development research. *Rural Sociology, 27*(1), 43–52.

Wilkinson, Kenneth P. (1986). In search of the community in the changing countryside. *Rural Sociology, 51*(1), 1–17.

Wilkinson, Kenneth P. (1991). *The community in rural America.* New York, NY: Greenwood.

Williams, F., & M. Moore. (1994). Telemedicine: A place on the information highway? Discussion paper, Center for Research on Communication Technology and Society, The University of Texas at Austin, cited in Zetzman, 1995.

Williams, Frederick. (1997). Telecommunications and economic development: A U.S. perspective. In Donald Alexander (Ed.), *Telecommunications policy: Have regulators dialed the wrong number?* (pp. 31–48). Westport, CT: Praeger Publishers.

Williams, Frederick, & J. Barnaby. (1992). *Telecommunications regulation and economic development: A view of the states.* Austin, TX: Center for Research on Communication Technology and Society, College of Communication, University of Texas.

Wilson, Bill. (1996). Wisconsin: Events signal promising future. *Library Hi-Tech, 14*(2–3), 344–347.

Wilson, Robert H. (1992). Rural telecommunications: A strategy for community development. *Policy Studies Journal, 20*(2), 289–300.

Wittig, C. R. (1991). Some characteristics of Mississippi adult library users. *Public Libraries, 30*(1), 25–32.

Wolfe, Gary D. (1996). Pennsylvania: Public-private partnership. *Library Hi-Tech, 14*(2–3), 263–267.

Wresch, William. (1996). *Disconnected: Haves and have-nots in the information age.* New Brunswick, NJ: Rutgers University Press.

Yarbrough, J. Paul. (1987, October 29). *Emerging generalizations on the diffusion of computers among farmers* (Special report (NCR-90) to North Central Regional Communication Research Committee annual meeting). Winrock International, Arkansas.

Yarbrough, J. Paul. (1990, June). *Information technology and rural economic development: Evidence from historical and contemporary research.* Report prepared for the Office of Technology Assessment, U.S. Congress, as part of its project, Information Age Technology and Rural Economic Development. Ithaca, NY: Cornell University, Department of Communication.

Yarbrough, J. Paul. (1997). Information technologies and rural community viability: Lessons from the past. In Eric A. Abbott (Ed.), *Telecommunications and rural community viability: Making wise choices* (pp. 11–19). Ames, IA: Department of Journalism and Mass Communication, Iowa State University.

Yates, JoAnne. (1989). *Control through communication: The rise of system in American management.* Baltimore, MD: Johns Hopkins University Press.

Yepsen, David. (1999, January 11). Up in the air: Should state sell fiber-optic network? *Des Moines Register*, p. 4A.

Yergin, Daniel, & Joseph Stanislaw. (1998). *The commanding heights: The battle between government and the marketplace that is remaking the modern world.* New York, NY: Simon & Schuster.

Zaltman, Gerald, & Robert Duncan. (1977). *Strategies for planned change.* New York, NY: John Wiley.

Zetzman, Marion. (1995). Telemedicine: POTS and PANS technology and rural health care in Texas. *Texas Journal of Rural Health XIV*, 1–4.

Zmud, Robert W., Mary R. Lind, & Forrest W. Young. (1990). An attribute space for organizational communication channels. *Information Systems Research, 1*(4), 440–457.

INDEX

About the Contributors

ERIC A. ABBOTT has been a professor of journalism and mass communication at the Greenlee School of Journalism and Communication, Iowa State University, for the past 25 years. He has also served seven years as chair of the University's interdisciplinary Technology and Social Change program. His research specialty is the study of implications of new communication technologies for rural development in the United States, Russia, and Nigeria. In Iowa, he has been a member of the Rural Development Initiative research team since 1994. Internationally, he has served as a consultant on technology transfer projects with the World Bank, Inter-American Development Bank, and the U.S. Agency for International Development.

DOM CARISTI is an associate professor of telecommunications at Ball State University. His research interest is in electronic media policy. He has published in Telecommunications Policy, Journal of Broadcasting & Electronic Media, and Suffolk University Law Review, and has contributed policy sections to a half-dozen different books. His research has been presented to the Broadcast Education Association, International Engineering Consortium, Telecommunications Policy Research Conference, and Alliance for Public Technology. In 1995, he served as a Fulbright Professor at the University of Ljubljana in Slovenia.

SAMI EL-GHAMRINI is currently affiliated with the Development Research Institute, Giza, Egypt. In addition to his research responsibilities, Dr. El-Ghamrini teaches courses in extension and sociology, and consults with numerous organizations on issues relating to extension education, the environment, and agricultural development. He recently was in Yemen to consult on an agricultural management support project with an extension training component. His master's thesis in rural sociology examined factors related to rural telephone company innovativeness.

JOY GIESEKE is the assistant director for the Kansas Center for Rural Initiatives at Kansas State University and has over 15 years experience as an advocate for rural development. She has experience as a small business owner and in nonprofit management, including administering community development and service programs. Joy has been active in a wide range of rural community boards including land use planning, economic development, community arts, faith-based community development, and leadership.

JENNIFER L. GREGG is a graduate assistant in the Department of Telecommunication who specializes in information technologies for older adults and persons with disabilities. Her most recent research activities include the Telewindows Project—an innovative program funded by the National Council on Aging that connects homebound seniors with their communities through video and audio conferencing.

BRENT HALES is assistant professor at Delta State University. In addition to his work with the Rural Development Initiative, his research interests include drug and alcohol use among Native Americans, network analyses of Iowa farm communities, gender stereotypes in the media, and use of Internet resources in small Iowa communities.

PATRICIA C. HIPPLE has a long career as an applied sociologist, with more than twenty years experience in social research, program evaluation, policy analysis, human service administration, and education. Dr. Hipple's areas of specialization include technology and social change, international development, social inequality, and communication and culture.

PETER F. KORSCHING, professor of soiciology at Iowa State University, has conducted extensive basic and applied research in adoption and diffusion of innovations, community and rural development, the changing structure of agriculture and the effects of planned social change. He has published articles and research reports in academic and popular periodicals, has several chapters in books, is co-author of an introductory rural sociology text titled Social Change in Rural Societies, and is coeditor of two other books. Organization affiliations include the Rural Sociological Society, the Community Development Society, and the Society for Applied Sociology. He served as director of the North Central Regional Center for Rural Development from 1984–1994. His experiences include both research and outreach in interdisciplinary programs.

BRIDGET MOSER PELLERIN has worked as a researcher at a marketing research firm for several years. Ms. Pellerin's areas of specialization include adoption and diffusion of new technologies, particularly those that are computer or telecommunications-based. She also has extensive domestic and international experiences researching audience preferences in television news.

WALTER E. NIEBAUER, JR., associate professor in Communication and Performance Studies at Northern Michigan University, has studied suburban and small town community newspapers for the past 20 years. He is a founding member of the Media Management and Economics Division of the Association for Education in Journalism and Mass Communication and a charter editorial board member of the Journal of Media Economics.

G. PREMKUMAR is an associate professor of information systems in the College of Business at Iowa State University. He has over nine years of industry experience in IS and related areas. His current research interest includes electronic commerce, telecommunications, inter-organizational systems/EDI, and adoption and diffusion of information technology. He has been actively involved in curriculum development for the MIS program and started an IS certificate program for working practitioners. He has given seminars in information systems to faculty in Russia. His research papers have been published in Information Systems Research, Decision Sciences, Journal of Management Information Systems, Information and Management, IEEE Transactions on Engineering Management, and other leading journals and conference proceedings.

DIANNE RAHM is associate professor of public administration and urban studies at Cleveland State University's Levin College of Urban Affairs. Professor Rahm's research interest is science, technology, and environmental policy. She has published numerous articles and book chapters on these subjects. She is the co-editor of Technology and American Competitiveness (Greenwood Press) and co-author of University-Industry R&D Collaboration in the United States, the United Kingdom, and Japan (Kluwer Academic Publishers). She is currently writing a book entitled American Public Policy (Peacock Press). Professor Rahm serves on the editorial board of Policy Studies Journal. She is past-president and current treasurer of the Section for Science and Technology in government of the American Society for Public Administration. She is a member of the steering committee for the Section on Science, Technology, and Environmental Policy for the American Political Science Association.

MELODY RAMSEY's interests include technology and social change and political processes. Ms. Ramsey currently serves as senior marketing analyst for Ethridge and Associates, L. L. C. in Memphis, Tennessee.

SCOTT SANDERS has conducted research on rural telecommunications, spoken at telecommunication conferences and town meetings, advised city councils and utility boards on the impacts of municipally owned telecommunication utilities, and developed a community telecommunications planning model as a thesis. Mr. Sanders currently works as a Consultant for Public Financial Management, Inc., in their Des Moines, Iowa office. He provides financial advice to cities, municipal utilities, and school districts. His experiences include debt structuring and sizing, refunding analysis, strategic financial planning, and credit analysis. Mr. Sanders was intimately involved with the issuance of the first ever Municipal Communication Revenue Bonds backed solely by municipal communication revenues.

ALLAN G. SCHMIDT has 13 years experience working with agricultural and Extension communications in the area of video, satellite, multimedia, and Internet technologies. Allan is currently Instructional Development Coordinator for the Instructional Technology Center at Iowa State University.

ERIN K. SCHRECK served as a research assistant for the Rural Development Initiative project and used her research and studies of telecommunication in Iowa local governments for her creative component topic. Ms. Schreck is a Human Resources professional for Eaton Corporation in Cleveland, Ohio.

MONTGOMERY VAN WART is an associate professor of Political Science and Director of the Center for Public Service at Texas Tech University. His areas of interest include organizational change, including the affects of technology innovation on structure and work process. Recent work has included the role of economic development and rural enterprise as related to contemporary public utilities. Currently he is completing research on the effects and effectiveness of distance education in political science curricula.

DELFINO VARGAS-CHANES specializes in the methodological aspects of sociological research. His major contributions are in structural equation modeling with latent variables, latent growth curves, multilevel modeling, and imputation methods for incomplete data.

J. PAUL YARBROUGH has studied patterns of adoption and impact of communication technologies in rural America. He authored a 1990 report for the Office of Technology Assessment, U.S. Congress, on the impacts of communication technologies in rural areas, and is continuing an 18-year study of adoption and impacts of computers and other information technologies on farmers in Iowa and New York. Yarbrough is professor emeritus of Communication at Cornell University.